Published for
**OXFORD INTERNATIONAL
AQA EXAMINATIONS**

International GCSE
PHYSICS

Jim Breithaupt
Editor: Lawrie Ryan

OXFORD
UNIVERSITY PRESS

Acknowledgements

The publishers would like to thank the following for permissions to use their photographs:

Cover: DAVID PARKER/SCIENCE PHOTO LIBRARY

p6: Topfoto; **p8:** Martyn F. Chillmaid/Science Photo Library; **p9:** Nasa/Science Photo Library; **p14:** Keith Kent/Science Photo Library; **p15:** Martyn Chillmaid; **p16:** iStockphoto; **p18t:** Data Harvest; **p18b:** Martyn Chillmaid; **p23:** Rob Melnychuk/Getty Images; **p25:** iStockphoto; **p27:** Shutterstock; **p30:** Getty Images; **p33:** AFP/Getty Images; **p34:** Copyright 2010 photolibrary.com; **p36t:** Fstop/Getty Images; **p36b:** iStockphoto; **p40l:** Getty Images; **p40r:** AFP/Getty Images; **p46:** cscredon/iStockphoto; **p47t:** Optare plc; **p50t:** Zoran Kolundzija/iStockphoto; **p50b:** AFP/Getty Images; **p52:** Getty Images; **p53:** AFP/Getty Images; **p56:** Fotolia; **p57 - p60r:** iStockphoto; **p60l:** Shutterstock; **p66:** Science Photo Library; **p68 & p69t:** Shutterstock; **p69b:** Science Photo Library; **p70:** Shutterstock; **p73 & p76:** istockphoto; **p82t:** NASA/ESA/STSCI/Hubble Heritage Team/Science Photo Library; **p82b - p90b:** iStockphoto; **p92:** Mauro Fermariello/Science Photo Library; **p93t:** Image100; **p93b:** Martyn F. Chillmaid/Science Photo Library; **p94t:** czardases/iStockphoto; **p94b:** emmy-images/iStockphoto; **p98:** Fotolia; **p100:** Photolibrary/Imagebroker.net; **p102:** iStockphoto; **p106:** Photolibrary; **p107:** Shout/Rex Features; **p109:** Pasieka/Science Photo Library; **p109:** DYK iStockphoto; **p110:** GIPhotostock/Science Photo Library; **p113:** David M.Martin, MD/Science Photo Library; **p117:** AlexTyum/Fotolia; **p122:** Ian Hooton/Science Photo Library; **p123:** Adam Gault/Science Photo Library; **p126:** Charles D. Winters/Science Photo Library; **p129:** G&D Images/Alamy; **p131:** Brian Stevenson/Getty Images; **p136:** Fotolia; **p137:** Gary Ombler/Getty Images; **p138:** iStockphoto; **p140:** Spohn Matthieu/Getty Images; **p141:** iStockphoto; **p142:** Ted Kinsman/Getty Images; **p143t:** Tony Craddock/Science Photo Library; **p143b:** Photolibrary/Tsuneo Nakamura; **p144:** AP/PA Photos; **p146tl:** Cordelia Molloy/Science Photo Library; **p146tr & p146l:** iStockphoto; **p147(a & b):** Fotolia; **p147:** iStockphoto; **p148:** David Taylor/Science Photo Library; **p149:** Mark Burnett/Science Photo Library; **p150:** ESQ2; **p152:** SSPL/Science Museum/Getty Images; **p159a:** Martyn F. Chillmaid/Science Photo Library; **p168t:** Peter Zijlstra/iStockphoto; **p168b:** Cordelia Molloy/Science Photo Library; **p174:** Boissonnet/Science Photo Library; **p177r:** Science Photo Library; **p180:** ImageBroker.net/Photolibrary; **p187 & p188:** iStockphoto; **p189:** ia_64/Fotolia.com; **p190:** SSPL/Science Museum/Getty Images; **p191:** Cordelia Molloy/Science Photo Library; **p194:** Jim Breithaupt; **p195:** iStockphoto; **p200:** Popperfoto/Getty Images; **p211:** PascalR/Fotolia.com; **p212:** SSPL/Science Museum/Getty Images; **p215:** Copyright 2010 photolibrary.com; **p216 & p217:** Getty Images; **p222:** Shutterstock; **p223:** Science Photo Library; **p225:** Shutterstock; **p226 & p228:** Science Photo Library.

Although we have made every effort to trace and contact all copyright holders before publication this has not been possible in all cases. If notified, the publisher will rectify any errors or omissions at the earliest opportunity.

Links to third party websites are provided by Oxford in good faith and for information only. Oxford disclaims any responsibility for the materials contained in any third party website referenced in this work.

Physics   Contents

How to use this book

Learning objectives

Each topic begins with statements of key content that you should know by the end of the lesson.

Study tip

Hints to give you important advice on things to remember and what to watch out for.

??? Did you know … ?

There are lots of interesting and often strange facts about science. This feature tells you about many of them.

∞ links

Links will tell you where you can find more information about what you are learning and how different topics link up.

Activity

An activity is linked to a main lesson and could be a discussion or task in pairs, in groups, or by yourself.

This book has been written for you by a team of very experienced teachers and subject experts. It covers everything you may need to know for your exams and is packed full of features to help you achieve the very best that you can.

Figure 1 *Many diagrams are as important for you to learn as the text, so make sure you revise them carefully*

Key words are highlighted in the text. You can look them up in the Glossary at the back of the book if you are not sure what they mean.

Demonstration or Required Practical

This feature helps you become familiar with key practicals. It may be a simple introduction, a reminder, or the basis for a practical in the classroom.

Summary questions

These questions give you the chance to test whether you have learnt and understood everything in the topic. If you get any wrong, go back and have another look. They are designed to be increasingly challenging.

And at the end of each chapter you will find …

Chapter summary questions

These will test you on what you have learnt throughout the whole chapter, helping you to work out what you have understood and where you need to go back and revise.

Practice questions

These questions are examples of the types of question you may encounter in your exam, so you can get lots of practice during your course.

Key points

At the end of the topic are the important points that you must remember. They can be used to help with revision and to summarise your knowledge.

Practical skills

During this course, you will develop your understanding of the scientific process and the skills associated with scientific enquiry. Practical work is an important part of the course, as it develops these skills and also reinforces concepts and knowledge you have learnt during the course.

As part of this course you will undertake practical work in many topics, and you are required to carry out five practicals as suggested below:

Required practicals

1 Investigating the relationship between force and extension for a spring. (3.1.1) [Topic 1.5]

2 Investigating refraction of light by different substances. (3.3.5) [Topic 10.2]

3 Investigating cooling curve for stearic acid. (3.4.1) [Topic 12.3]

4 Investigating the potential difference–current characteristics of a filament lamp, a diode, and a resistor at constant temperature. (3.5.1) [Topic 14.5]

5 Investigating the factors that determine the strength of an electromagnet. (3.5.2) [Topic 15.2]

In Paper 2, you will be assessed on aspects of the practical skills listed below, and may be required to read and interpret information from scales given in diagrams and charts, present data in appropriate formats, design investigations, and evaluate information that is presented to you.

Designing a practical procedure

- Design a practical procedure to answer a question, solve a problem, or test a hypothesis.
- Comment on/evaluate plans for practical procedures.
- Select suitable apparatus for carrying out experiments accurately and safely.

Control

- Appreciate that, unless certain variables are controlled, experimental results may not be valid.
- Recognise the need to choose appropriate sample sizes, and study control groups where necessary.

Risk assessment

- Identify possible hazards in practical situations, the risks associated with these hazards, and methods of minimising the risks.

Collecting data

- Make and record observations and measurements with appropriate precision and record data collected in an appropriate format (such as a table, chart, or graph).

Analysing data

- Recognise and identify the cause of anomalous results and suggest what should be done about them.
- Appreciate when it is appropriate to calculate a mean, calculate a mean from a set of at least three results, and recognise when it is appropriate to ignore anomalous results in calculating a mean.
- Recognise and identify the causes of random errors and systematic errors.
- Recognise patterns in data, form hypotheses, and deduce relationships.
- Use and interpret tabular and graphical representations of data.

Making conclusions

- Draw conclusions that are consistent with the evidence obtained and support them with scientific explanations.

Evaluation

- Evaluate data, considering its repeatability, reproducibility, and validity in presenting and justifying conclusions.
- Evaluate methods of data collection and appreciate that the evidence obtained may not allow a conclusion to be made with confidence.
- Suggest ways of improving an investigation or practical procedure to obtain extra evidence to allow a conclusion to be made.

1.1

Forces between objects

When you apply a **force** to a tube of toothpaste, be careful not to apply too much force. The force you apply to squeeze the tube changes its shape and pushes toothpaste out of the tube. If you apply too much force, the toothpaste might come out too fast.

Forces can change the shape of an object or change its state of rest or its motion.

A force is a push or pull that acts on an object because of an interaction with another object. If two objects must touch each other to interact, the forces are called **contact forces**. Examples include friction, air resistance, tension (or stretching forces), and normal contact forces when an object is supported by or strikes another object. **Non-contact forces** include magnetic force, electrostatic force, and the force of gravity.

Equal and opposite forces

When two objects push or pull on each other, they exert equal and opposite forces on one another.

The unit of force is the **newton** (abbreviated as N). The rule stated above is called **Newton's Third Law**. (You will meet Newton's First Law in Topic 1.2, and Newton's Second Law in Chapter 2, when you study forces and motion.) Here are some examples of Newton's Third Law:

- A boxer who punches an opponent with a force of 100 N experiences an equal and opposite force of 100 N from his opponent.

- Two roller skaters pull on opposite ends of a rope. The skaters move towards each other because they pull on each other with equal and opposite forces. Two newtonmeters could be used to show this. A newtonmeter is a spring balance calibrated in newtons (see Topic 1.4).

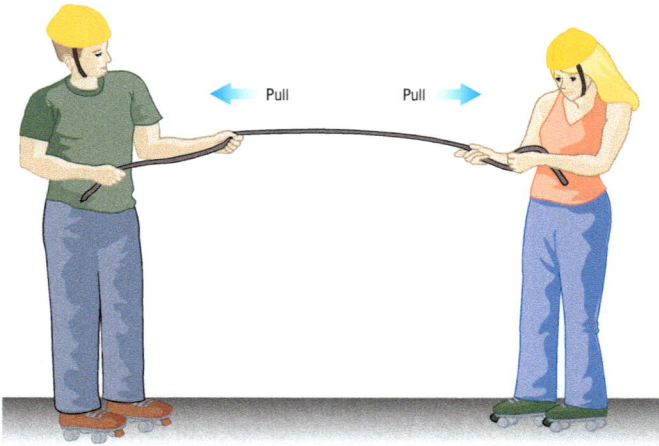

Figure 1 *Equal and opposite forces*

Practical

Action and reaction

Test Newton's Third Law as shown in Figure 1 with a partner if you can, using roller skates and two newtonmeters. Don't forget to wear protective head gear.

- What did you find out?
- Evaluate the precision of your repeat readings.

Safety: You might want someone to help support you.

In the mud

A car stuck in mud can be difficult to shift. A tractor can be very useful here. Figure 2 shows the idea. At any stage, the force of the rope on the car is equal and opposite to the force of the car on the rope.

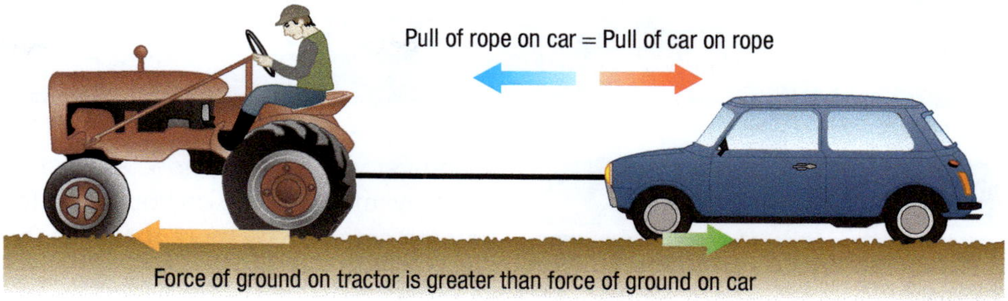

To pull the car out of the mud, the force of the ground on the tractor needs to be greater than the force of the ground on the car. Note that these two forces are not equal and opposite to each other. The 'equal and opposite force' to the force of the mud on the tractor is the force of the tractor on the ground. The 'equal and opposite force' to the force of the mud on the car is the force of the car on the ground.

Pull of rope on car = Pull of car on rope

Force of ground on tractor is greater than force of ground on car

Figure 2 *In the mud*

Friction in action

The driving force on a car is the force that makes it move. This is sometimes called the engine force or the **motive force**. This force is caused by friction between the road and the tyre of each drive wheel. Friction is a force that opposes the motion of any two surfaces in contact with each other when they slide or try to slide across each other. In Figure 3, friction acts between the tyre and the road where they are in contact with each other.

When the car moves forwards:

* the force of friction of the road on the tyre is in the forward direction
* the force of friction of the tyre on the road is in the reverse direction.

The two forces are equal and opposite to one another.

Direction of car

Force of tyre on road   Force of road on tyre

Figure 3 *Driving force*

Summary questions

1 **a** When the brakes of a moving car are applied, what is the effect of the braking force on the car?
   b When you sit on a cushion, what forces act on you?
   c When you kick a football, what is the effect of the force of your foot on the ball?

2 **a** **i** A hammer hits a nail with a downward force of 50 N. What is the size and direction of the force of the nail on the hammer?
      ii A lorry tows a broken-down car. When the force of the lorry on the tow rope is 200 N, what is the force of the tow rope on the lorry?
   b Copy and complete **i–iii** using the words below:

   downwards   equal   opposite   upwards

   i The force on a ladder resting against a wall is .... and .... to the force of the wall on the ladder.
   ii A book is at rest on a table. The force of the book on the table is .... The force of the table on the book is ....
   iii When a ball is dropped onto the floor, the force of the floor on the ball is ....

3 When a student is standing at rest on bathroom scales, the scales read 500 N.
   a What is the size and direction of the force of the student on the scales?
   b What is the size and direction of the force of the scales on the student?
   c What is the size and direction of the force of the floor on the scales?

Key points

* A force can change the shape of an object, or change its motion, or its state of rest.

* The unit of force is the newton (N).

* When two objects interact, they always exert equal and opposite forces on each other.

* Friction opposes the motion of any two surfaces in contact with each other when they slide or try to slide across each other.

3

1.2 Resultant force

Resultant force

Learning objectives

After this topic, you should know:

- what a resultant force is

- what happens if the resultant force on an object is:
  - zero
  - not zero

- how to calculate the resultant force when an object is acted on by two forces acting along the same line.

Wherever you are right now, at least two forces are acting on you. These are the gravitational force on you and a force supporting you. Most objects around you are acted on by more than one force. You can work out the effect of all the forces acting on an object by replacing them with a single force, the **resultant force**. This is a single force that has the same effect as all the forces acting on the object. If the resultant force is zero, you know that the forces acting on the object are **balanced**.

Zero resultant force

Newton's First Law of motion states that if **the resultant force on an object is zero**, the object:

- remains stationary if it was at rest (i.e., it was in equilibrium), or
- continues to move at the same speed and in the same direction if it was already moving.

If the resultant force is zero, and only two forces act on the object, they must be equal to each other and act in opposite directions. Here are two examples.

1 **A glider on a linear air track** floats on a cushion of air. As long as the track stays level, the glider moves at constant velocity (i.e., with no change of speed or direction) along the track. This is because friction is absent. Newton's First Law tells you that the resultant force on the glider is therefore zero.

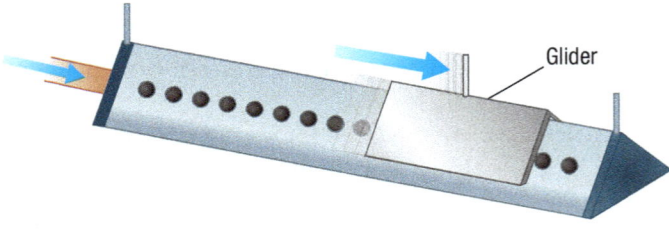

Figure 1 *The linear air track*

2 **When a heavy crate is pushed across a rough floor at a constant velocity**, Newton's First Law tells you that the resultant force on the crate is zero. The push force on the crate is equal in size but acts in the opposite direction to the force of friction of the floor on the crate.

Figure 2 *Overcoming friction*

Practical

Investigating forces

Use a glider on an air track to investigate the relationship between force and acceleration.

- Record your observations and explain what you observed.

Alternatively:

Make and test a model hovercraft floating on a cushion of air provided by an inflated balloon.

Non-zero resultant force

When the resultant force on an object is not zero, the forces acting on the object are not balanced. The movement of the object depends on the size and direction of the resultant force.

1 **When a jet plane is taking off**, the thrust force of its engines is greater than the force of air resistance on it. The resultant force on it is the difference between the thrust force and the force of air resistance on it. The resultant force is therefore non-zero. The greater the resultant force, the quicker the take-off is.

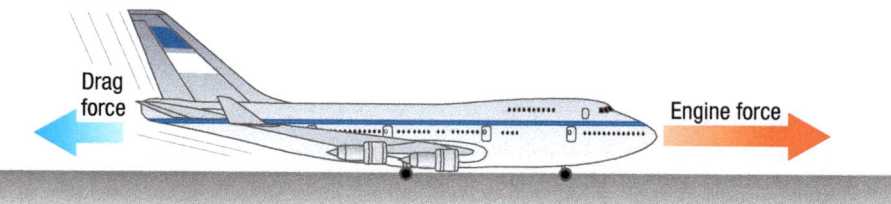

Figure 3 *A passenger jet on take-off*

2 **When a car driver applies the brakes**, the braking force is greater than the force from the engine. The resultant force is the difference between the braking force and the engine force. It acts in the opposite direction to the car's direction so it slows the car down.

The examples above show that if an object is acted on by two unequal forces acting in opposite directions, the resultant force is:

- equal to the difference between the two forces
- in the direction of the larger force.

For example, suppose two forces A and B act on an object in opposite directions. If A = 5 N and B = 9 N, the resultant force on the object is 4 N (= 9 N – 5 N) in the direction of B.

What happens if the two forces act in the same direction? The resultant force is equal to the sum of the two forces and acts in the same direction as the two forces.

Figure 4 *Braking*

Summary questions

1 a What happens to the glider in Figure 1 if the air track blower is switched off, and why?
  b When a jet plane is moving at constant velocity at a constant height, what can be said about the thrust force and the force of air resistance?

2 A jet plane lands on a runway and stops.
  a What can you say about the direction of the resultant force on the plane as it lands?
  b What can you say about the resultant force on the plane when it has stopped?

3 A car is stuck in the mud. A tractor tries to pull it out.
  a The tractor pulls the car with a force of 250 N. The car does not move. Explain in terms of forces why not.
  b Increasing the driving force of the tractor to 300 N pulls the car steadily out of the mud at constant velocity. What is the force of the ground on the car now?

Key points

- The resultant force is a single force that has the same effect as all the forces acting on an object.
- If the resultant force on an object is zero, the object stays at rest or at constant velocity.
- If the resultant force on an object is not zero, the velocity of the object will change.
- If two forces act on an object along the same line, the resultant force is:
  1 their sum if the forces act in the same direction
  2 their difference if the forces act in opposite directions.

1.3

Force as a vector

Learning objectives

After this topic, you should know:

- how to represent a force accurately on a diagram

- how to find the resultant of two forces at any angle to each other

- what the parallelogram of forces is.

Force diagrams

When an object is acted on by more than one force, you can draw a **force diagram** to work out the resultant force on the object. A force diagram shows the forces acting on the object. Each force can be represented accurately on the diagram by a **vector**, which is shown as an arrow. The length of the arrow is proportional to the size (i.e., magnitude) of the force and the arrow points in the direction of the force.

In Topic 1.2 'Resultant force', you saw that when two forces act along the same line on an object, the resultant force is equal to:

- the sum of the two forces if the forces are in the same direction
- the difference between the two forces if the forces are in opposite directions.

Figure 1 shows a tug-of-war in which the pull force of each team is represented by a vector. A scale of 10 mm to 200 N is used. Team A pulls with a force of 1000 N and team B pulls with a force of 800 N. So the resultant force is 200 N in team A's direction.

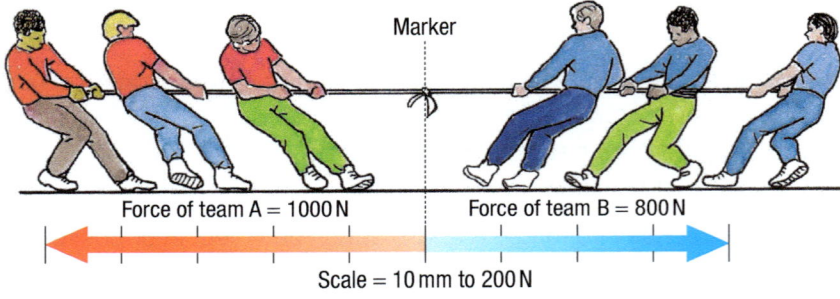

Figure 1 *A tug-of-war*

Figure 2 *In tow*

The parallelogram of forces

What if the two forces do not act along the same line, as shown in Figure 2? Here you can see a ship being towed by cables from two tugboats. The tension force in each cable pulls on the ship. The combined effect of these tension forces is to pull the ship forwards. This is the resultant force.

Figure 3 shows how the two tension forces T_1 and T_2, represented as vectors, combine to produce the resultant force. The tension forces are drawn as adjacent sides of a parallelogram – the resultant force is the length of the diagonal of the parallelogram from the origin of T_1 and T_2. This geometrical method of finding the resultant force is called the **parallelogram of forces**.

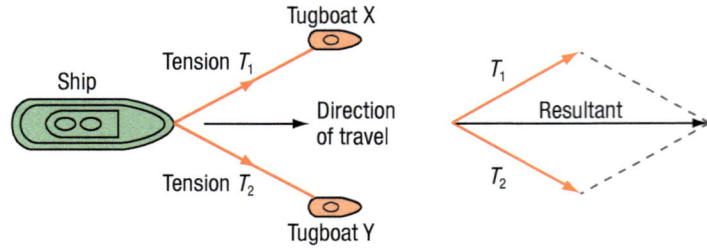

Figure 3 *Combining forces*

Investigating the parallelogram of forces

You can use weights and pulleys to demonstrate the parallelogram of forces, as shown in Figure 4. The tension in each string is equal to the weight it supports, either directly or over a pulley.

The point where the three strings meet is in equilibrium (i.e., at rest). The string supporting the middle weight (W_3) is vertical. Using a protractor, you can measure angles θ_1 and θ_2. If you know the values of the three weights, you can draw a parallelogram to scale such that:

*   the line down the centre of the diagram represents a vertical line
*   adjacent sides of the parallelogram at angles θ_1 and θ_2 to the vertical line represent the tensions in the strings supporting W_1 and W_2.

The resultant of W_1 and W_2 represented by the diagonal line should be equal and opposite in direction to the vector representing W_3.

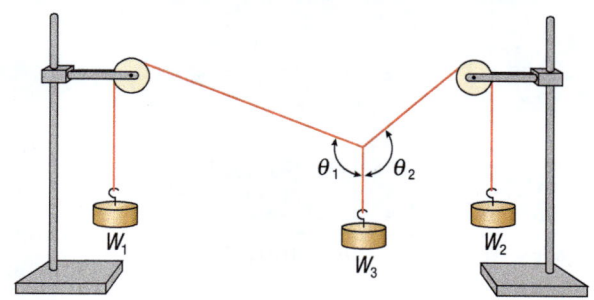

Figure 4 *The parallelogram of forces*

Worked example

A tow rope is attached to a car at two points 0.80 m apart. The two sections of rope joined to the car are of the same length and are at 30° to each other (see Figure 5). The pull on each attachment should not exceed 3000 N. Use the parallelogram of forces to determine the maximum tension in the main tow rope.

Solution

The maximum tension T in the main tow rope is equal and opposite to the resultant of the two 3000 N forces at 30° to each other.

Drawing the parallelogram of forces as shown in Figure 5 and measuring the diagonal gives:

$$T = 5800 \text{ N}$$

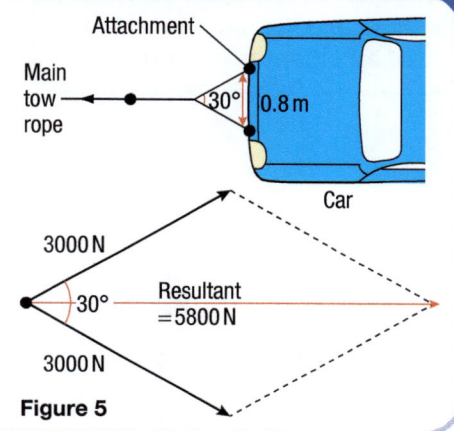

Figure 5

Summary questions

1   The diagrams below show several examples where two forces act on an object X. In each case, work out the magnitude and direction of the resultant force on X.

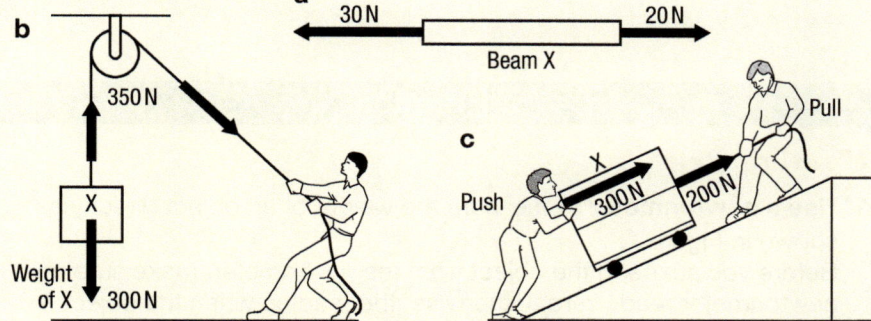

2   A force of 3.0 N and a force of 4.0 N act on a point. Determine the magnitude of the resultant of these two forces if the angle between their lines of action is:

a   90°   b   60°   c   45°.

3   In Figure 5, suppose the angle between the two sections of rope joined to the car had been 50° instead of 30°. Use the parallelogram of forces to find the maximum tension in the main tow rope.

Study tip

Remember you cannot always use arithmetic to add and subtract forces. When the two forces act at an angle, you will have to use method – draw a parallelogram of forces and measure the diagonal.

Key points

*   On a diagram, an arrow is used to show a force as a vector.

*   The parallelogram of forces is used to find the resultant of two forces that do not act along the same line.

1.4

Mass and weight

Learning objectives

After this topic, you should know:

- the difference between mass and weight

- what causes the weight of an object on or near the Earth

- what is meant by gravitational field strength

- how to calculate the weight of an object from its mass.

Your weight is caused by the gravitational force of attraction between you and the Earth. This force is very slightly weaker at the equator than at the poles. So your weight would be slightly less at the equator. However, your mass will be the same no matter where you are.

- The **weight** of an object is the force of gravity on it. Weight is measured in newtons, or N.

- The **mass** of an object depends on the quantity of matter in it. Mass is measured in kilograms, or kg.

You can measure the weight of an object using a newtonmeter.

The weight of an object:

- of mass 1 kg is 10 N
- of mass 5 kg is 50 N.

The gravitational force acting on each kilogram of an object's mass is the **gravitational field strength** at the place where the object is. The unit of gravitational field strength is the newton per kilogram (N/kg).

The value of the Earth's gravitational field strength at its surface is about 10 N/kg.

If you know the mass of an object, you can calculate the gravitational force on it (i.e., its weight) using the equation:

$$\text{weight} = \text{mass} \times \text{gravitational field strength}$$
(newtons, N)   (kilograms, kg)   (newtons per kilogram, N/kg)

You can write this word equation using symbols as:

$$W = mg$$

where:
W = weight in N
m = mass in kg
g = gravitational field strength in N/kg.

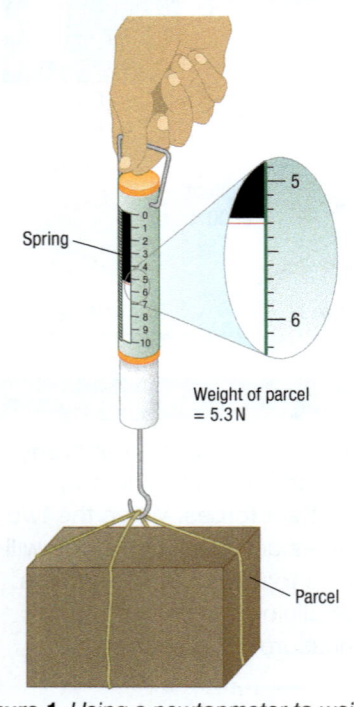

Figure 1 *Using a newtonmeter to weigh an object*

Spring

Weight of parcel = 5.3 N

Parcel

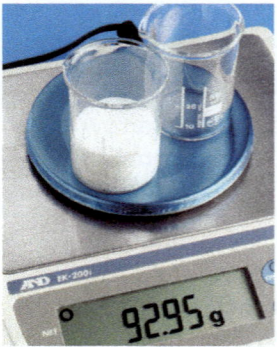

Figure 2 *Using a top pan balance*

Practical

Measuring weight

1 **Use a newtonmeter** to measure the weight of an object directly as shown in Figure 1.
Before you suspend the object from the newtometer, make sure the newtonmeter reads zero. Otherwise, the reading when the object is suspended from the newtonmeter will be inaccurate.

2 **Use a top pan balance** to measure the mass of the object in kilograms. Then multiply by the mass by g (= 10 N/kg) to find the weight of the object. Check the balance reads zero before you put the object on the balance pan.

More about the force of gravity

Objects would weigh less on the Moon's surface than they do on the Earth's surface. The force of gravity on an object near a large body depends on the mass of the large body, the mass of the object, and the distance between them.

Because the Moon has a much smaller mass than the Earth, the force of gravity on an object at the Moon's surface is less than the force of gravity on it at the Earth's surface. The gravitational field strength on the Moon's surface is about 1.6 N/kg, compared with 10 N/kg at the Earth's surface.

Figure 3 *On the Moon's surface*

Summary questions

1 The gravitational field strength at the surface of the Earth is 10 N/kg. Explain what this statement means.

2 The gravitational field strength at the surface of the Moon is 1.6 N/kg.

  a Calculate the weight of a person of mass 50 kg on the Earth.

  b Calculate the weight of the same person if she was on the Moon.

3 A lunar vehicle weighs 190 N on the Moon. Calculate its weight on the Earth.

4 A space probe travelling directly along a line between the Earth and the Moon experiences a force towards the Earth and a force in the opposite direction towards the Moon.

  a Describe how these forces change as the probe moves from the Earth to the Moon.

  b Describe how the resultant force on the space probe changes as it moves from the Earth to the Moon.

Key points

- The weight of an object is the gravitational force acting on it. Its mass is the quantity of matter in the object.

- The weight of an object is due to the gravitational attraction between the object and the Earth.

- The gravitational field strength, *g*, at a point is the gravitational force acting on each kilogram of an object's mass at that point.

- The weight of an object = its mass × *g*.

Forces and elasticity

Learning objectives

After this topic, you should know:

- what is meant when an object is called elastic

- how to measure the extension of an object when it is stretched

- how the extension of a spring varies with the force applied to it.

Table 1 *Weight versus length measurements for a rubber strip*

Weight (N)	Length (mm)	Extension (mm)
0	120	0
1.0	124	4
2.0	135	15
3.0	152	
4.0		

??? Did you know …?

If you freeze rubber or another soft material, such as a flower, by dipping it in liquid nitrogen, it becomes as brittle as glass. It will then shatter when struck with a hammer, or explode when hit with a projectile.

Figure 2 *A flower dipped in nitrogen and then shattered*

Squash players know that hitting a squash ball changes the ball's shape briefly. A squash ball is **elastic** because it goes back to its original shape. A rubber band is also elastic as it returns to its original length after it is stretched and then released. Rubber is an example of an elastic material.

An object is elastic if it returns to its original shape when the forces deforming it are removed.

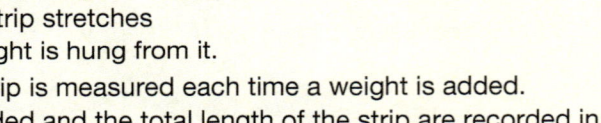

Required Practical

Stretch tests

You can investigate how easily a material stretches by hanging weights from it, as shown in Figure 1.

- The strip of material to be tested is clamped at its upper end. The material is held to keep it straight.
- The length of the strip is measured using a metre ruler. This is its original length.
- The weight hung from the material is increased by adding weights one at a time. The strip stretches each time more weight is hung from it.

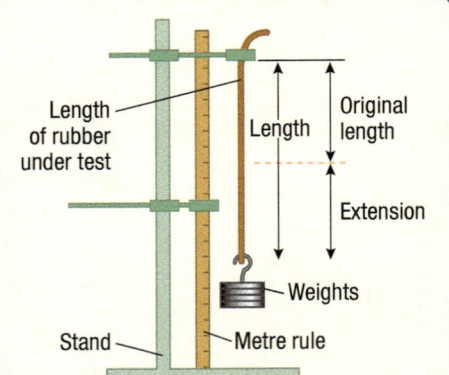

Figure 1 *Investigating stretching*

- The length of the strip is measured each time a weight is added. The total weight added and the total length of the strip are recorded in a table.

Safety: Protect the bench and your feet from falling objects.

The increase in length of a stretched material from its original length is called the **extension**. This is calculated each time a weight is added, as shown in Table 1.

extension = stretched length – original length

The measurements may be plotted on a graph of extension on the vertical axis against weight on the horizontal axis. Figure 3 shows the results for strips of different materials and a steel spring plotted on the same axes.

- The steel spring gives a straight line through the origin. This shows that the extension of the steel spring is **directly proportional** to the weight hung on it. For example, doubling the weight from 2.0 N to 4.0 N doubles the extension of the spring.
- The rubber band does not give a straight line. When the weight on the rubber band is doubled from 2.0 N to 4.0 N, the extension more than doubles.
- The polythene strip does not give a straight line either. As the weight is increased from zero, the polythene strip stretches very little at first, then it 'gives' and stretches easily.

Hooke's law

In the tests above, the extension of a steel spring is directly proportional to the force applied to it. You can use the graph to predict what the extension would be for any given force. But if the force is too large, the spring stretches more than predicted. This is because the spring has been stretched beyond its **limit of proportionality**.

The extension of a spring is directly proportional to the force applied, as long as its limit of proportionality is not exceeded.

The above statement is known as **Hooke's law**. If the extension of any stretched object or material is directly proportional to the stretching force, then it obeys Hooke's law.

1 Hooke's law may be written as an equation:

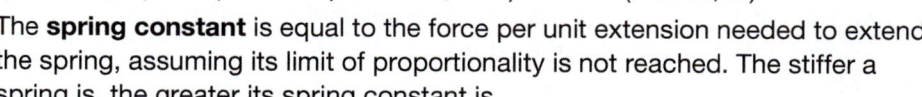

force applied =	spring constant	×	extension
(newtons, N)	(newtons per metre, N/m)		(metres, m)

The **spring constant** is equal to the force per unit extension needed to extend the spring, assuming its limit of proportionality is not reached. The stiffer a spring is, the greater its spring constant is.

2 The line on the graph in Figure 3 for the spring shows that the extension is directly proportional to the weight so the spring obeys Hooke's law. The other lines on the graph show that rubber and polythene have a low limit of proportionality. Beyond this limit, they do not obey Hooke's law. A steel spring has a much higher limit of proportionality.

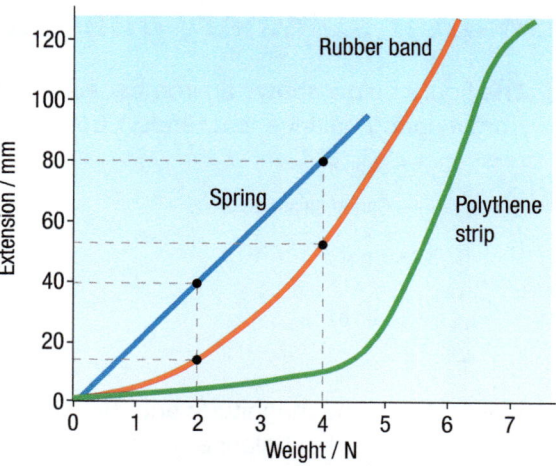

Figure 3 *Extension versus weight for different materials*

Elastic energy

When an elastic object is stretched, elastic potential energy is stored in the object. This is because work is done on the object by the stretching force.

When the stretching force is removed, the elastic energy stored in the object is released. Some of this energy may be transferred into kinetic energy of the object or may make its atoms vibrate more so it becomes warmer.

Summary questions

1 a i State Hooke's law.
ii A spring has a spring constant of 25 N/m. How much force is needed to make the spring extend by 0.10 m?
b i What happens to a strip of polythene if it is stretched beyond its elastic limit of proportionality?
ii How does the result of stretching then releasing a rubber band differ from that of stretching a strip of polythene?

2 What is meant by:
a the limit of proportionality of a spring?
b the spring constant of a spring?
c the extension of a stretched spring?

3 a In Figure 3, when the weight is 4.0 N, what is the extension of:
i the spring?
ii the rubber band?
iii the polythene strip?
b i What is the extension of the spring when the weight is 3.0 N?
ii Calculate the spring constant of the spring.
iii What does the gradient of the line for the spring in Figure 3 represent?

Chapter summary questions

1 The figure below shows an iron bar suspended at rest from a spring balance that reads 1.6 N.

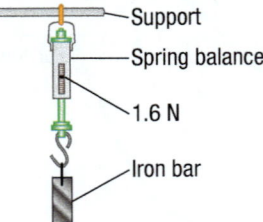

Support
Spring balance
1.6 N
Iron bar

a i What is the magnitude and the direction of the force on the spring balance due to the iron bar?

ii What is the weight of the bar in newtons?

b When a magnet is held under the iron bar, the spring balance reading increases to 2.0 N.
What is the magnitude and the direction of:

i the force on the iron bar due to the magnet?

ii the force on the magnet due to the iron bar?

2 The figure below shows a stationary helium-filled balloon attached to a vertical thread. Because the helium-filled balloon is less dense than air, an upward force, or upthrust, acts on the balloon. The lower end of the string is attached to a weight on a table.

a i What can you say about the resultant force on the balloon?

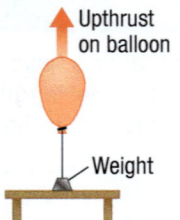

Upthrust on balloon

Weight

ii Which force is greater: the gravitational force on the balloon or the upthrust? Give a reason for your answer.

b Describe and explain what would happen to the balloon if the thread was cut.

3 A space vehicle of mass 200 kg rests on its four wheels on a flat area of the lunar surface. The gravitational field strength at the surface of the Moon is 1.6 N/kg.

a Calculate the weight of the space vehicle on the lunar surface.

b Calculate the force that each wheel exerts on the lunar surface.

4 In a Hooke's law test on a spring, the following results were obtained.

Weight (N)	Length (mm)	Extension (mm)
0	245	0
1.0	285	40
2.0	324	
3.0	366	
4.0	405	
5.0	446	
6.0	484	

a Copy and complete the third column of the table.

b Plot a graph of the extension on the vertical axis against the weight on the horizontal axis.

c If a weight of 7.0 N is suspended on the spring, work out what the extension of the spring would be.

d i Calculate the spring constant of the spring.

ii An object suspended on the spring gives an extension of 140 mm. Calculate the weight of the object.

5 a A tugboat is towing a ship steadily into a port. The tugboat cable exerts a horizontal force of 7200 N on the ship, which is moving at a constant velocity. A resistive force acts on the ship because of water flowing past it as it moves through the water. The resistive force acts in the opposite direction to the force of the cable.

i What is the magnitude of the **resistive** force on the ship?

ii What is the resultant force on the ship?

b Near the port, the ship stops and two tugboats are used to pull the ship towards the quay. Each tugboat exerts a force of 7200 N on the ship at an angle of 45° between their cables, as shown in the figure. Use the parallelogram of forces to find the magnitude of the resultant of the tugboat forces on the ship.

7200 N
45°
7200 N

Practice questions

1 Some quantities are vector quantities – others are scalar quantities.

  a Which three of the following are vector quantities? (3)

    acceleration force mass speed velocity

  b A toy train travels round a circular track at a steady 50 cm/s.

    Discuss whether the train is accelerating. (4)

  c Two tugboats tow a ship along a river.

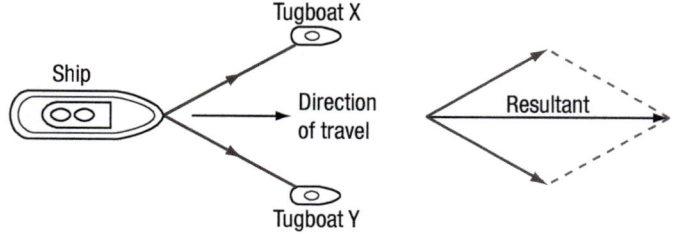

    Use the parallelogram of forces to explain why it is better to use long tow ropes rather than short ones. (3)

2 The diagram shows a car of mass 1200 kg travelling along a level road.

The diagram also shows the two forces, **A** and **B**, acting on the car.

The car is travelling at constant speed.

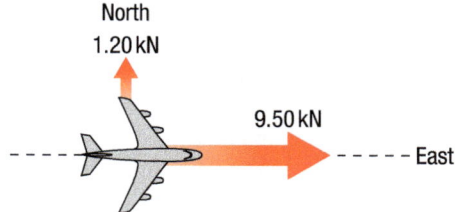

  a Force **B** is 1600 N.

    What is the size of force **A**? (1)

  b Force **B** is increased to 4000 N. Describe the effect on the car. (3)

3 An aircraft in level flight is travelling at a constant velocity due east with an engine force of 9.50 kN when it experiences a horizontal crosswind acting due north with a force of 1.20 kN.

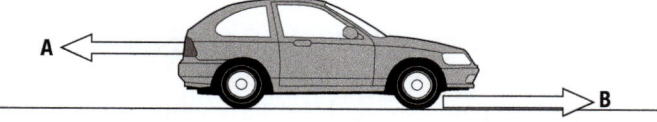

  a Use the parallelogram of forces to show that the aircraft is pushed off course by about 7°. (3)

  b Calculate the magnitude of the resultant force on the aircraft. (2)

4 Two boys pull a sledge through snow. The diagram shows the forces they exert on the sledge.

The dotted line shows the direction in which the sledge moves.

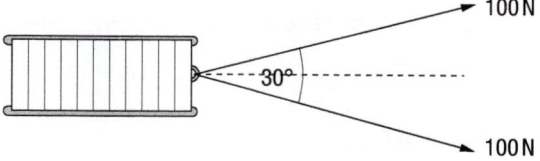

  a Describe how you would determine the resultant force acting on the sledge. (6)

  b One of the boys is stronger than the other and increases his force to 150 N.

    Explain what he would have to do to keep the sledge moving forward in the same straight line. (2)

5 A car of mass 1500 kg is pulling a trailer of mass 300 kg.

  a The force of the car pulling the trailer forward is labelled **A**. The force of the trailer on the car is labelled **B**. If force **A** is 120 N, what can be said about force **B** in comparison? (2)

  b The car and the trailer are moving at a constant velocity.

    i Calculate the weight of the trailer. Assume $g = 10$ N/kg. (1)

    ii Identify two other forces acting on the trailer as well as force of gravity on it and force **A**? (2)

    iii What is the magnitude of the resultant force on the trailer? (1)

  c If the velocity of the car increases, describe how the forces acting on trailer change during this time. (3)

Chapter 2 Forces and motion

Speed

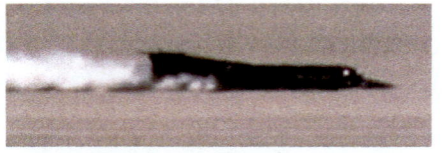

Figure 2 Capturing the land speed record

Some motorways have marker posts every kilometre. If you are a passenger in a car on a motorway, you can use these posts to check the speed of the car. You need to record the time as the car passes each post. Table 1 shows some measurements made on a car journey.

Table 1 *Measurements made on a car journey*

Distance (metres, m)	0	1000	2000	3000	4000	5000	6000
Time (seconds, s)	0	40	80	120	160	200	240

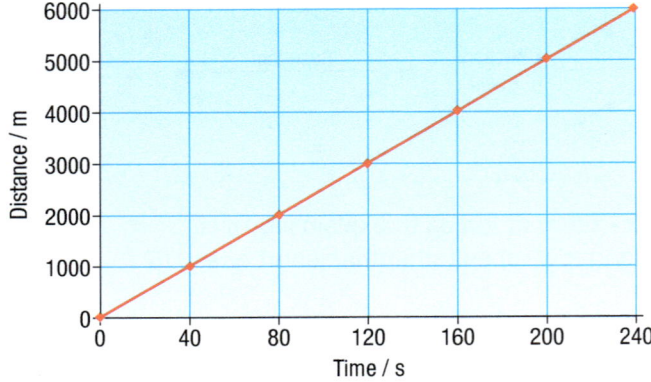

Figure 1 *A distance–time graph*

Look at the readings plotted on a graph of distance against time in Figure 1.

The graph shows that:

- the car took 40 s to go from each marker post to the next. So its speed was **constant** (or uniform).
- the car went a distance of 25 metres every second (= 1000 metres ÷ 40 seconds). So its speed was 25 metres per second.

If the car had travelled faster, it would have gone further than 1000 metres every 40 seconds. So the line on the graph would have been **steeper**. In other words, the **gradient** of the line would have been greater.

The gradient of a line on a distance–time graph represents speed.

Equation for constant speed

For an object moving at **constant speed**, you can calculate its speed using the equation:

$$\textbf{speed in (metres per second, m/s)} = \frac{\textbf{distance travelled (metres, m)}}{\textbf{time taken (seconds, s)}}$$

The scientific unit of speed is the metre per second, usually written as metre/second or m/s.

Average speed

The equation above can also be used to calculate the **average speed** of an object whose speed varies. For example, if a motorist in a traffic queue took 50 s to travel a distance of 300 m, the car's average speed was 6.0 m/s (= 300 m ÷ 50 s).

Speed in action

Long-distance vehicles are fitted with recorders called **tachographs.** These can check that their drivers don't drive for too long. Look at the distance–time graphs in Figure 3 for three lorries, X, Y, and Z, on the same motorway.

- X went fastest – it travelled furthest in the same time.
- Y travelled more slowly than X. From the graph, you can see it travelled 30 000 metres in 1250 seconds. So its average speed was:

  distance ÷ time = 30 000 m ÷ 1250 s = 24 m/s.

- Z travelled the shortest distance. It stopped for some of the time. Its speed was zero during this time. When it was moving its speed was also less than that of X or Y.

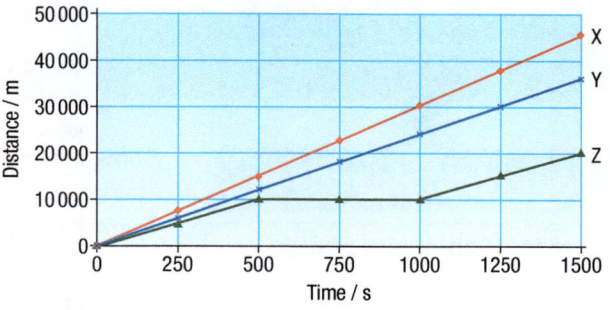

Figure 3 *Comparing distance–time graphs*

Practical

Be a distance recorder!

Take measurements to plot distance–time graphs for a person:

- walking
- running
- riding a bike.

Remember to label the graph axes and include the correct units.

- Work out the average speeds.

Figure 4 *Measuring distance*

Summary questions

1. **a** For an object travelling at constant speed:
   - **i** What can you say about the distance it travels every second?
   - **ii** What can you say about the gradient of its distance–time graph?
   b Look at the distance–time graphs in Figure 3.
   - **i** Calculate the speed of X.
   - **ii** How long did Z stop for?
   - **iii** Calculate the **average** speed of Z, using the total distance Z travels in its journey.

2. A vehicle on a motorway travels 1800 m in 60 seconds. Calculate:
   a the average speed of the vehicle in m/s
   b how far it would travel if it travelled at this speed for 300 seconds
   c how long it would take to travel a distance of 3300 m at this speed.

3. A car on a motorway travels a certain distance d in 6 minutes at a speed of 21 m/s. A coach takes 7 minutes to travel the same distance. Calculate the distance d and the speed of the coach.

4. A train takes 2 hours and 40 minutes to travel a distance of 360 kilometres.
   a Calculate the average speed of the train in metres per second on this journey.
   b The train travelled at a constant speed of 40 m/s for a distance of 180 km. Calculate the time taken in minutes for this section of the journey.

Maths skills

Rearranging the speed formula

The word equation for speed may be written as:

$$v = \frac{s}{t}$$

where v is the speed, s is the distance travelled, and t is the time taken. If two of the three quantities are known, the third can be found by using the equation above, or rearranging it to give $s = vt$ or $t = \frac{s}{v}$.

Study tip

Always convert time into seconds in these calculations if it is given in minutes or hours.

Key points

- The distance–time graph for any object that is
  - stationary is a horizontal line
  - moving at constant speed is a straight line that slopes upwards.

- The gradient of a distance–time graph for an object represents the object's speed.

- Speed (m/s) = $\dfrac{\text{distance travelled (m)}}{\text{time taken (s)}}$.

- Rearrange the equation to give distance = speed × time or

  time = $\dfrac{\text{distance}}{\text{speed}}$.

2.2 Velocity and acceleration

Learning objectives

After this topic, you should know:

- the difference between speed and velocity

- the difference between vectors and scalars

- how to calculate the acceleration of an object

- the difference between acceleration and deceleration.

Some fairground rides spin you round and round. Your direction of motion keeps changing. The word **velocity** means speed in a given direction. An exciting ride would be one that changes your velocity often and unexpectedly!

Velocity is speed in a given direction.

- Two moving objects can have the same speed but different velocities. For example, a car travelling north at 30 m/s on a motorway has the same speed as a car travelling south at 30 m/s. But their velocities are not the same because they are moving in opposite directions.

- An object moving steadily round in a circle has a constant speed. It moves in a circular motion. Its direction of motion changes continuously as it goes round so its velocity is not constant. For example,

  - The Earth moves round the Sun at a constant speed but its direction of motion continually changes so its velocity is continually changing even though its speed is constant.

  - A car travelling round a roundabout at constant speed has a continually changing velocity. This is because the direction in which it is moving is continually changing.

An object that travels at constant velocity travels at a constant speed without changing its direction. It therefore travels in a straight line in a certain direction. The word **displacement** means the distance moved in a certain direction.

Displacement is distance in a certain direction.

For example, the displacement of a car that travels 20 km on a straight motorway due north is 20 km due north. A car travelling the same distance in the opposite direction would have a displacement of 20 km due south.

Vectors and scalars

Physical quantities that are directional are called **vectors**. In Topic 1.3, you met force as a vector. Other examples of vectors in this book include displacement, acceleration, force, momentum, weight, and gravitational field strength.

Physical quantities that are not directional are called **scalars**. Examples include speed, distance, time, mass, energy, and power.

A vector has **magnitude** (i.e., size) and a direction. A scalar has magnitude only.

Acceleration

A car maker claims their new car 'accelerates more quickly than any other new car'. A rival car maker is not pleased by this claim and issues a challenge. Each car in turn is tested on a straight track with a velocity recorder fitted.

The results are shown in Table 1.

Table 1 *Acceleration*

Time from a standing start (s)	0	2	4	6	8	10
Velocity of car X (m/s)	0	5	10	15	20	25
Velocity of car Y (m/s)	0	6	12	18	18	18

Which car has a greater **acceleration**? The results are plotted on the velocity–time graph in Figure 2. You can see the velocity of Y goes up from zero faster than the velocity of X does. So Y accelerates more in the first 6 seconds then its acceleration is zero.

Figure 1 *You experience plenty of changes in velocity on a corkscrew ride!*

Figure 2 *Velocity–time graph*

The acceleration of an object is its change of velocity per second. The unit of acceleration is the metre per second squared, usually written as m/s².

Any object with a changing velocity is accelerating. You can work out its acceleration using the equation:

$$\text{acceleration (metres per second squared, m/s}^2) = \frac{\text{change in velocity (m/s)}}{\text{time taken for the change (s)}}$$

For an object that accelerates steadily from an initial velocity u to a final velocity v,

its change of velocity = final velocity – initial velocity = $v - u$.

Therefore, you can write the equation for acceleration as:

$$\text{acceleration, } a = \frac{v - u}{t}$$

Worked example

In Figure 2, the velocity of Y increases from 0 to 18 m/s in 6 seconds. Calculate its acceleration.

Solution

Change of velocity = $v - u$ = 18 m/s – 0 m/s = 18 m/s
Time taken, t = 6 s

$$\text{Acceleration, } a = \frac{\text{change in velocity (m/s)}}{\text{time taken for the change (s)}}$$

$$= \frac{v - u}{t} = \frac{18 \text{ m/s}}{6 \text{ s}} = \textbf{3 m/s}^2$$

Deceleration

A car decelerates when the driver brakes. The term **deceleration** or **negative acceleration** describes a situation where an object slows down.

Maths skills

You can write the word equation for acceleration in symbols as:
$a = \dfrac{\Delta v}{t}$,
where a is the average acceleration, Δv is the change of velocity in metres per second, and t is the time taken in seconds.

For constant acceleration a, the change of velocity
$\Delta v = v - u,$
where u is the initial velocity and v is the final velocity.
Therefore: $a = \dfrac{(v - u)}{t}$,
for constant (i.e., steady) acceleration.

Note: Speed and velocity may also be described in kilometres per hour (km/h).

As 1000 m = 1 km and 3600 s = 1 hour, then a speed of 1 km/h is equal to 1000 m ÷ 3600 s = 0.278 m/s.

Summary questions

1 **a** What is the difference between speed and velocity?
   b A car on a motorway is travelling at a constant speed of 30 m/s when it overtakes a lorry travelling at a speed of 22 m/s. If both vehicles maintain their speeds, how far ahead of the lorry will the car be after 300 s?

2 The velocity of a car increased from 8 m/s to 28 m/s in 16 s without change of direction. Calculate its acceleration.

3 The driver of a car increased the speed of the car as it joined the motorway. It then travelled at constant velocity before slowing down as it left the motorway at the next junction.
   a i When did the car decelerate?
      ii When was the acceleration of the car zero?
   b When the car joined the motorway, it accelerated from a speed of 7.0 m/s for 10 s at an acceleration of 2.0 m/s². Calculate its speed at the end of this time.

4 A sprinter in a 100 m race accelerated from rest and reached a speed of 9.2 m/s in the first 3.1 seconds.
   a Calculate the acceleration of the sprinter in this time.
   b The sprinter continued to accelerate to top speed and completed the race in 10.4 s. Calculate the sprinter's average speed.

Key points

- Velocity is speed in a given direction.
- A vector is a physical quantity that has a direction as well as a magnitude.
- A scalar is a physical quantity that has a magnitude only and does not have a direction.
- Displacement is distance in a given direction.
- Acceleration is change of velocity per second. The unit of acceleration is the metre per second squared (m/s²).
- Acceleration = change of velocity ÷ time taken.
- Deceleration is the change of velocity per second when an object slows down.

2.3 More about velocity–time graphs

Learning objectives

After this topic, you should know:

- what a horizontal line on a velocity–time graph shows

- how to tell from a velocity–time graph if an object is accelerating or decelerating

- what the area under a velocity–time graph shows

- how to measure velocity changes.

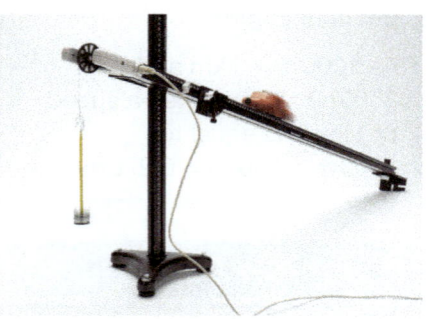

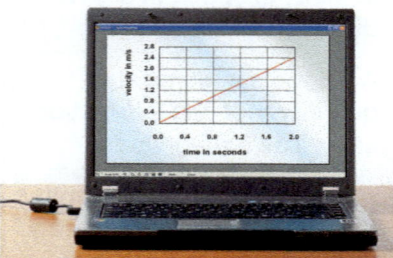

Figure 2 *Measuring motion using a computer*

⚲ links

For more information on variables and relationships between them, see 'Experimental data handling' on Pages 232–7.

Investigating acceleration

You can use a motion sensor linked to a computer to record how the velocity of an object changes. Figure 1 shows how you can do this, using a trolley as the moving object. The computer can also display the measurements as a velocity–time graph.

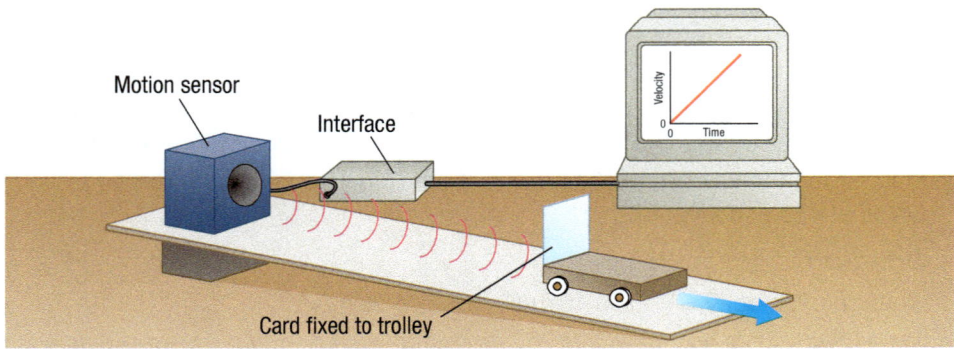

Motion sensor
Interface
Velocity
0 Time
Card fixed to trolley

Figure 1 *A velocity–time graph on a computer*

Test A: If you let the trolley accelerate down the runway, its velocity increases with time. Look at the velocity–time graph from a test run in Figure 1.
- The line goes up because the velocity increases with time. It shows that the trolley was accelerating as it ran down the runway.
- The line is straight, which tells us that the increase in velocity was the same every second. In other words, the acceleration of the trolley was constant.

Test B: If you make the runway steeper, the trolley accelerates faster. This would make the line on the graph in Figure 1 steeper than for test A. The acceleration in test B would be greater.

The tests show that:

the gradient of the line on a velocity–time graph represents acceleration.

Practical

Investigating acceleration

Use a motion sensor and a computer to find out how the slope of a runway affects a trolley's acceleration.
- In this investigation, name:
  i   the independent variable
  ii the dependent variable.
- What relationship do you find between the variables?

Safety: Use foam or an empty cardboard box to stop the trolley falling off the bench. Protect the bench and your feet from falling objects.

Braking

Braking reduces the velocity of a vehicle. Figure 3 shows the velocity–time graph for a vehicle that brakes and stops at a set of traffic lights. The velocity is constant until the driver applies the brakes.

Using the gradient of the line:

- The section of the graph showing constant velocity is horizontal. The gradient of the line is zero, so the acceleration in this section is zero.
- When the brakes are applied, the vehicle decelerates and its velocity decreases to zero. The gradient of the line is negative in this section. So the acceleration is negative.

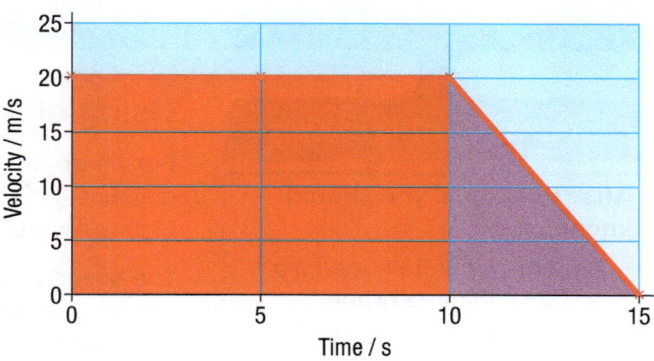

Figure 3 *Braking*

Look at Figure 3 again.

Using the area under the line:

- Before the brakes are applied, the vehicle moves at a velocity of 20 m/s for 10 s. It therefore travels 200 m in this time (= 20 m/s × 10 s). This distance is represented on the graph by the area under the line from 0 s to 10 s. This is the rectangle shaded red on the graph.
- When the vehicle decelerates in Figure 3, its velocity drops from 20 m/s to 0 m/s in 5 s. You can work out the distance travelled in this time from the area of the purple triangle. This area is ½ × the height × the base of the triangle. So, the vehicle travelled a distance of 50 m when it was decelerating.

The area under the line on a velocity–time graph represents distance travelled.

Summary questions

1 Match each of the following descriptions to one of the lines, labelled **A**, **B**, **C**, and **D**, on the velocity–time graph.
   a accelerated motion throughout
   b zero acceleration
   c accelerated motion, then decelerated motion
   d deceleration

2 Look at the graph in Question **1**.
   a Which line represents the object that travelled:
      i the longest distance?
      ii the shortest distance?
   b Which object, **B** or **D**, travelled further?

3 Look again at the graph in Question **1**.
   a Show that the object that produced the data for line **A** (the horizontal line) travelled a distance of 160 m.
   b Which one of the other three lines represents the motion of an object that decelerated throughout its journey?

4 **a** Calculate the distance travelled by object **C** in Question **1**.
   b Calculate the difference in the distances travelled by **A** and **D**.

Study tip

If you are drawing a straight-line graph, always use a ruler.

Key points

- If a velocity–time graph is a horizontal line, the acceleration is zero.

- The gradient of the line on a velocity–time graph represents acceleration.

- The area under the line on a velocity–time graph represents distance travelled.

- A motion sensor linked to a computer can be used to measure velocity changes.

2.4 Using graphs

Learning objectives

After this topic, you should know:

- how to calculate speed from a distance–time graph
  - where the speed is constant
  - where the speed is changing
- how to calculate acceleration from a velocity–time graph
- how to calculate distance from a velocity–time graph.

Using distance–time graphs

For an object moving at constant speed, you saw in Topic 2.1 that the distance–time graph is a straight line sloping upwards, as shown in Figure 1 in the margin.

The speed of the object is represented by the gradient of the line. To find the gradient, you can draw a triangle under the line. The height of the triangle represents the distance travelled and the base represents the time taken. So:

$$\text{the gradient of the line} = \frac{\text{the height of the triangle}}{\text{the base of the triangle}}$$

and this represents the object's speed.

For a moving object with a changing speed, the distance–time graph is not a straight line. The red line in Figure 2 shows an example.

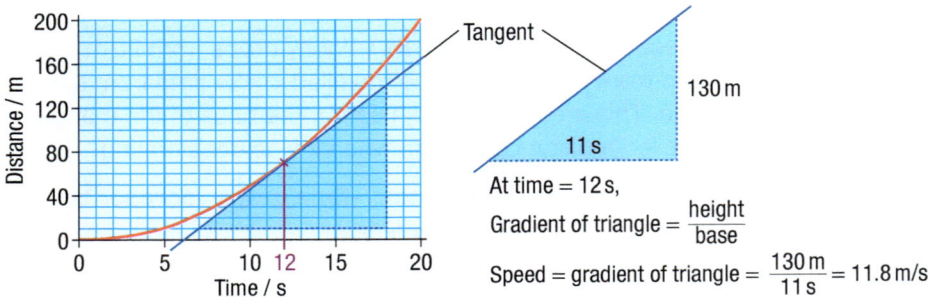

At time = 12 s,

Gradient of triangle = $\frac{\text{height}}{\text{base}}$

Speed = gradient of triangle = $\frac{130\,\text{m}}{11\,\text{s}}$ = 11.8 m/s

Figure 2 *A distance–time graph for changing speed*

In Figure 2, the gradient of the line increases gradually, so the object's speed must have increased gradually. You can find the speed at any point on the line by drawing a tangent to the line at that point, as shown in Figure 2. The **tangent** to the curve is a straight line that touches the curve at a single point without cutting through it. The gradient of the tangent is equal to the speed at that point.

Using velocity–time graphs

Look at the graph in Figure 3. It shows the velocity–time graph of an object X moving with a constant acceleration. Its velocity increases at a steady rate. So the graph shows a straight line that has a constant gradient.

To find the acceleration from the graph, remember the gradient of the line on a velocity–time graph represents the acceleration.

In Figure 3, the gradient is given by the height divided by the base of the triangle under the line.

The height of the triangle represents the change of velocity and the base of the triangle represents the time taken for this change.

Therefore, the gradient represents the acceleration, because:

$$\text{acceleration} = \frac{\text{change of velocity}}{\text{time taken}}$$

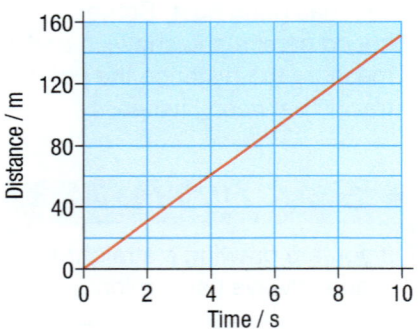

Figure 1 *A distance–time graph for constant speed*

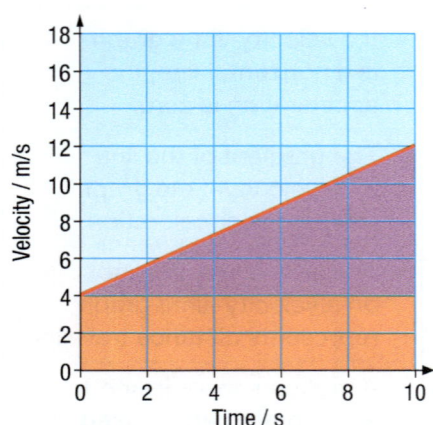

Figure 3 *A velocity–time graph for constant acceleration*

Worked example

Use the graph in Figure 3 to find the acceleration of object X.

Solution

The height of the triangle represents an increase of velocity of 8 m/s (= 12 m/s – 4 m/s).

The base of the triangle represents a time of 10 s.

Therefore, the acceleration = $\dfrac{\text{change of velocity}}{\text{time taken}}$

$= \dfrac{8\,\text{m/s}}{10\,\text{s}} = \mathbf{0.8\,\text{m/s}^2}$

To find the distance travelled from the graph, remember the area under a line on a velocity–time graph represents the distance travelled. The shape under the line in Figure 3 is a triangle on top of a rectangle. So the distance travelled is represented by the area of the triangle plus the area of the rectangle under it. Prove for yourself that the triangle represents a distance travelled of 40 m and the rectangle also represents a distance of 40 m. Then look at the worked example in the margin.

Worked example

Use the graph in Figure 3 to calculate the distance moved by object X.

Solution

The area of the purple triangle = ½ × height × base.

Therefore, the distance represented by the area of the triangle = ½ × 8 m/s × 10 s
= 40 m

The area of the red rectangle = height × base

Therefore, the distance represented by the area of the rectangle = 4 m/s × 10 s = 40 m

So the distance travelled by X = 40 m + 40 m = **80 m**

Summary questions

1 **a** Find the speed of the object in the graph in Figure 1.
   b **i** What does the gradient of the line at the origin of the graph in Figure 2 tell you about the speed at time = 0?
   ii What can you say about the speed in the graph of Figure 2?

2 The graph opposite shows how the velocity of a cyclist on a straight road changes with time.
   a Describe the motion of the cyclist.
   b Use the graph to work out the acceleration of the cyclist and the distance travelled in:
   i the first 40 s
   ii the next 20 s.
   c Calculate the average speed of the cyclist over the journey.

3 In a motorcycle test, the speed from rest was recorded at intervals.

Time (s)	0	5	10	15	20	25	30
Velocity (m/s)	0	10	20	30	40	40	40

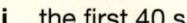

   a Plot a velocity–time graph of these results.
   b What was the initial acceleration?
   c How far did the motorcycle move in:
   i the first 20 s?
   ii the next 10 s?

Key points

- The speed of an object is given by the gradient of the line on its distance–time graph if the speed is constant or the gradient of the tangent to the line if the speed is changing.

- The acceleration of an object is given by the gradient of the line on its velocity–time graph.

- The distance travelled by an object is given by the area under the line of its velocity–time graph.

2.5 Forces and acceleration

After this topic, you should know:

- how the acceleration of an object depends on the size of the resultant force

- the effect that the mass of an object has on its acceleration

- how to calculate the resultant force on an object from its acceleration and its mass.

Practical

Investigating force and acceleration

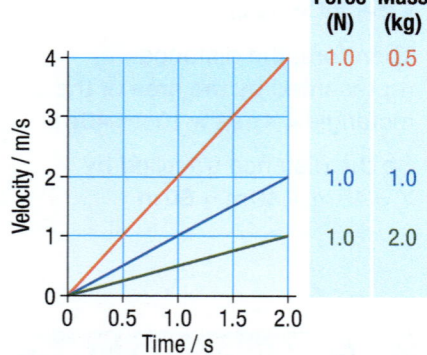

Figure 1 *Investigating the link between force and motion*

You can use the apparatus above to accelerate a trolley with a constant force.

Use the newtonmeter to pull the trolley along with a constant force.

You can double or treble the total moving mass by using double-deck and triple-deck trolleys.

A motion sensor and a computer can be used to record the velocity of the trolley as it accelerates.

- What are the advantages of using a data logger and computer in this investigation?

Safety: Protect the bench and your feet from falling trolleys.

You can display the results as a velocity–time graph on the computer screen.

Figure 2 shows velocity–time graphs for different masses. You can work out the acceleration from the gradient of the line, as explained in Topic 2.4 'Using graphs'.

Look at some typical results in Table 1.

Table 1 *Investigating force and acceleration*

Resultant force (N)	0.5	1.0	1.5	2.0	4.0	6.0
Mass (kg)	1.0	1.0	1.0	2.0	2.0	2.0
Acceleration (m/s²)	0.5	1.0	1.5	1.0	2.0	3.0
Mass × acceleration (kg m/s²)	0.5	1.0	1.5	2.0	4.0	6.0

The results show that the resultant force, the mass, and the acceleration are linked by the equation:

$$\text{resultant force} = \text{mass} \times \text{acceleration}$$
$$\text{(N)} \qquad \text{(kg)} \qquad \text{(m/s}^2\text{)}$$

You can write the word equation above using symbols as follows:

$$F = m \times a,$$

where:

F = resultant force in N
m = mass in kg
a = acceleration in m/s².

Figure 2 *Velocity–time graphs for the same force acting on different masses*

Force (N)	Mass (kg)
1.0	0.5
1.0	1.0
1.0	2.0

You can rearrange the equation $F = m \times a$ to give

$$a = \frac{F}{m} \text{ or } m = \frac{F}{a}.$$

Newton's Second Law

Newton's Second Law of motion says that the acceleration of an object is:

- proportional to the resultant force on the object
- inversely proportional to the mass of the object.

So the acceleration a is proportional to the resultant force F divided by the object's mass m. You can write this as the equation $a = \dfrac{F}{m}$ where the force is in newtons, the mass is in kilograms, and the acceleration is in metres per second squared (m/s²).

You can see from this equation that 1 N is the force that gives a 1 kg mass an acceleration of 1 m/s².

Worked example

Calculate the acceleration of an object of mass 5 kg acted on by a resultant force of 40 N.

Solution

Rearranging $F = m \times a$ gives $a = \dfrac{F}{m} = \dfrac{40\,\text{N}}{5\,\text{kg}} = \mathbf{8\,m/s^2}$

Speeding up or slowing down

If the velocity of an object changes, it must be acted on by a resultant force. Its acceleration is always in the same direction as the resultant force.

- The velocity of the object increases if the resultant force is in the **same** direction as the velocity. Its acceleration is described as positive because it is in the same direction as its velocity.
- The velocity of the object decreases (i.e., it decelerates) if the resultant force is **opposite** in direction to its velocity. Its acceleration is described as negative because it is opposite in direction to its velocity.

Summary questions

1 **a** Calculate the resultant force on a sprinter of mass 80 kg who accelerates at 8 m/s².
   b Calculate the acceleration of a car of mass 800 kg acted on by a resultant force of 3200 N.

2 Copy and complete the following table:

	a	b	c	d	e
Force (N)		200	840		5000
Mass (kg)	20		70	0.40	
Acceleration (m/s²)	0.80	5.0		6.0	0.20

3 A car and a trailer have a total mass of 1500 kg.
   a Find the force needed to accelerate the car and the trailer at 2.0 m/s².
   b The mass of the trailer is 300 kg. Calculate:
   i the force of the tow bar on the trailer
   ii the resultant force on the car.

If you're in a car that suddenly brakes, your neck pulls on your head and slows it down. The equal and opposite force of your head on your neck can injure your neck.

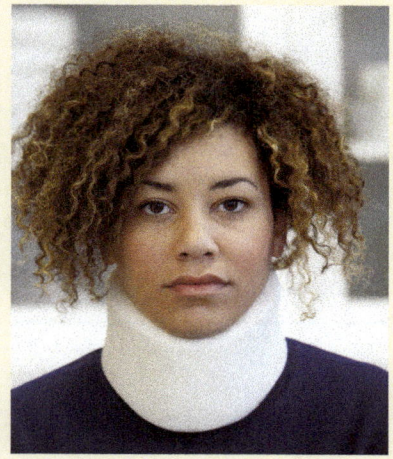

Figure 3 A 'whiplash' injury

Study tip

- If an object is accelerating, it can be speeding up or changing direction. If it is decelerating, it is slowing down.
- If an object is accelerating or decelerating, there must be a resultant force acting on it.

Key points

- The bigger the resultant force on an object, the greater the object's acceleration.
- The greater the mass of an object, the smaller its acceleration for a given force.
- Resultant force (N) = mass (kg) × acceleration (m/s²).

2.6

Forces and braking

Learning objectives

After this topic, you should know:

- the forces that oppose the driving force of a vehicle

- what the stopping distance of a vehicle depends on

- the factors that can increase the stopping distance of a vehicle.

Did you know ...?

The mass of a BMW Mini Cooper car is just over 1000 kg.

Did you know ...?

When the brakes of a car are applied, friction between the brake pads and the car wheels causes kinetic energy to be transferred from the vehicle by heating the brakes and the brake pads. If the brake pads wear away too much, they need to be replaced.

Practical

Reaction times

Use an electronic stopwatch to test your own reaction time. Ask a friend to start the stopwatch when you are looking at it with your finger on the stop button. The read-out from the watch will give your reaction time.

- How can you make your data as precise as possible?

- What conclusions can you draw?

Forces on the road

For any car travelling at constant velocity, the resultant force on it is zero. This is because the **driving force** of its engine is balanced by the **resistive forces**, which are mostly caused by air resistance. Friction between parts of the car that move against each other also contributes to the resistive forces.

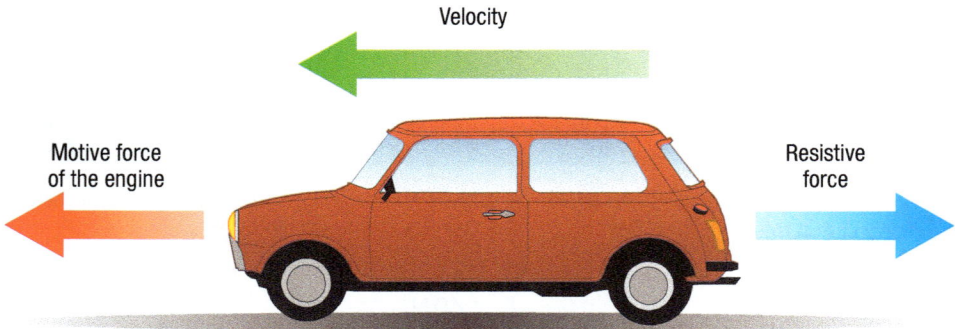

Figure 1 *Constant velocity*

A car driver uses the accelerator pedal (also called the gas pedal) to vary the driving force of the engine.

The **braking force needed to stop a vehicle** in a certain distance depends on:

- the speed of the vehicle when the brakes are first applied
- the mass of the vehicle.

You can see this using the equation: resultant force = mass × acceleration, in which the braking force is the resultant force.

1 The greater the speed, the greater the deceleration needed to stop the vehicle within a certain distance. So, the braking force must be greater than it would be at lower speeds.

2 The greater the mass, the greater the braking force needed for a given deceleration.

Stopping distances

Driving tests always ask about **stopping distances**. This is the shortest distance a vehicle can safely stop in, and is in two parts:

The **thinking distance**: the distance travelled by the vehicle in the time it takes the driver to react (i.e., during the driver's reaction time).

The **braking distance**: the distance travelled by the vehicle during the time the braking force acts.

stopping distance = thinking distance + braking distance

Figure 2 shows the stopping distance for a vehicle on a dry flat road travelling at different speeds. Check for yourself that the stopping distance at 31 m/s (70 miles per hour) is 96 m.

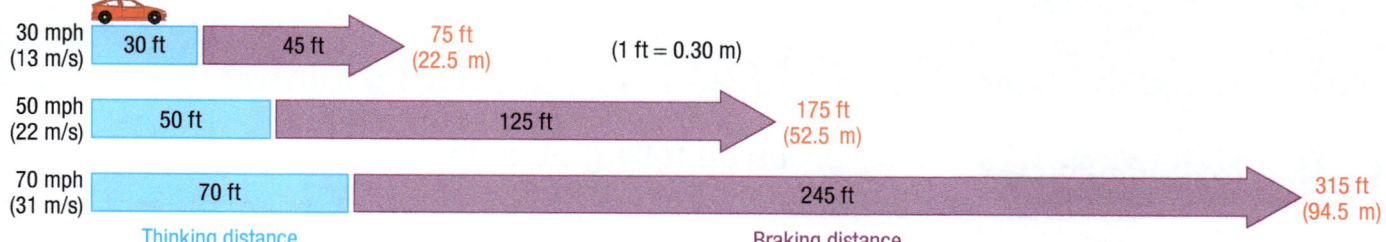

30 mph (13 m/s)	30 ft	45 ft		75 ft (22.5 m)	(1 ft = 0.30 m)
50 mph (22 m/s)	50 ft	125 ft		175 ft (52.5 m)	
70 mph (31 m/s)	70 ft	245 ft		315 ft (94.5 m)	

Thinking distance   Braking distance

Figure 2 *Stopping distances*

Factors affecting stopping distances

1 **Tiredness, alcohol, and drugs** all increase reaction times. Distractions such as using a mobile phone can also increase reaction time. All these factors increase the thinking distance (because thinking distance = speed × reaction time). Therefore, the stopping distance is greater.

2 **The faster a vehicle is travelling**, the further it travels before it stops. This is because the thinking distance and the braking distance both increase with increased speed.

3 **In adverse road conditions**, for example on wet or icy roads, drivers have to brake with less force to avoid skidding. Stopping distances are therefore greater in poor road conditions.

4 **Poorly maintained vehicles**, for example with worn brakes or tyres, take longer to stop because the brakes and tyres are less effective.

Figure 3 *Stopping distances are further than you might think!*

Summary questions

1 For each of the following factors, which distance is affected: the thinking distance or the braking distance of a vehicle?
   a the road surface
   b the tiredness of the driver
   c poorly maintained brakes

2 a Use the chart in Figure 2 to work out, in metres, what is the effect of the increase from 13 m/s (30 mph) to 22 m/s (50 mph) on the following:
      i   the thinking distance
      ii   the braking distance
      iii the stopping distance.
   b A driver has a reaction time of 0.8 s. Calculate the change in her thinking distance if she travels at 15 m/s instead of 30 m/s.

3 a When the speed of a car is doubled:
      i   explain why the thinking distance of the driver is doubled, assuming that the driver's reaction time is unchanged
      ii   explain why the braking distance is more than doubled.
   b A student states that braking distance is proportional to the square of the speed. Use the chart in Figure 2 to decide whether this is a valid claim.

2.7 Forces and terminal velocity

Learning objectives

After this topic, you should know:

- about the motion of a falling object acted on only by gravity

- what terminal velocity means

- what can be said about the resultant force acting on an object that is falling at terminal velocity.

The forces on falling objects

If you release an object above the ground, it falls because of its weight (i.e., the gravitational force on it).

If the object falls with no other forces acting on it, the resultant force on it is its weight. It accelerates downwards at a constant acceleration of $10 \, \text{m/s}^2$. This is called the acceleration due to gravity, or the acceleration of free fall. For example, if you release an object of mass 1 kg above the ground:

- the gravitational force on it is 10 N (see Topic 1.4 'Mass and weight')

- its acceleration $\left(= \dfrac{\text{force}}{\text{mass}} = \dfrac{10 \, \text{N}}{1 \, \text{kg}} \right) = 10 \, \text{m/s}^2$.

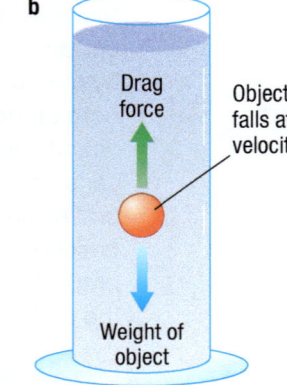

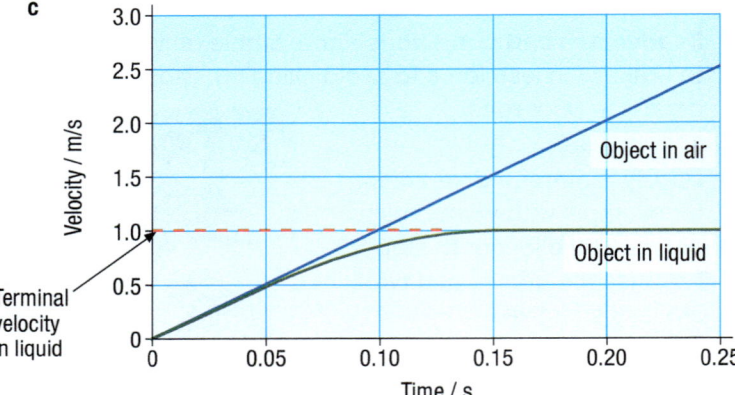

Figure 1 *Falling objects* **a** *Falling in air* **b** *Falling in a liquid* **c** *Velocity–time graph for* **a** *and* **b**

If the object falls in a fluid (a liquid or a gas), the fluid drags on the object because of friction between the fluid and the surface of the moving object. This frictional force or drag force increases with speed. At any instant, the resultant force on the object is its weight minus the frictional force on it.

- The acceleration of the object decreases as it falls. This is because the frictional force increases as it speeds up. So the resultant force on it decreases and therefore its acceleration decreases.

- The object reaches a constant velocity when the frictional force on it is equal and opposite to its weight. This velocity is its **terminal velocity**. The resultant force is then zero, so its acceleration is zero.

The frictional force is higher in a liquid than in a gas such as air. When an object moves through the air, the frictional force is called **air resistance**. This is not shown in Figure 1a because air resistance is very small in a short descent. In Figure 1b, the drag force is much greater in the liquid than for an object of the same weight falling through air.

Velocity–time graphs for a falling object

Figure 1c shows how the velocity of an object changes with time with and without drag. As explained in Topic 2.3, the gradient of a velocity–time graph gives acceleration.

- Without drag, the object falls freely and its velocity increases at a constant rate. The velocity–time graph is a straight line. The gradient of the line is positive and is equal to the acceleration of free fall, g, which is $10 \, \text{m/s}^2$.

Practical

Investigating falling

Release an object with and without a parachute.

Make suitable measurements to compare the two situations.

- Why does the object fall at constant speed when the parachute is open?

- Evaluate the quality of the data you collected. How could you improve your data?

Safety: Do not climb on stools and benches. Find a safe area to release parachutes.

Figure 2 *Using a parachute*

- With drag, the velocity increases at a decreasing rate. The velocity–time graph is a curve which has a gradient that decreases with time.
  - The initial gradient is equal to g. This is because its initial velocity is zero so the initial drag force is zero. So the resultant force is initially due to gravity only.
  - The gradient decreases because as the velocity increases, the drag force increases from zero. So the resultant force and therefore the acceleration decreases.
  - The gradient gradually decreases to zero so the velocity becomes constant (equal to the terminal velocity). As explained above, the drag force is equal to the weight of the object when it is moving at its terminal velocity.

Drag forces and streamlining

The drag force on an object moving through a fluid is due to friction between the fluid and the surface of the object it flows past. The drag force opposes the motion of the object. The object reaches its terminal speed when the drag force is equal and opposite to the force that makes the object move forwards.

Streamlining by design

The maximum speed of a powered vehicle is its speed when the drag force acting on it is equal and opposite to the driving force of the vehicle engine. This is its terminal speed. The maximum speed of a vehicle can be increased by streamlining its shape. This reduces the drag force on it, enabling it to reach a greater terminal speed.

Most vehicles are designed with streamlined shapes in order to increase their fuel efficiency. Because streamlining reduces the drag force at any speed, the engine force to maintain a constant speed is reduced, so less fuel is used over any given distance.

Figure 3 *Streamlining by design*

Figure 4 *A shark can move very fast through water because of its streamlined shape, which has developed naturally through evolution and adaptation. Its top speed is its terminal speed which is reached when the drag force on it is equal and opposite to the force of its muscles propelling it forwards*

Summary questions

1 When an object is released in a fluid, what can be said about:
  a the resultant force on it initially?
  b the weight of the object and the frictional force on it before it reaches its terminal velocity?
  c its acceleration after it reaches its terminal velocity?
  d the resultant force on it when it moves at its terminal velocity?

2 Look at Figure 1c.
  a Write down the acceleration of the object 0.20 s after being released:
    i in air   ii in the liquid.
  b Describe and explain how the acceleration of the object changes with time.

3 A parachutist of mass 70 kg supported by a parachute of mass 20 kg reaches a constant speed.
  a Explain why the parachutist reaches a constant speed.
  b Calculate:
    i the total weight of the parachutist and the parachute
    ii the size and direction of the force of air resistance on the parachute when the parachutist falls at constant speed.

4 a Use Figure 1c to determine the acceleration of the object in liquid 0.10 s after it was released.
  b Show that the ratio of the drag force to the weight at 0.10 s is about 0.5.

Key points

- The terminal velocity of an object is the velocity it eventually reaches when it is falling in a fluid. The weight of the object is then equal to the frictional force on the object.

- When an object is moving at terminal velocity, the resultant force on it is zero.

Chapter summary questions

1 A model car travels round a circular track at constant speed.

If you were given a stopwatch, a marker, and a tape measure, how would you measure the speed of the car?

2 The figure shows a distance–time graph for a motorcycle approaching a speed limit sign.

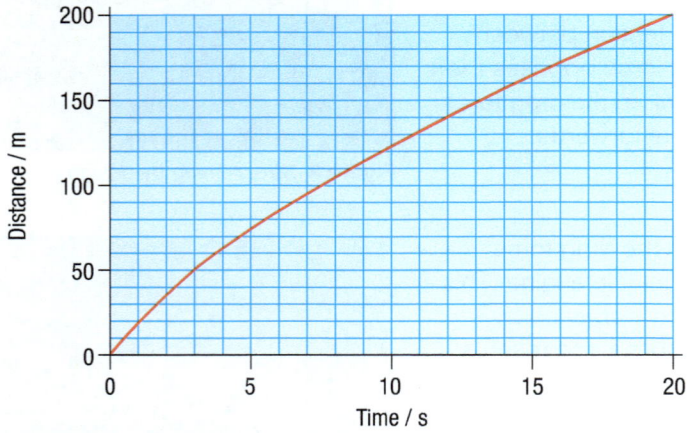

a Describe how the speed of the motorcycle changed with time.

b Use the graph to determine the speed of the motorcycle:
  i initially
  ii 10 seconds later.

3 The table shows how the velocity of a train changed as it travelled from one station to the next.

Time (s)	0	20	40	60	80	100	120	140	160
Velocity (m/s)	0	5	10	15	20	20	20	10	0

a Plot a velocity–time graph using this data.

b Calculate the acceleration in each of the three parts of the journey.

c Calculate the total distance travelled by the train.

d Show that the average speed for the train's journey was 12.5 m/s.

4 A car accelerates from rest with an initial acceleration of 1.2 m/s². The total mass of the car and its occupants is 800 kg.

a Assuming the resultant force is initially due to the driving force of the car engine only, calculate the initial driving force on the car.

b i If the car was used to pull a trailer of mass 70 kg, what would be its initial acceleration for the same driving force?
  ii Calculate the force on the car and trailer at this acceleration.

5
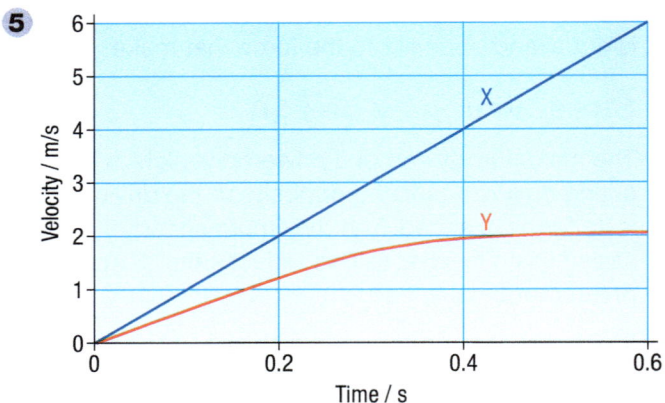

a The gravitational field strength *g* at the surface of the Earth is 10 N/kg. Explain why a freely falling object released near the Earth's surface has a constant acceleration of 10 m/s² as it falls.

b A stone dropped from the top of a water well hits the water in the well 1.7 s later. Calculate the speed of the object just before it hits the surface of the water.

c The figure above shows the velocity–time graphs for a metal object **X** dropped in air and a similar object **Y** dropped in a tank of water.
  i What does the graph for **X** tell you about its acceleration?
  ii In terms of the forces acting on **Y**, explain why it reached a constant velocity.

6 a A racing cyclist accelerates at 5.0 m/s² when she starts from rest. The total mass of the cyclist and her bicycle is 45 kg. Calculate:
  i the resultant force that produces this acceleration
  ii the total weight of the cyclist and the bicycle.

b Explain why she can reach a higher speed by crouching than by staying upright.

c When the cyclist was travelling at a velocity of 6.6 m/s on a level road, she used her brakes to slow down and stop in a time of 3.7s. Calculate the braking force acting during this time.

Practice questions

1 A van has a fault and leaks one drop of oil every second.

The diagram shows the oil drops left on the road as the van moves from **A** to **D**.

●   ●   ●   ●●●●●   ●   ●   ●
A   **B**   **C**   **D**

a Describe the motion of the van as it moves from:
  i   **A** to **B**
  ii   **B** to **C**
  iii   **C** to **D**.   (3)

b The distance from **B** to **C** is 100 metres.

Calculate the average speed of the van between **B** and **C**.   (3)

c Later in the journey, the van slows down from a speed of 25 m/s to 5 m/s in 10 s.

Calculate the acceleration of the van.   (5)

2 Graphs can give useful information.

a Use words from the list to complete the sentences.

accelerating   travelling at constant speed
stationary

  i   In a distance–time graph, a horizontal line shows that the vehicle was ....   (1)
  ii   In a velocity–time graph, a horizontal line shows that the vehicle was ....   (1)

b A car driver sees a dog on the road ahead and has to make an emergency stop.

The graph shows how the speed of the car changes with time after the driver first sees the dog.

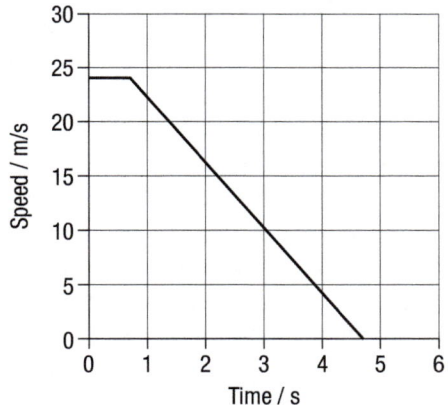

  i   What is the time interval between the driver seeing the dog and applying the brakes?   (1)
  ii   How far does the car travel between the driver seeing the dog and applying the brakes?   (3)
c i   Calculate the deceleration of the car, in m/s², after the brakes are applied.   (4)
  ii   How far does the car travel after the brakes are applied?   (3)

3 A headteacher wants the local council to put a 20 mph speed limit on the road outside the school.

She asks some students to carry out a survey of vehicles passing the school.

a She wants one group of students to investigate the average speed of vehicles passing the school.
  i   Describe how the students would obtain the data needed. Your description should include the equipment they would use.   (6)
  ii   Outline how they could make the result as accurate as possible.   (3)

b The headteacher wants another group of students to produce a graph showing the number of vehicles travelling at different speeds along the road.
  i   Which two of the following could the students use to display their results?
  Give reasons for your choice.   (5)

  bar chart   line graph   pie chart   scattergram

  ii   Discuss how long the students should spend collecting their data.   (2)

4 a Complete the following sentences about graphs.
  i   The gradient of a velocity–time graph represents ....   (1)
  ii   The gradient of a distance–time graph represents ....   (1)
  iii   Distance is represented by the area under a .... graph.   (1)

b The graphs describe the motion of two runners in a race.

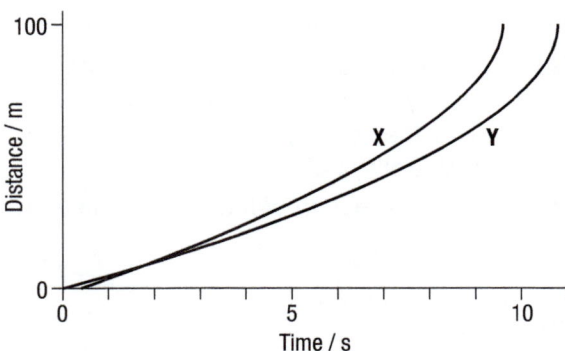

Compare and evaluate the information shown in the two graphs.   (6)

3.1 Momentum

Learning objectives

After this topic, you should know:

- how to calculate momentum
- the unit of momentum
- what happens to the total momentum of two objects when they collide.

Momentum is important to anyone who plays a contact sport. In a game of rugby, a player with a lot of momentum is very difficult to stop.

The momentum of a moving object = mass × velocity.

So momentum has a size and a direction.

The unit of momentum is the kilogram metre/second (kg m/s).

You can write the word equation above using symbols: $\boldsymbol{p = m \times v}$
where:
p = momentum in kg m/s
m = mass in kg
v = speed in m/s.

Figure 1 *A contact sport*

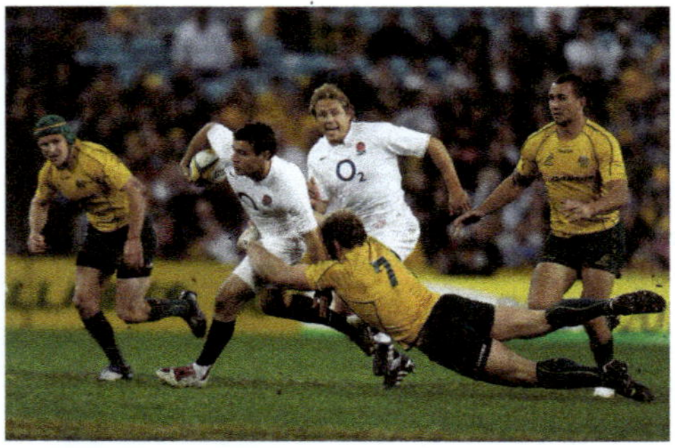

Worked example

Calculate the momentum of a sprinter of mass 50 kg running at a velocity of 10 m/s.

Solution

Momentum = mass × velocity
= 50 kg × 10 m/s
= **500 kg m/s**

Study tip

Make sure you know the unit of momentum – kg m/s.

Practical

Investigating collisions

When two objects collide, the momentum of each object changes. Figure 2 shows how to use a computer and a motion sensor to investigate a collision between two trolleys.

Trolley A is given a push so it collides with stationary trolley B. The two trolleys stick together after the collision. The computer gives the velocity of A before the collision and the velocity of both trolleys afterwards.

- What does each section of the velocity–time graph show?

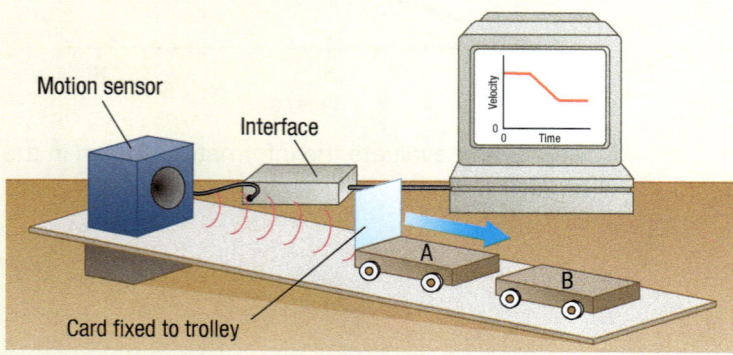

Figure 2 *Investigating collisions*

1. **For two trolleys of the same mass**, the velocity of trolley A is halved after the impact. The combined mass after the collision is twice the moving mass before the collision. So the momentum (= mass × velocity) after the collision is the same as before the collision.
2. **For a single moving trolley colliding with a stationary double trolley**, the velocity of A is reduced to one-third. The combined mass after the collision is three times the initial mass. So again, the momentum after the collision is the same as the momentum before the collision.

In both tests, the total momentum is unchanged (i.e., is conserved) by the collision. This is an example of the **conservation of momentum**. It applies to any system of objects as long as the system is **closed**, which means that the resultant force acting on the system is zero.

Safety: Use foam or an empty cardboard box to stop trolleys. Protect the bench and your feet from falling trolleys.

Figure 3 A 'shunt' collision

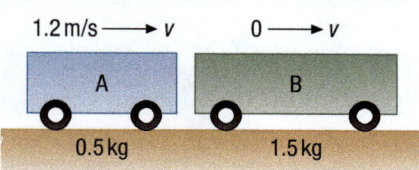

Figure 4 Worked example

In general, the **law of conservation of momentum** states that:

In a closed system, the total momentum before an event is equal to the total momentum after the event.

You can use this law to predict what happens whenever objects collide, or when they push each other apart in an explosion. Momentum is conserved in any collision or explosion as long as no external forces act on the objects.

Worked example

Trolley A of mass 0.5 kg is pushed at a velocity of 1.2 m/s into a stationary trolley B of mass 1.5 kg as shown in Figure 4. The two trolleys stick together after the impact.

Calculate:

a the momentum of trolley A before the collision

b the velocity of the two trolleys immediately after the impact.

Solution

a Momentum = mass × velocity = 0.5 kg × 1.2 m/s = **0.6 kg m/s**

b The momentum after the impact = the momentum before the impact
= 0.6 kg m/s

$(1.5 \text{ kg} + 0.5 \text{ kg}) \times$ velocity after the impact = 0.6 kg m/s

the velocity after the impact = $\dfrac{0.6 \text{ kg m/s}}{2 \text{ kg}}$ = **0.3 m/s**

Worked example

A truck of mass 3000 kg moving at a velocity of 16 m/s crashes into the back of a stationary car of mass 1000 kg. The two vehicles move together immediately after the impact. Calculate their velocity.

Solution

Let v represent the velocity of the vehicles after the impact.

momentum of truck before impact = 48 000 kg m/s

momentum of car before impact = 0 m/s

momentum of truck after impact = 3000 kg × v

momentum of car after impact = 1000 kg × v

$(3000 \text{ kg} \times v) + (1000 \text{ kg} \times v) = (48\,000 + 0) \text{ kg m/s}$

$v = \dfrac{48\,000 \text{ kg m/s}}{4000 \text{ kg}}$

= **12 m/s**

Summary questions

1 a Define momentum and state its unit.

  b Calculate the momentum of a person of mass 40 kg running at 6 m/s.

  c In the worked example above, calculate the speed after the collision if trolley A had a mass of 1.0 kg.

2 a Calculate the momentum of a rugby player of mass 80 kg running at a velocity of 5 m/s.

  b A car of mass 800 kg moves with the same momentum as the rugby player in **a**. Calculate the velocity of the car.

  c Calculate the velocity of a ball of mass 0.40 kg that has the same momentum as the rugby player in **a**.

3 A rail wagon of mass 1000 kg moving at a velocity of 5.0 m/s on a level track collides with a stationary wagon of mass 1500 kg. The two wagons move together after the collision.

    5 m/s ⟶ v   0 ⟶ v
    1000 kg   1500 kg

  a Calculate the momentum of the 1000 kg wagon before the collision.

  b Calculate the velocity of the wagons after the collision.

Key points

- Momentum = mass × velocity.

- The unit of momentum is kg m/s.

- Momentum is conserved whenever objects interact, as long as the objects are in a closed system so that no external forces act on them.

3.2 Explosions

Learning objectives

After this topic, you should know:

- how momentum can be described as having direction as well as size

- why two objects that push each other apart:
  - move away at different speeds
  - have zero total momentum.

If you are a skateboarder, you will know that the skateboard can shoot away from you when you jump off it. Its momentum is in the opposite direction to your own momentum. What does this say about the total momentum of objects when they fly apart from each other?

Practical

Investigating a controlled explosion

Figure 1 shows a controlled explosion using trolleys. When the trigger rod is tapped, a bolt springs out and the trolleys recoil (spring back) from each other.

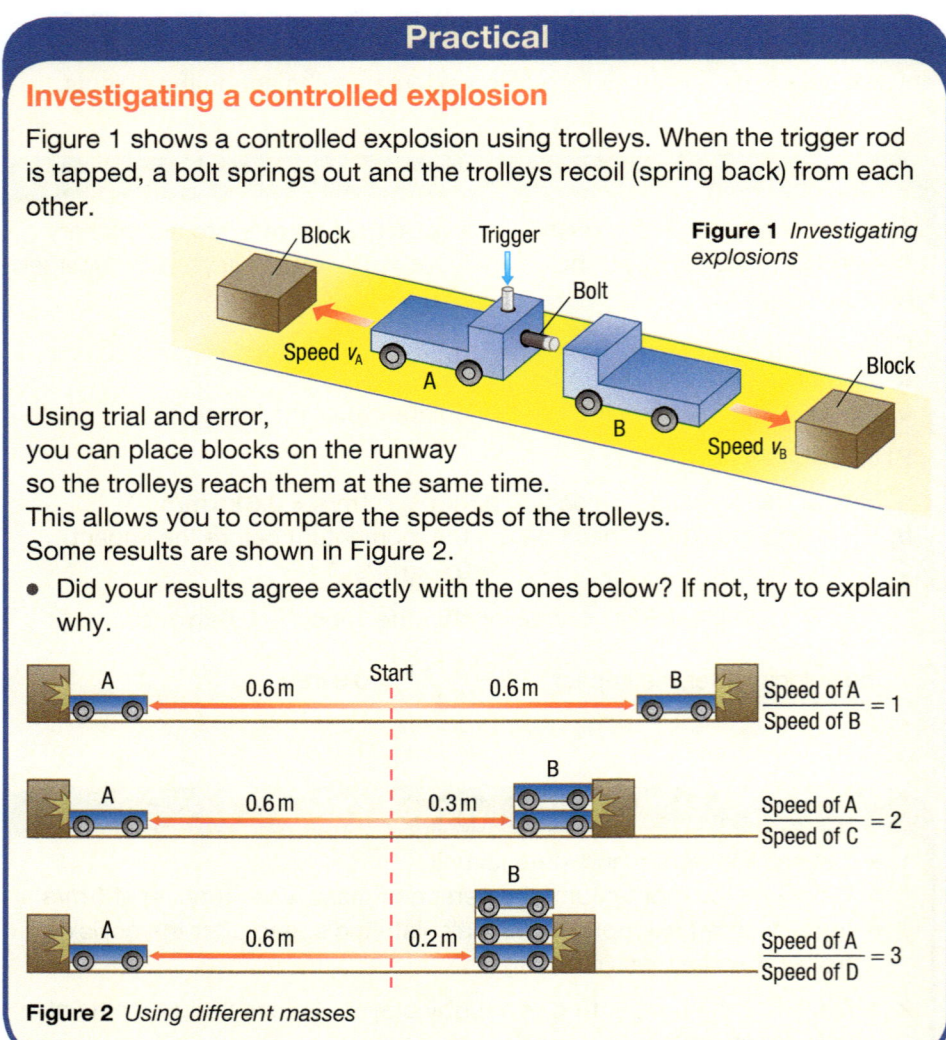

Figure 1 *Investigating explosions*

Using trial and error,
you can place blocks on the runway
so the trolleys reach them at the same time.
This allows you to compare the speeds of the trolleys.
Some results are shown in Figure 2.

- Did your results agree exactly with the ones below? If not, try to explain why.

Figure 2 *Using different masses*

- Two single trolleys travel equal distances in the same time. This shows that they recoil at equal speeds.
- A double trolley travels only half the distance that a single trolley does. Its speed is half that of the single trolley.

In each test:

1. the mass of the trolley × the speed of the trolley is the same, and
2. they recoil in opposite directions.

So momentum has size and direction. The results show that the trolleys recoil with equal and opposite momentum.

Study tip

Be careful in calculations – momentum is a vector quantity, so if two objects are travelling in opposite directions, one has positive momentum, and the other has negative momentum.

Conservation of momentum

In the trolley examples:

- momentum of A after the explosion = (mass of A × velocity of A)
- momentum of B after the explosion = (mass of B × velocity of B)
- total momentum before the explosion = 0 (because both trolleys were at rest).

Using conservation of momentum gives:

(mass of A × velocity of A) + (mass of B × velocity of B) = 0

Therefore:

(mass of A × velocity of A) = – (mass of B × velocity of B)

The minus sign after the equals sign shows that the momentum of B is in the opposite direction to the momentum of A. The equation shows that A and B move apart with equal and opposite amounts of momentum. So, the total momentum after the explosion is the same as before it.

Momentum in action

When a shell is fired from an artillery gun, the gun barrel recoils backwards. The recoil of the gun barrel is slowed down by a spring. This lessens the backwards motion of the gun.

Figure 3 An artillery gun in action

Summary questions

1 A skater of mass 60 kg and a skater of mass 80 kg are standing in the middle of an ice rink. They push each other away.

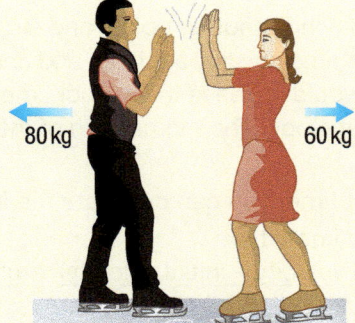

80 kg   60 kg

What can be said about:
a the force they exert on each other when they push apart?
b the momentum each skater has just after they separate?
c each of their velocities just after they separate?
d their total momentum just after they separate?

2 In Question 1, the skater of mass 60 kg moves away at 2.0 m/s. Calculate:
a her momentum
b the velocity of the other skater.

3 A cannon of mass 600 kg recoils at a speed of 0.5 m/s when a cannon ball of mass 12 kg is fired from it.
a Calculate the velocity of the cannon ball when it leaves the cannon.
b How would the recoil velocity of the cannon have been different if a cannon ball of mass 4 kg had been used instead?

Key points

- Momentum is mass × velocity and has direction.

- When two objects push each other apart, they move:
  - with different speeds if they have unequal masses
  - with equal and opposite momentum, so their total momentum is zero.

3.3 Impact forces

Learning objectives

After this topic, you should know:

- what affects the force of impact when two vehicles collide

- how the impact force depends on the impact time

- about the impact forces and the total momentum when two vehicles collide.

Figure 2 *A crash test. Car makers test the design of a crumple zone by driving a remotely controlled car into an obstacle*

Crumple zones at the front end and rear end of a car are designed to lessen the force of an impact. In a collision, the force changes the momentum of the car.

- In a front-end impact, the momentum of the car is reduced.
- In a rear-end impact (when a vehicle is struck from behind by another vehicle), the momentum of the car is increased.

In both cases the effect of a crumple zone is to increase the impact time and so lessen the impact force.

Practical

Investigating impacts

You can test an impact using a trolley and a brick, as shown in Figure 1. When the trolley hits the brick, the plasticine flattens on impact, making the impact time longer. This is the key factor that reduces the impact force.

Safety: Protect the bench and your feet from falling objects.

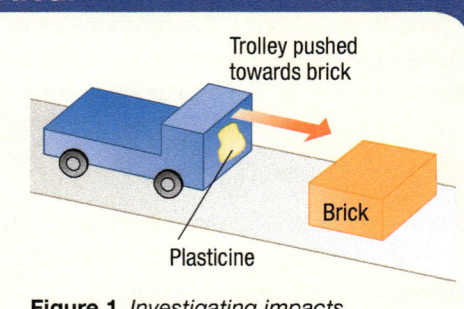

Figure 1 *Investigating impacts*

Impact time

Let's see why making the impact time longer reduces the impact force.

Suppose a moving trolley hits another object and stops. The impact force on the trolley acts for a certain time (the impact time) and causes it to stop. A soft pad on the front of the trolley would increase the impact time and would allow the trolley to travel further before it stops. The momentum of the trolley would be lost over a longer time.

If you know the impact time, you can calculate the impact force as follows:

- From Topic 2.2, you know that:

$$\text{acceleration} = \frac{\text{(final velocity − initial velocity)}}{\text{time taken}} = \frac{\text{change of velocity}}{\text{time taken}}$$

- From Topic 2.5, you know that: force = mass × acceleration

Therefore, because mass × change of velocity = change of momentum:

$$\text{force} = \frac{\text{mass × change of velocity}}{\text{time taken}} = \frac{\text{change of momentum}}{\text{time taken}}$$

For a change of momentum Δp in time t, you can write this as:

$$\textbf{force } F = \frac{\Delta p}{t}$$

This shows that if the impact time t is longer, the force F is smaller for the same change of momentum Δp. Car safety features such as crumple zones and side bars increase the impact time and so reduce the impact force.

The longer the impact time, the more the impact force is reduced.

Study tip

Remember that the duration of the impact is important. To make the force as large as possible – for example when hitting a ball – the impact should be as short as possible. To make the force smaller – for example in a crash – the impact should last as long as possible.

Worked example

A ball of mass 0.060 kg moving at a velocity of 20 m/s is stopped by a fielder in 0.25 s.

Calculate **a** the deceleration, **b** the change of momentum, and **c** the impact force.

Solution

a Initial velocity of ball = 20 m/s

Final velocity of ball = 0

Change of velocity = final velocity − initial velocity = 0 − 20 m/s = −20 m/s
(the minus sign shows that the change of velocity is a decrease)

$$\text{Deceleration} = \frac{\text{change of velocity}}{\text{impact time}} = \frac{-20\,\text{m/s}}{0.25\,\text{s}} = \mathbf{-80\,m/s^2}$$

b Change of momentum = mass × change of velocity
$$= 0.060\,\text{kg} \times (0 - 20\,\text{m/s}) = \mathbf{-1.2\,kg\,m/s}$$

c $\text{Force} = \dfrac{\text{change of momentum}}{\text{time taken}} = \dfrac{-1.2\,\text{kg\,m/s}}{0.25\,\text{s}} = \mathbf{-4.8\,N}$
(the minus sign shows that the force decelerates the ball).

Two-vehicle collisions

When two vehicles collide, they exert equal and opposite impact forces on each other for the same length of time. The change of momentum of one vehicle is therefore equal and opposite to the change of momentum of the other vehicle. The total momentum of the two vehicles is the same after the impact as it was before the impact, so momentum is conserved – assuming no external forces act.

For example, suppose a fast-moving truck runs into the back of a stationary car. The impact decelerates the truck and accelerates the car. The truck loses momentum and the car gains momentum.

Summary questions

1 a In a car crash, why would a seat belt reduce the impact force on a passenger?

b A ball of mass 0.12 kg moving at a velocity of 18 m/s is caught by a person in 0.0003 s. Calculate the impact force.

2 a A car of mass 800 kg travelling at 30 m/s is stopped safely when the brakes are applied. What braking force is required to stop it in:

    i 6.0 s?   **ii** 30 s?

b If the vehicle in **a** had been stopped in a collision lasting less than a second, explain by referring to momentum why the force on it would have been much greater.

3 A van of mass 2000 kg moving at a velocity of 12 m/s crashes into the back of a stationary truck of mass 10 000 kg. Immediately after the impact, the two vehicles move together.

a Show that the velocity of the van and the truck immediately after the impact was 2 m/s.

b The impact lasted for 0.3 seconds. Calculate:

    i the deceleration of the van

    ii the change of momentum of the van

    iii the force of the impact on the van.

Did you know … ?

Scientists at Oxford University have developed lightweight material for bulletproof vests. The material is so strong and elastic that bullets bounce off it.

Maths skills

The equation $F = \dfrac{\Delta p}{t}$ tells us that force F is **inversely proportional** to time t. For example, if t is doubled, F is halved. See the 'Using data' section for more about inverse proportion.

Did you know … ?

The effect of an impact on an object or person may be described as the **g-force**. For example, a g-force of $2g$ means that the force on an object is twice its weight. You would experience a g-force of:

– about 3–4g on a fairground ride that whirls you round

– about 10g in a low-speed car crash

– more than 50g in a high-speed car crash.

Key points

- When vehicles collide, the force of the impact depends on changes of momentum and the duration of the impact.

- The longer the impact time, the more the impact force is reduced.

- When two vehicles collide:
  – they exert equal and opposite forces on each other
  – their total momentum is unchanged.

3.4 Safety first

Figure 1 *An air bag in action*

Figure 2 *A child car seat*

When you go on a cycle ride or travel in a car, you want to feel safe if you crash. Even falling over in a playground can cause serious injury. In this topic, you'll look at different safety features that are designed to keep you safe when you are travelling around.

A **seat belt** stops its wearer from continuing forwards when the car suddenly stops. Someone without a seat belt would hit the windscreen.

- The time taken to stop someone moving in a car that suddenly stops is longer if they are wearing a seat belt than if they are not. So the decelerating force is reduced by wearing a seat belt.

- The seat belt acts across the chest so it spreads the force out. Without the seat belt, the force would act on the head when it hit the windscreen.

Air bags

Air bags in a car are designed to protect the driver and the front passenger. Some cars also have side air bags. These bags protect people in the car from an impact on the side of the car. In a car crash, an inflated air bag spreads the force of an impact across the upper part of the body. It also increases the duration of the impact time. So the effect of the force is lessened even further than with just a seat belt. Crumple zones, side impact bars, and collapsible steering wheels also increase the impact times to lessen the force of an impact.

Child car seats

A baby or a child in a car should be strapped in a child car seat. This applies to children up to 12 years old or up to 1.35 metres in height. Different types of child car seat should be used for babies up to 9 months old, infants up to about 4 years old, and children over 4.

- Baby seats should face backwards.

- Children under 4 years old should usually be in a child car seat fitted to a back seat.

These measures are intended to reduce deaths and serious injuries of children in cars. Before the measures were introduced, many children were killed and seriously injured each year in car accidents. Many such accidents happened whilst children were travelling to school in cars. The driver is responsible for making sure every child in their car is seated safely in a correct type of seat.

> ### Study tip
>
> Make sure that you can explain how seat belts, air bags, and crumple zones make cars safer.

Playground safety

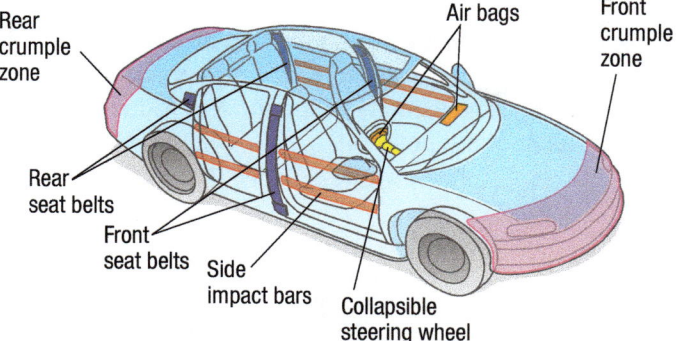

Figure 3 *Car safety features*

Playgrounds may have cushioned surfaces underneath swings and slides in case children fall off. When a child falls on such a soft surface, the duration of the impact is longer than it would be if the surface had been hard, such as concrete. So the child's momentum is reduced to zero over a longer time by the cushioning effect of the softer surface. The change of momentum per second is therefore reduced so the impact force is reduced compared with an impact at same momentum on a hard surface.

Gymnasium crash mats contain cushioning material so they have the same effect as a cushioned playground surface when someone falls on a crash mat.

Summary questions

1 Explain why a cyclist should wear a safety helmet when riding a bicycle.

2 a Explain why an inflated air bag in front of a car user reduces the force on the driver in a head-on crash.

 b Why are rear-facing car seats for babies safer than front-facing seats?

3 A car crashed into a lorry that was crossing a busy road. The speed limit on the road was 30 m/s.

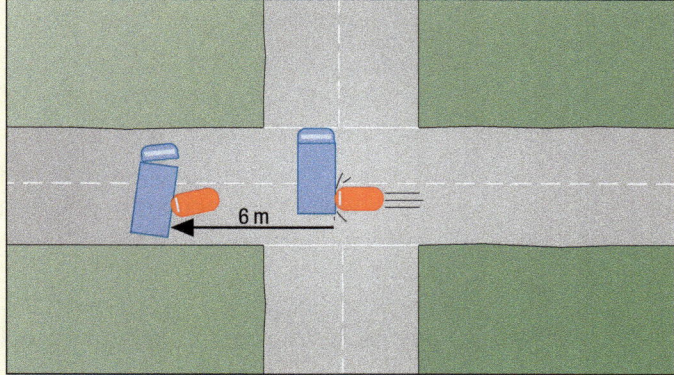

The following measurements were made by police officers at the scene of the crash:

- The car and lorry ended up 6 m from the point of impact.
- The car's mass was 750 kg and the lorry's mass was 2150 kg.
- The speed of a vehicle for a braking distance of 6 m is 9 m/s.

 a Use this speed to calculate the momentum of the car and the lorry immediately after the impact.

 b Use conservation of momentum to calculate the velocity of the car immediately before the collision.

 c Was the car travelling over the 30 m/s speed limit before the crash?

Momentum at work

When a vehicle crashes into another vehicle, their velocities immediately after the crash can be worked out separately from the distances they stopped in from the point of impact. So their total momentum after the collision can be worked out. Their total momentum immediately after the crash is the same as before the crash. So if the velocity of one of the vehicles before the collision is known, the velocity of the other vehicle before the collision can be worked out to tell if it was 'over the speed limit'. See Q3 in the Summary questions.

Key points

- Seat belts and air bags spread the force across the chest and increase the impact time.

- Side impact bars and crumple zones deform in an impact, so increasing the impact time.

- You can use the conservation of momentum to find the speed of a car before an impact.

Chapter summary questions

1 A car of mass 1500 kg is moving at a speed of 30 m/s on a horizontal road when the driver applies the brakes and the car stops 12 seconds later.

30 m/s

a i Calculate the initial momentum of the car before the brakes are applied.

  ii Calculate the braking force.

b Describe how the momentum of the car changes when the brakes are applied.

c Discuss the effect on the motion of the car if the brakes had been applied with much greater force.

2 A child of mass 14 kg is strapped into a forward-facing car seat in a car. When the car was moving at a speed of 8.2 m/s the driver suddenly braked and stopped the car.

a i Calculate the momentum of the child before the driver applied the brakes.

  ii What was the direction of the force on the child that stopped her continuing to move forward when the brakes were applied?

b Explain why the child would have suffered an injury if she had not been strapped into the car seat.

3 A student of mass 50 kg standing at rest on a boat of mass 75 kg jumps off the boat onto a pier. The boat recoils and moves away at a speed of 0.5 m/s.

a Explain why the boat recoiled when the student jumped off it.

b Calculate:

  i the momentum of the boat when it recoiled

  ii the speed of the student when he jumped off the boat.

4 A truck of mass 2000 kg moving at a velocity of 18.0 m/s on a level road collides with a stationary van of mass 1200 kg. The velocity of the truck is reduced to 10 m/s as a result of the collision.

18 m/s

a Calculate the momentum of the truck:

  i before the collision

  ii after the collision.

b For the van after the collision, calculate:

  i the momentum

  ii the velocity.

5 Safety footwear is designed to protect the wearer's feet if a heavy weight falls on a foot. In a test of a safety boot, an object of mass 7.0 kg was dropped from a certain height above the boot.

a i The time taken for the object to fall was measured electronically and found to be 0.63 s. Show that the speed of the object just before the impact was 6.3 m/s. As explained in Topic 2.7, assume that the falling object had an acceleration of 10 m/s².

  ii Calculate the momentum of the object just before the impact.

b An electronic sensor attached to the toecap of the boot recorded that the impact lasted for 0.0022 seconds. Calculate the impact force.

c Discuss how the impact force would have differed if the same weight had been dropped from a greater height without damaging the boot.

6 When a stationary football of mass 0.44 kg was kicked, it gained a velocity of 19 m/s as a result of the impact.

a Calculate the gain of momentum of the football due to the impact.

b The impact lasted 0.0384 s. Show that the impact force was 220 N.

Practice questions

1 Momentum is a vector quantity.

  a Explain why momentum is a vector quantity. (3)

  b Two lorries, **P** and **Q**, are travelling in the same direction along a motorway.

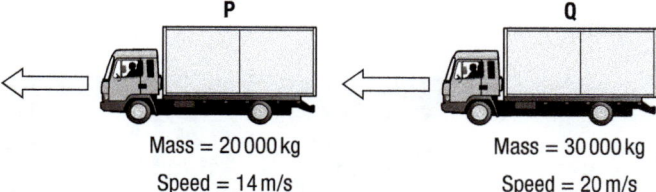

  i Calculate the momentum of lorry **P**.
  ii Calculate the momentum of lorry **Q**. (4)
  iii Lorry **Q** collides with lorry **P** and they stick together. Calculate the speed of the lorries immediately after the collision. (4)

2 The picture shows an ice dancer and her coach.

  They are standing still on the ice facing each other.

  The dancer pushes her coach and they move away from each other.

Coach   Dancer

  a When the dancer pushes her coach, the momentum of each of them changes.
  i What are the units of momentum?

  Choose from the list below. (1)

  kg m s   kg m/s   kg m/s² kg/m/s

  ii Outline the similarities and the differences between the momentum of the coach and the momentum of the dancer after the push. (3)

  b The mass of the dancer is 50 kg. The mass of her coach is 90 kg.

  As they move away from each other, the speed of the dancer is 1.5 m/s.

  Calculate the speed of her coach. (3)

3 Modern cars have many safety features.

  a Explain how seat belts and air bags keep passengers safe. (6)

  b Discuss the importance of duration of the impact when a player uses a cricket bat to hit a cricket ball and when a fielder catches the cricket ball. (4)

4 Most cars have crumple zones at the front and back of the car.

  a Explain how crumple zones reduce impact forces. (4)

  b A snooker player uses a cue to hit the stationary white ball. The cue hits the ball of mass 0.2 kg with a force of 60 N. The cue is in contact with the ball for 0.008 s.

  Calculate the speed of the ball immediately after impact. (5)

5 In a vehicle test, the velocity of a car was recorded when it collided with a fixed concrete block. The graph shows how the velocity v of the car varied with time t during the impact.

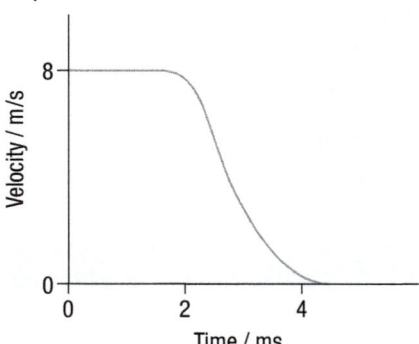

  a The mass of the car was 1200 kg.
  i Use the graph to determine the velocity of the car before the impact and therefore determine the car's momentum before the impact. (2)
  ii Estimate the impact time from the graph and therefore calculate the impact force. (4)

  b Explain how the graph shows that the impact force varied with time during the impact. (4)

4.1

Centre of mass

Learning objectives

After this topic, you should know:

- what the centre of mass of an object is

- about the centre of mass of an object that is suspended from a fixed point

- how to find the centre of mass of a symmetrical object.

??? Did you know …?

Tightrope walkers carry a long pole to keep balanced. They use the pole to keep their centre of mass directly above the rope. A slight body movement one way is counterbalanced by shifting the pole slightly the other way.

The design of racing cars has changed a lot since the first models. But one thing that has not changed is the need to keep the car near the ground. The weight of the car must be as low as possible. Otherwise the car would overturn when cornering at high speeds.

Figure 1 *Racing cars from the 1920s to modern day*

You can think of the mass of an object as if it is concentrated at a single point. This point is called the **centre of mass** (or the centre of gravity) of the object.

The centre of mass of an object is the point at which its mass may be thought to be concentrated.

Practical

Suspended equilibrium

If you suspend an object and then release it, it will sooner or later come to rest with its centre of mass directly below the point of suspension, as shown in Figure 2a. The object is then in equilibrium, which means it is at rest. Its weight does not exert a turning effect on the object because its centre of mass is directly below the point of suspension.

If the object is moved from this position and then released, it will swing back to its equilibrium position. This is because its weight has a turning effect that returns the object to equilibrium, as shown in Figure 2b. A *freely suspended* object returns to its equilibrium position.

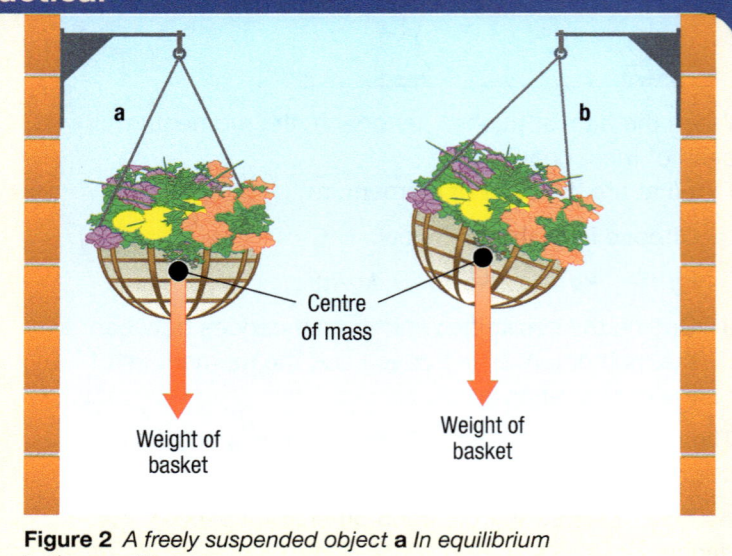

Figure 2 *A freely suspended object* **a** *In equilibrium* **b** *Non-equilibrium*

The centre of mass of a symmetrical object

For a symmetrical object, its centre of mass is along the axis of symmetry. You can see this in Figure 3.

If the object has more than one axis of symmetry, its centre of mass is where the axes of symmetry meet.

- A rectangle has two axes of symmetry, as shown Figure 3a. The centre of mass is where the axes meet.
- The equilateral triangle in Figure 3b has three axes of symmetry, each bisecting one of the angles of the triangle. The three axes meet at the same point. This is where the centre of mass of the triangle is.

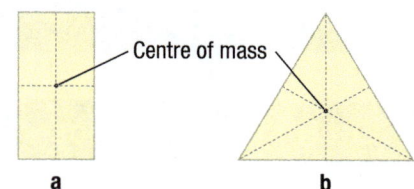

Figure 3 *Symmetrical objects*

Practical

A centre of mass test

Figure 4 shows how to find the centre of mass of an irregularly shaped flat card. The card is at rest, freely suspended from a rod.

Its centre of mass is directly below the rod. You can use a plumbline – a string with a weight on the end – to draw a vertical line on the card from the rod downwards.

The procedure is repeated with the card suspended from a second point to give another similar line. The centre of mass of the card is where the two lines meet.

Try drawing a third line to see if all three cross at the same point.

- What can you say about the accuracy of your experiment?

Test your results to see if you can balance the card at this point on the end of a pencil.

- Now find the centre of mass of a semicircular card of radius 100 mm.

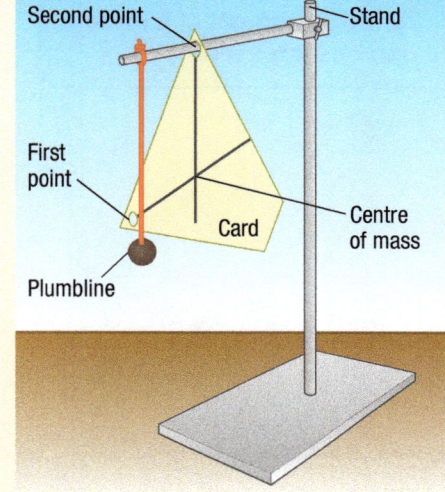

Figure 4 *Finding the centre of mass of a card*

Study tip

Make sure you can describe all the steps in the practical above to find the centre of mass of a thin sheet of material.

Summary questions

1 Describe what is meant by an object's 'centre of mass'.

2 Sketch each of the objects shown and mark its centre of mass.

a

b

c

3 Explain why a child on a swing comes to rest directly below the top of the swing.

4 Describe how you would find the centre of mass of a flat semicircular card.

Key points

- **The centre of mass of an object is the point where its mass may be thought to be concentrated.**

- **When an object is freely suspended, it comes to rest with its centre of mass directly beneath the point of suspension.**

- **The centre of mass of a symmetrical object is along the axis of symmetry.**

4.2 Moments at work

Learning objectives

After this topic, you should know:

- what the moment of a force measures
- how to calculate the moment of a force
- how the moment of a force can be increased.

Figure 1 *A turning effect*

To undo a very tight wheel-nut on a bicycle, you need a spanner. The force you apply to the spanner has a turning effect on the nut. You couldn't undo a tight nut with your fingers but you can with the spanner. The spanner exerts a much larger force on the nut than the force you apply to the spanner.

If you had a choice between a long-handled spanner and a short-handled one, which would you choose? The longer the spanner handle, the less force you need to exert on it to loosen the nut.

In this example, the turning effect of the force, called the **moment** of the force, can be increased by:

- increasing the size of the force
- using a spanner with a longer handle.

Levers

A crowbar is a lever that can be used to raise one edge of a heavy object. Look at Figure 2.

The weight of the object is called the **load**, and the force the person applies to the crowbar is called the **effort**. The point about which the crowbar turns is called the **pivot** or the **fulcrum**. Using the crowbar, the effort needed to lift the same object is only a small fraction of its weight. The lever used in this way is an example of a **force multiplier** because the effort moves a much bigger load.

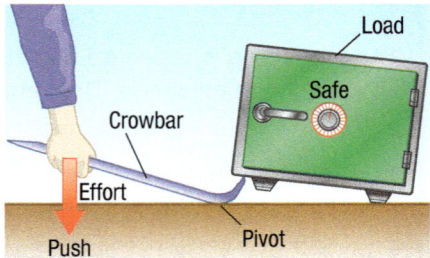

Figure 2 *Using a crowbar*

The line along which a force acts is called its **line of action**.

Did you know …?

A patient fitted with a replacement hip joint has to be very careful at first. A slight movement can cause a turning effect that could pull the hip joint apart.

Practical

Investigating the turning effect of a force

The diagram in Figure 3 shows one way to investigate the turning effect of a force.

- How do you think the reading on the newtonmeter compares with the weight?

You should find that the newtonmeter reading (i.e., the force F needed to support the ruler) increases if the weight W is increased.

- How does this reading change as the weight is moved away from the pivot?

You should find that the newtonmeter reading increases as the weight is moved away from the pivot.

Safety: Protect the bench and your feet from falling weights.

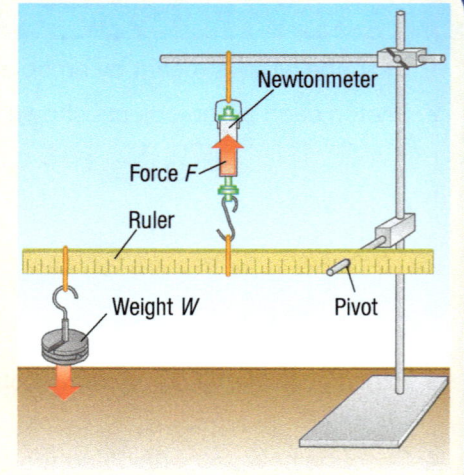

Figure 3 *Investigating turning forces*

You can work out the moment of a force using this equation:

moment
(newton metres, N m) **=** **force**
(newtons, N) **×** **perpendicular distance from the line of action of the force to the pivot**
(metres, m)

 Maths skills

The word equation can be written using symbols, as follows:

$$M = F \times d$$

where:

M = moment in N m

F = force in N

d = perpendicular distance from the line of action of the force to the pivot, in m.

Look at Figure 4. The claw hammer is being used to remove a nail from a wooden beam.

- The applied force F on the claw hammer tries to turn it clockwise about the pivot.
- The moment of force F about the pivot is $F \times d$, where d is the perpendicular distance from the pivot to the line of action of the force.
- The effect of the moment is to cause a much larger force to be exerted on the nail than you could exert without the claw hammer.

Summary questions

1 a A force acts on an object and makes it turn about a fixed point. State the effect on the moment of the force if:
  i the force is increased without changing its line of action
  ii the force is doubled and the perpendicular distance from its line of action to the pivot is halved
  iii the force is halved and the perpendicular distance from its line of action to the pivot is halved.
  b A force of 72 N is exerted on a claw hammer of length 0.25 m, as shown in Figure 4. Calculate the moment of the force.

2 In Figure 1, a force is applied to a spanner to undo a nut. State whether the moment of the force is:
  a clockwise or anticlockwise
  b increased or decreased by:
    i increasing the force
    ii exerting the force nearer the nut.

3 Explain each of the following statements:
  a it is easier to remove a nail with a claw hammer if the hammer has a long handle
  b a door with rusty hinges is more difficult to open than a door of the same size with lubricated hinges.

4 A spanner of length 0.25 m is used to turn a nut as in Figure 1. Calculate the force that needs to be applied to the end of the spanner if the moment it exerts is to be 18 N m.

Worked example

A force of 50 N is exerted on a claw hammer of length 0.30 m, as shown in Figure 4. Calculate the moment of the force.

Solution

Moment = 50 N × 0.30 m
= **15 N m**

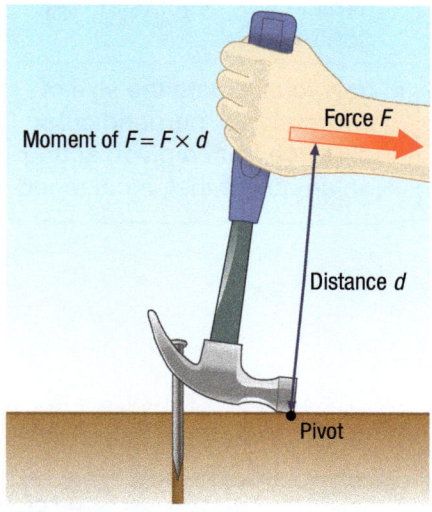

Figure 4 *Using a claw hammer*

Study tip

Remember when calculating moments, you use the *perpendicular* distance from the pivot.

Be careful with units. You need to be consistent – N m or N cm, but not a mixture of both.

Key points

- The moment of a force is a measure of the turning effect of the force on an object.
- The moment of a force M about a pivot = $F \times d$, where d is the perpendicular distance from the line of action of the force, F, to the pivot.
- To increase the moment of a force, increase F or increase d.

4.3

Moments in balance

Learning objectives

After this topic, you should know:

- how to use our knowledge of forces and moments to explain why objects at rest don't turn

- how to calculate the size of a force (or its perpendicular distance from a pivot) acting on an object that is balanced.

A see-saw is an example in which clockwise and anticlockwise moments might balance each other out. The girl in Figure 1 balances her younger brother at the far end of the see-saw. Her brother is not as heavy as his big sister, so she sits nearer the pivot than he does. That means her anticlockwise moment about the pivot balances his clockwise moment.

A model see-saw

Look at the model see-saw in Figure 2. The ruler is balanced horizontally by adjusting the positions of the two weights. When it is balanced:
- the anticlockwise moment due to W_1 about the pivot = W_1d_1
- the clockwise moment due to W_2 about the pivot = W_2d_2.

When the ruler is balanced, the anticlockwise moment due to W_1 = the clockwise moment due to W_2 and therefore:

$$W_1d_1 = W_2d_2$$

Study tip

Make sure the units in your calculations are consistent.

Figure 1 *The see-saw*

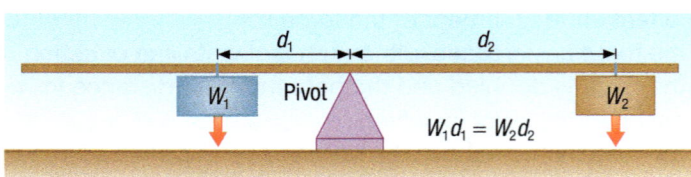

Figure 2 *The principle of moments*

Practical

Measuring an unknown weight

You can use the arrangement in Figure 2 to find an unknown weight, W_1, if you know the other weight, W_2, and you measure the distances d_1 and d_2. Then you can calculate the unknown weight using the equation: $W_1d_1 = W_2d_2$.

Safety: Protect the bench and your feet from falling weights.

The see-saw is an example of the **principle of moments**. This states that, for an object that is not turning:

$$\text{the sum of all the clockwise moments about any point} = \text{the sum of all the anticlockwise moments about that point}$$

Worked example

Calculate W_1 in Figure 2, if W_2 = 4.0 N, d_1 = 0.25 m, and d_2 = 0.20 m.

Solution
Rearranging $W_1d_1 = W_2d_2$ gives:

$$W_1 = \frac{W_2d_2}{d_1} = \frac{4.0\,\text{N} \times 0.20\,\text{m}}{0.25\,\text{m}} = \textbf{3.2 N}$$

Practical

Measuring the weight of a beam

Figure 3 shows how you can measure the weight of a beam by balancing it off-centre using a known weight. The weight of the beam acts at its centre of mass, which is at distance d_0 from the pivot.

- The moment of the beam about the pivot = $W_0 d_0$ clockwise, where W_0 is the weight of the beam.
- The moment of W_1 about the pivot = $W_1 d_1$ anticlockwise, where d_1 is the perpendicular distance from the pivot to the line of action of W_1.

Applying the principle of moments gives $W_1 d_1 = W_0 d_0$.

So you can calculate W_0 if you know W_1 and distances d_1 and d_0.

Safety: Protect the bench and your feet from falling weights.

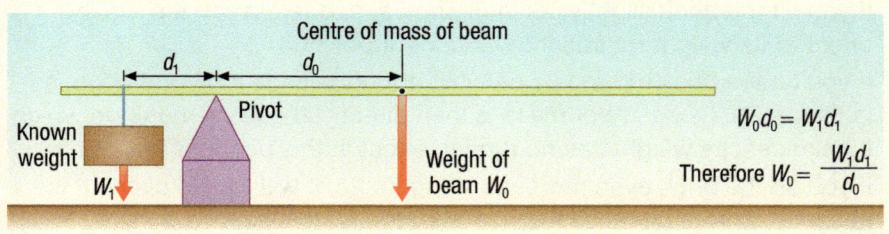

$$W_0 d_0 = W_1 d_1$$

$$\text{Therefore } W_0 = \frac{W_1 d_1}{d_0}$$

Figure 3 *Finding the weight of a beam*

Lifting a load

If you have to move a heavy load, think beforehand about how to make the job easier. Figure 4 shows a wheelbarrow and a trolley being used to move a load. The load (weight W_0) is lifted using a much smaller effort (force F_1). Once the load has been lifted, it can easily be moved by pushing the wheelbarrow or the trolley forwards.

Summary questions

1. **a** **i** In Figure 2, calculate W_1 if $W_2 = 6.0\,\text{N}$, $d_1 = 0.30\,\text{m}$, and $d_2 = 0.15\,\text{m}$.
   ii In Figure 3, calculate the weight of the beam if $W_1 = 2.0\,\text{N}$, $d_1 = 0.15\,\text{m}$, and $d_0 = 0.25\,\text{m}$.
   b In Figure 4 explain why the effort is smaller than the load.

2. Dawn sits on a see-saw 2.50 m from the pivot. Jasmin balances the see-saw by sitting 2.00 m on the other side of the pivot.
   a Who is lighter, Dawn or Jasmin?
   b Jasmin weighs 425 N. What is Dawn's weight?
   c John now sits on the see-saw on the same side as Dawn at a distance of 0.50 m from the pivot. Jasmin stays in the same position as before. John's weight is 450 N. How far and in which direction should Dawn move to rebalance the see-saw?

3. For the balanced beam in the figure, work out its weight, W.

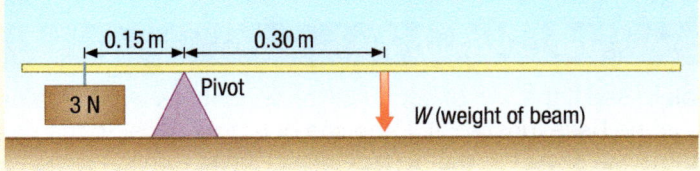

Study tip

In calculations, be systematic – write down the clockwise moments, then the anticlockwise moments, and then equate them.

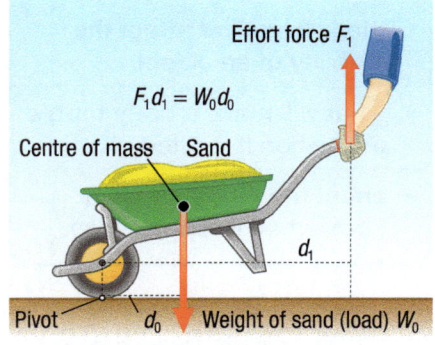

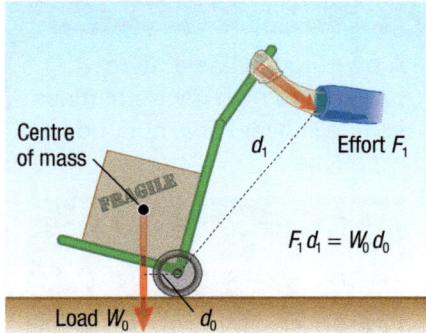

Figure 4 *Using moments*

Key points

- If an object at rest doesn't turn, the sum of the anticlockwise moments about any point = the sum of the clockwise moments about that point.

- To calculate the force needed to stop an object turning, use the equation: $W_1 d_1 = W_2 d_2$. You need to know all the forces (apart from any acting through the pivot) and their perpendicular distances from the line of action to the pivot.

4.4

Stability

After this topic, you should know:

- the factors that affect the stability of an object
- what will make a body topple over when it is tilted
- about the moments on a body when it topples over.

??? Did you know ...?

A bowling pin has a narrow base and a high centre of mass so it falls over if it is nudged slightly.

Figure 1 *At the bowling alley*

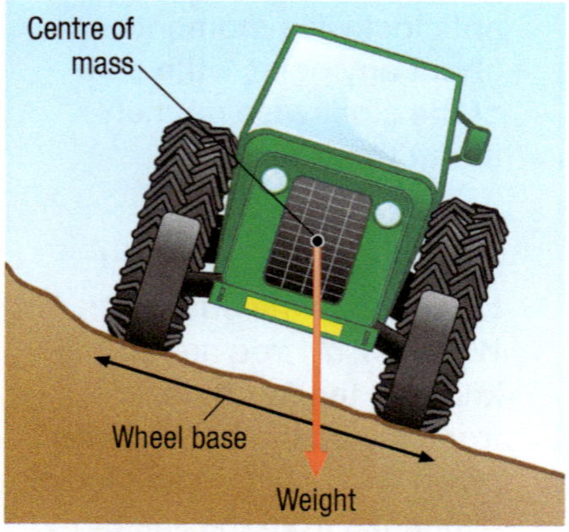

Stability and safety

Look around you and see how many objects could topple over. Bottles, table lamps, and floor-standing bookcases are just a few objects that can easily topple over. Lots of objects are designed to be stable, so they can't topple over easily.

Practical

Tilting and toppling tests

How far can you tilt something before it topples over? Figure 2 shows how you can test your ideas using a tall box or a brick on its end.

- If you tilt the brick slightly, as in Figure 2a, and release it, the turning effect of its weight returns it to its upright position.
- If you tilt the brick more, you can just about balance it on one edge, as in Figure 2b. Its centre of mass is then directly above the edge on which it balances. Its weight has no turning effect in this position.
- If you tilt the brick even more, as in Figure 2c, it will topple over if it is released. This is because the line of action of its weight is outside its base. So its weight has a turning effect that makes it topple over.

Investigate the factors that affect the stability of an object.

Safety: Protect the bench and your feet from falling bricks.

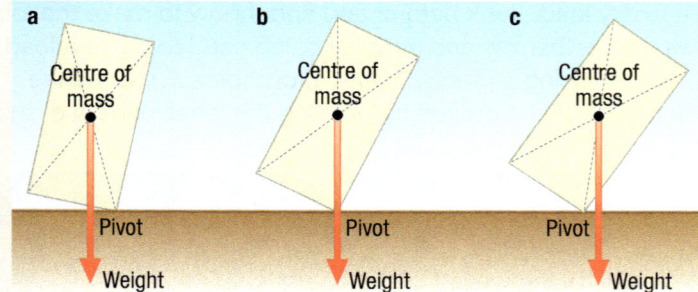

Figure 2 *Tilting and toppling* **a** *Tilted* **b** *At balance* **c** *Toppled over*

1 Tractor safety

Look at the tractor on a hillside in Figure 3. It doesn't topple over because the line of action of its weight acts within its wheel base.

If it is tilted more, it will topple over when the line of action of its weight acts outside its wheel base. This is because the moment of its weight about the lower wheel is clockwise. The moment of the support force from the ground on the upper wheel is zero (because the upper wheel lifts off the ground). So the resultant moment about the lower wheel makes it topple over.

Study tip

In diagrams, look to see if the line of action of the weight lies within the base (stable) or outside the base (unstable). If the line of action is inside the base, the moment returns the object to the horizontal.

Figure 3 *Forces on a tilting tractor*

2 Bus tests

Look at the double-decker bus in Figure 4. It is being tested on a platform to see how far it can tilt without toppling. Such tests are important to make sure buses are safe to travel on, especially when they go round bends and on hilly roads. The explanation of why it topples over when the platform is tilted too much is the same as for the tractor.

3 High chairs

A high chair for a young infant needs to have a wide base. When the child is sitting in it, the centre of mass is above the seat. If the base were narrow, the chair would topple over if the child leant sideways too much. The wide base prevents this (see Figure 5).

The chair topples over if the child's weight acts outside the chair base on one side and the chair is not heavy enough to stop it toppling over. The high chair will turn about the position where the chair legs on that side are in contact with the floor.

Toppling happens if the moment of the child's weight about this position is greater than the moment of the chair's weight.

How to stop an object toppling over

To prevent toppling, the centre of mass of the object should be as low as possible and:
- either the base should be wide enough to prevent toppling when the object is tilted or knocked sideways
- or the base should be bolted or clamped down.

Suppose an object is not clamped or bolted down. If the line of action of its weight lies outside its base, the object will topple over. This is because there is a **resultant moment** on the object. In other words, the object topples over because the sum of the clockwise moments about any point is **not** equal to the sum of the anticlockwise moments about that point.

Figure 4 *A toppling test*

Figure 5 *A high chair needs a wide base*

Summary questions

1 a i Would a double-decker bus be more stable or less stable if everyone on it sat upstairs?
   ii Why are stabiliser wheels fitted to bicycles that are designed for young children?
   b Think of an object that needs to be redesigned because it is knocked over too easily. Sketch the object and explain how it could be redesigned to make it more stable.

2 A well-designed baby chair has a wide base and a low seat.
   a If the base of a baby chair was too narrow, why would the chair be unsafe?
   b Why is a baby chair with a low seat safer than one with a high seat?

3 a Explain why a tall plastic bottle is less stable when it is empty than when it is half-full of water, using the idea of moments.
   b Explain why a traffic cone is difficult to knock over.

Key points

- The stability of an object is increased by making its base as wide as possible and its centre of mass as low as possible.

- An object will tend to topple over if the line of action of its weight is outside its base.

- An object topples over if the resultant moment about its point of turning is not zero and acts away from equilibrium.

Chapter summary questions

1 The bottle opener in the figure is being used to force the cap off a bottle.

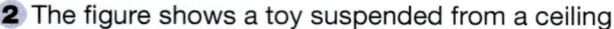
Cap
Bottle opener
Bottle

a Explain why the force of the bottle opener on the cap is much larger than the force applied to the bottle opener by the person opening the bottle.

b In the figure, a force of 25 N had to be applied to force the cap off the bottle. By measuring the appropriate distances in the figure, estimate the force exerted by the bottle opener on the cap. Explain your working.

2 The figure shows a toy suspended from a ceiling.

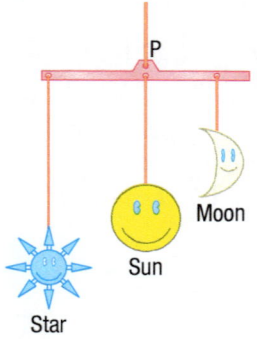
P
Moon
Sun
Star

a How would the stability of the toy be affected if the Sun were removed from it?

b The star on the toy has a weight of 0.04 N and is a distance of 0.30 m from the point P where the thread is attached to the toy. The moon attached to the toy is at a distance of 0.20 m from P.
Calculate the weight of the moon.

3 The figure shows a wheelbarrow being used to move a bag of sand.

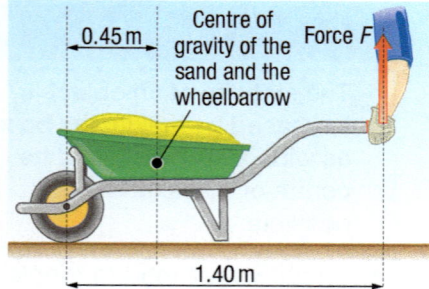
0.45 m
Centre of gravity of the sand and the wheelbarrow
Force F
1.40 m

a Explain why the vertical force F needed to lift the wheelbarrow's legs off the ground is much less than the combined weight of the sand and the wheelbarrow.

b A vertical force of 48 N was needed to lift the wheelbarrow's legs off the ground. The force was applied to the handles at a horizontal distance of 1.40 m from the wheel axle. The centre of mass of the bag of sand and the wheelbarrow was a horizontal distance of 0.45 m from the wheel axle.
  i Calculate the combined weight of the bag of sand and the wheelbarrow.
  ii The weight of the wheelbarrow was 65 N. Calculate the weight of the bag of sand.

4 The figure below shows a simple model of a see-saw in which a ruler pivoted at its centre supports two weights A and B. The forces acting on the ruler are shown.

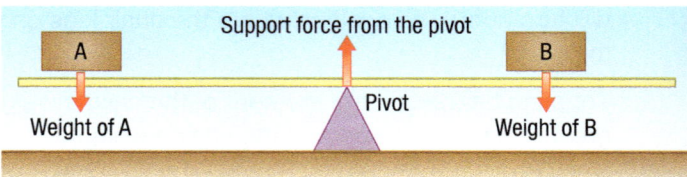
Support force from the pivot
A
Pivot
B
Weight of A
Weight of B

a i The weight of A is 2.2 N and its distance from the pivot is 0.38 m. Calculate the moment of A about the pivot.
  ii The distance from B to the pivot is 0.24 m. Calculate the weight of B.

b Describe and explain what would happen if A were moved towards the pivot.

5 The figure shows a metre ruler balanced on a pivot which is not at the centre of mass of the ruler. A 2.5 N weight W_1 is suspended from a thread which is attached to the ruler.

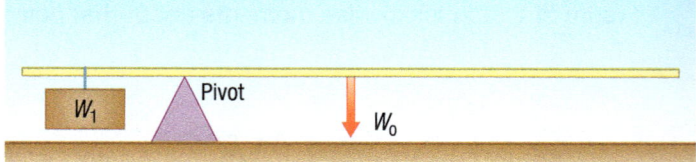
W_1
Pivot
W_0

a Explain why the ruler does not tip over.

b The arrangement shown in the figure is used to measure the weight W_0 of the ruler.
  i Describe the measurements you would need to make to determine W_0.
  ii Explain how you would use the measurements to determine W_0.

Practice questions

1 a The centre of mass of an object is important to its stability.

The diagram shows a thin sheet of metal with two holes drilled in it.

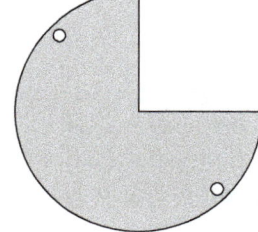

Describe how you would carry out an experiment to find the centre of mass of the sheet of metal. (6)

b The diagrams show a table lamp. The lamp is at the end of an arm that can be rotated about the pivot.

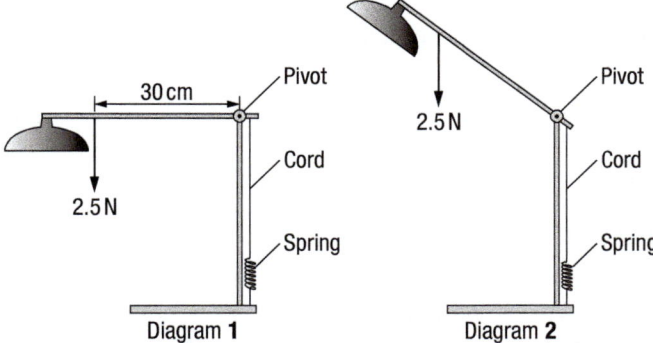

Diagram **1**   Diagram **2**

A cord is fixed near the right-hand end of the arm. The cord passes vertically downwards to a stretched spring. The centre of mass of the lamp and arm is 30 cm from the pivot. The weight of the lamp and arm is 2.5 N.

i From Diagram **1**, calculate the moment of the weight about the pivot. (2)

ii The arm is now raised, as in Diagram **2**. Explain what has happened to the tension in the spring. (4)

2 a The following article appeared in a newspaper.

> **Four-wheel-drive vehicle fails tilt test**
>
> A four-wheel-drive vehicle rolled over yesterday when travelling at 40 mph during stability tests carried out by the government. A motoring organisation had complained that the narrow-track, short-wheel-base vehicles are prone to rolling over. The government tests emphasised how passengers raise the centre of mass of the vehicle.

i Name **two** factors mentioned in the article that affect the stability of the vehicle. (2)

ii Why should stability tests be carried out by the government and not by the manufacturers? (1)

b The diagram shows a tilted vehicle.

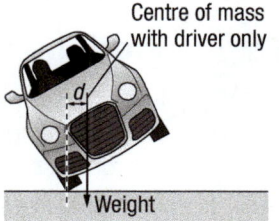

i The distance *d* in the diagram, is 50 cm. The mass of the vehicle is 1200 kg.
Calculate the moment of the weight about the point of contact with the road. (4)

ii Explain why the vehicle is stable. (3)

iii Explain how passengers make the vehicle less stable. (3)

3 The diagram shows a car hand brake.

a When the brake is used, explain why the applied force pulling on the cable is much greater than the force the driver applies to the hand brake handle. (3)

b Use the diagram to estimate the force pulling on the cable when a force of 35 N is applied by the driver to the brake. (3)

c The pivot is supported in a steel bracket. State and explain the approximate direction of the force on the pivot of the steel bracket. (3)

4 A student wanted to measure the weight of a small object X. She balanced a metre ruler at its centre of mass on a pivot. She suspended X and an object W of known weight from the ruler on opposite sides of the pivot. She then adjusted the positions of X and W until the ruler was balanced horizontally on the pivot.

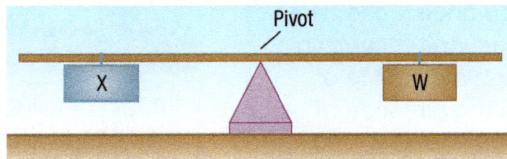

a The student measured the distances of W and X from the pivot P as follows: WP = 400 mm, XP = 260 mm. The weight of W was 3.0 N. Calculate the weight of X. (3)

b To check her results, she moved W away from P by a distance of 80 mm. Calculate how far she would need to have moved X and in which direction to balance the ruler again. (3)

5.1   Energy and work

Learning objectives

After this topic, you should know:

- what is meant by 'work' in science

- the relationship between work and energy

- how to calculate the work done by a force

- what happens to the work done to overcome friction.

Figure 1 *Working out*

Study tip

One joule of work is done when a force of 1 newton moves an object through a distance of one metre in the direction of the force.

Working out

In a fitness centre or a gym, you have to work hard to keep fit. Lifting weights and pedalling on an exercise bike are just two ways to keep fit. Whichever way you choose to keep fit, you have to apply a force to move something. So the work you do causes a **transfer** of energy.

The work done by a force depends on the size of the force and the distance moved. You can use the following equation to calculate the work done by a force when it moves an object:

work done = force applied × distance moved in the direction of the force
(joules, J)   (newtons, N)   (metres, m)

You can write the above word equation using symbols:

$$W = F \times d$$

where:
W = work done in J
F = force in N
d = distance moved in the direction of the force, in m.

When an object is moved by a force, **work** is done on the object by the force. The force therefore transfers energy to the object. The amount of energy transferred to the object is equal to the work done on it. For example, to raise an object, you need to apply a force to it to overcome the force of gravity on it. If the work you do on the object is 20 J, the energy transferred to it must be 20 J. So its gravitational potential energy (its energy due to its position) increases by 20 J.

energy transferred = work done

Worked example

A builder pushed a wheelbarrow a distance of 5 m across flat ground with a force of 50 N. How much work was done by the builder?

Solution

Work done = force applied × distance moved = 50 N × 5 m = **250 J**

 Did you know ... ?

Imagine pulling a truck a distance of 40 metres. On level ground, a pull force of about 2000 N is needed to do the necessary work of 80 000 N (= 2000 N × 40 m). Very few people can manage to pull with such force. Don't try it yourself – the people who can do it are very, very strong and have trained specially for it.

Figure 2 *Pulling a truck – hard work*

Practical

Doing work

Carry out a series of experiments to calculate the work done in performing the tasks below. Use a newtonmeter to measure the force applied and a metre ruler to measure the distance moved.

1 Drag a small box a measured distance across a rough surface.

2 Repeat the test above with two rubber bands wrapped around the box as shown in Figure 3.

• What is the resolution of your measuring instruments? Repeat your tests and comment on the precision of your repeat measurements. Can you be confident about the accuracy of your results?

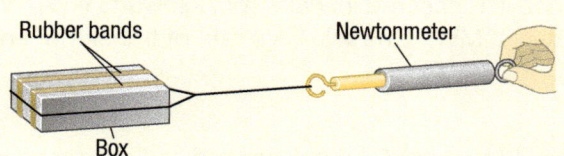

Figure 3 *At work*

Friction at work

Work done to overcome friction causes energy to be transferred, mainly by heating.

1 If you rub your hands together vigorously, they become warm. Your muscles do work to overcome the friction between your hands. The work you do is transferred to energy that warms your hands.

2 Brake pads become hot if the brakes are applied for too long a time. Friction between the brake pads and the wheel discs opposes the motion of the wheel. The kinetic energy of the vehicle (its energy due to its motion) is transferred to energy that heats the brake pads and the wheel discs, as well as the surrounding air. A small proportion of the energy will be transferred to the surroundings by sound waves if the brakes 'squeal'.

Summary questions

1 a State what happens to the energy transferred:
   i   by a rower rowing a boat
   ii by an electric motor used to raise a car park barrier.
   b How much work is done when a force of 2000 N pulls a truck through a distance of 40 m in the direction of the force?

2 A car is brought to a standstill when the driver applies the brakes.
   a Explain why the brake pads become warm.
   b The car travelled a distance of 20 metres after the brakes were applied. The braking force on the car during this time was 7000 N. Calculate the work done by the braking force.

3 a Calculate the work done when:
   i   a force of 20 N makes an object move 4.8 m in the direction of the force
   ii an object of weight 80 N is raised through a height of 1.2 m.
   b When a cyclist brakes, his kinetic energy is reduced from 1400 J to zero in a distance of 7.0 m. Calculate the braking force.

??? Did you know ...?

Meteorites are small objects from space that enter the Earth's atmosphere and fall to the ground. As they pass through the atmosphere, friction caused by air resistance acts on them and kinetic energy is transferred by heating. If a meteorite becomes hot enough, it glows and becomes visible as a 'shooting star'. Very small objects may burn up completely. The surface of a space vehicle is designed to withstand the very high temperatures caused by such friction when it re-enters the Earth's atmosphere.

Key points

• Work is done on an object when a force makes the object move or stop.

• Energy transferred = work done.

• Work done (J) = force (N) × distance moved in the direction of the force (m).

• Work done to overcome friction is transferred as energy that heats the objects that rub together and heats the surroundings.

5.2 Power

Learning objectives

After this topic, you should know:

- what is meant by 'power' in science
- how power is measured
- the relationship between power and energy transferred.

Figure 1 *Rocket power for launching into space*

??? Did you know … ?

The unit of power is named after James Watt, who invented the steam engine. Before this invention, horse-drawn carriages and barges were used to transport people and goods overland. When steam engines first replaced horses, each engine was rated in terms of the number of horses it could replace. The output power of a road vehicle is still sometimes expressed in 'horsepower', where 1 horsepower is equal to 756 watts.

Powerful machines

When you use a lift (elevator), a powerful electric motor pulls you and the lift up. The work done by the lift motor transfers energy from electricity to gravitational potential energy. Sound and energy that heats the motor and its surroundings are also transferred.

- The work done per second by the motor is the output **power** of the motor.
- The more powerful the lift motor, the faster it takes you up.

The power of an appliance is measured in watts (W), kilowatts (kW), or megawatts (millions of watts, MW). One **watt** is a rate of transfer of energy of 1 joule per second (J/s).

For example:

- a 5 W electric torch would transfer 5 J every second as light energy and energy that heats the surroundings
- a lift motor with an output power of 6000 W would transfer 6000 J to the lift as gravitational potential energy every second.

Here are typical values of power levels for some energy transfer 'mechanisms':

- a torch   1 W
- an electric light bulb   100 W
- an electric cooker   10 000 W = 10 kW (where 1 kW = 1000 watts)
- a railway engine   1 000 000 W = 1 megawatt (MW) = 1 million watts
- a Saturn V rocket   100 MW
- a very large power station   10 000 MW
- the Sun   100 000 000 000 000 000 000 MW.

Energy and power

Machines are labour-saving devices that do work for us. The faster a machine can do work, the more powerful it is.

Whenever a machine does work on an object, energy is transferred to the object. The useful energy transferred is equal to the work done by the machine. This is the form of energy needed that is transferred to the object.

The output power of a machine is the rate at which it does work. This is the same as the rate at which it transfers useful energy.

$$\text{power } P \text{ (watts, W)} = \frac{\text{work done (joules, J)}}{\text{time taken (seconds, s)}} = \frac{\text{useful energy transferred (J)}}{\text{time taken (s)}}$$

If work W is done (or energy E is transferred) in time t:

$$P = \frac{W}{t} = \frac{E}{t}$$

where:
P = power in W
W = work in J
E = energy in J
t = time in s.

Worked example

A crane lifts an object of weight 4000 N through a vertical distance of 2.5 m in 5.0 s. Calculate:

a the force needed to lift the object steadily

b how much gravitational potential energy the object gains

c the output power of the crane.

Solution

a Force needed = weight of object = **4000 N**

b Work done on the object = force needed × distance moved
$$= 4000 \text{ N} \times 2.5 \text{ m} = \textbf{10 000 J}$$

c Output power = $\dfrac{\text{work done}}{\text{time taken}}$

$$= \dfrac{10\,000 \text{ J}}{5.0 \text{ s}}$$

$$= \textbf{2000 W}$$

Figure 2 *A crane at work*

??? Did you know …?

How powerful is a weightlifter?

A dumbbell of mass 30 kg has a weight of 300 N. Raising it by 1 m would give it 300 J of gravitational potential energy. A weightlifter could lift it in about 0.5 seconds.

The rate of transfer of energy would be:

$$P = \frac{E}{t} = \frac{300 \text{ J}}{0.5 \text{ s}} = 600 \text{ W}$$

So, the weightlifter's output power would be about 600 W.

Study tip

Be careful with units – power is sometimes expressed in kilowatts.

Summary questions

1 a Which is more powerful:

    i a torch bulb or a mains-powered filament lamp?

    ii a 3 kW electric kettle or a 10 000 W electric cooker?

b There are about 2 million homes in a certain city. If a 3 kW electric kettle was switched on in 1 out of 10 homes in the city at the same time, how much extra power would need to be supplied?

2 An electric motor raises an object of weight 200 N by a height of 4.0 m in 5.0 s. Calculate:

    a how much gravitational potential energy the object gains

    b the work done by the motor on the object

    c the output power of the motor.

3 The engine of a goods vehicle has an output power of 150 kW when the vehicle is travelling at a constant velocity of 30 m/s on a level road.

    a Calculate:

        i the distance travelled by the vehicle in 60 seconds

        ii the useful energy transferred by the engine in this time.

    b Show that the driving force of the engine is 5000 N.

Key points

- Power is the rate at which energy is transferred.

- The unit of power is the watt (W), which is equal to 1 J/s.

- Power (W) = $\dfrac{\text{energy transferred (J)}}{\text{time taken (s)}}$

5.3 Gravitational potential energy

Gravitational potential energy transfers

Every time you lift an object up, you do some work. Some of your muscles transfer chemical energy from your muscles into **gravitational potential energy** of the object. Gravitational potential (sometimes called GPE or E_p) is energy stored in an object because of its position in the Earth's gravitational field.

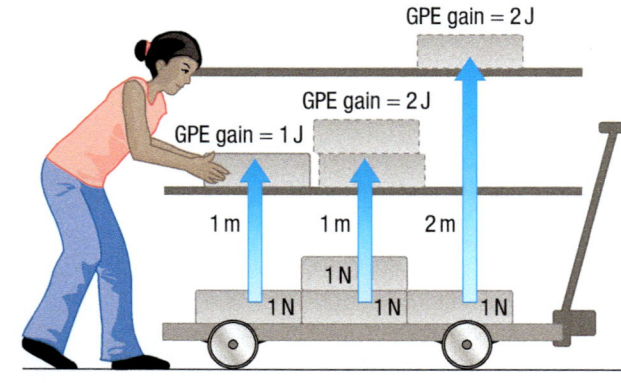

Figure 1 *GPE transfers*

The force you need to lift an object steadily is equal and opposite to the gravitational force on the object. Therefore, the upward force you need to apply to it is equal to the object's weight. For example, a force of 80 N is needed to lift a box of weight 80 N.

- **When an object is moved up**, its gravitational potential energy increases. The increase in its gravitational potential energy is equal to the work done on it by the lifting force to overcome the gravitational force on the object.
- **When an object moves down**, its gravitational potential energy decreases. The decrease in its gravitational potential energy is equal to the work done by the gravitational force acting on it as it moves down.

The work done when an object moves up or down depends on:

1 how far it is moved vertically (its change of height)
2 its weight.

You know from Topic 5.1 'Energy and work' that $W = F \times d$, or work done = force applied × distance moved in the direction of the force. Therefore:

$$\text{change of gravitational potential energy (J)} = \text{weight (N)} \times \text{change of height (m)}$$

Gravitational potential energy and mass

Astronauts on the Moon can lift objects much more easily than they can on the Earth. This is because, at their surfaces, the gravitational field strength of the Moon is only about a sixth of the Earth's gravitational field strength.

In Topic 1.4 'Mass and weight', you saw that the weight of an object in newtons is equal to its mass × the gravitational field strength.

Therefore, when an object is lifted or lowered, because its change of gravitational potential energy is equal to its weight × its change of height:

$$\text{change of gravitational potential energy (J)} = \text{mass (kg)} \times \text{gravitational field strength (N/kg)} \times \text{change of height (m)}$$

Worked example

A student of weight 300 N climbs on a platform that is 1.2 m higher than the floor. Calculate the increase of her gravitational potential energy.

Solution

Increase of GPE = 300 N × 1.2 m
= **360 J**

You can write the equation at the bottom of the previous page using symbols:

$$E_p = m \times g \times h$$

where:
E_p = change of gravitational potential energy in J
m = mass in kg
g = gravitational field strength in N/kg
h = change of height in m.

Worked example

An object of mass 2.0 kg is raised through a height of 0.4 m. Calculate the gain of gravitational potential energy of the object. The gravitational field strength of the Earth at its surface is 10 N/kg.

Solution

E_p = mass × gravitational field strength × height gain
= 2.0 kg × 10 N/kg × 0.4 m
= **8.0 J**

Summary questions

1 a Describe the energy changes of a ball when it falls and rebounds without regaining its initial height.
  b When a ball of weight 1.4 N is dropped from rest from a height of 2.5 m above a flat surface, it rebounds to a height of 1.7 m above the surface.
    i   Calculate the total loss of energy of the ball on reaching this maximum rebound height.
    ii State two causes of the energy loss.

2 A student of weight 450 N steps on a box of height 0.20 m.
  a Calculate the gain of gravitational potential energy of the student.
  b Calculate the work done by the student if she steps on and off the box 50 times.

3 a A weightlifter raises a steel bar of mass 25 kg through a height of 1.8 m. Calculate the change of gravitational potential energy of the bar. The gravitational field strength at the surface of the Earth is 10 N/kg.
  b The weightlifter then lowers the bar by 0.3 m and throws it so it falls to the ground. Assume that air resistance is unimportant. What is the change of its gravitational potential energy in this fall?

4 Read the 'Did you know?' box on the previous page. Explain what happens to the energy supplied to the muscles to keep them contracted.

Key points

- The gravitational potential energy of an object increases when it moves up and decreases when the object moves down.

- An object gains gravitational potential energy when it is lifted up because work is done on it to overcome the gravitational force.

- The change of gravitational potential energy of an object is equal to its mass × the gravitational field strength × its change of height.

5.4 Conservation of energy

Figure 1 *Energy transfers on a roller coaster*

At the funfair

Funfairs are very exciting places because lots of energy transfers happen quickly. A roller coaster gains gravitational potential energy when it climbs. This energy is then transferred as the roller coaster races downwards.

As it descends:

its gravitational → kinetic + sound + energy transfer by heating due
potential energy   energy   to air resistance and friction

The energy transferred by heating is 'wasted' energy, which you will learn more about in Topic 5.6 'Useful energy'.

Practical

Investigating energy changes

Pendulum swinging

When energy changes happen, does the total amount of energy stay the same? You can investigate this question with a simple pendulum.

Figure 2 shows a pendulum bob swinging from side to side.

As it moves towards the middle, its gravitational potential energy is transferred to kinetic energy.

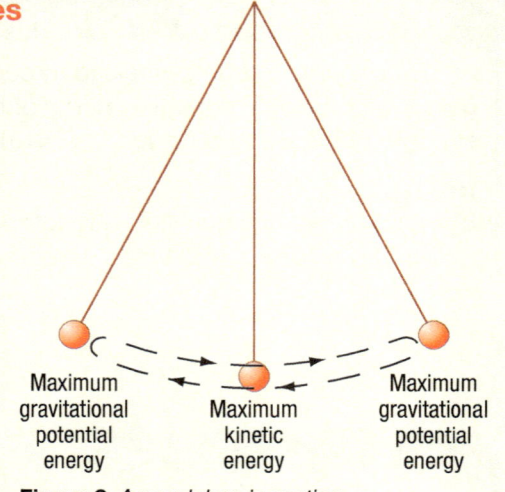

Maximum gravitational potential energy — Maximum kinetic energy — Maximum gravitational potential energy

Figure 2 *A pendulum in motion*

As it moves away from the middle, its kinetic energy transfers back to gravitational potential energy. If the air resistance on the bob is very small, you should find that the bob reaches the same height on each side.

- What does this tell you about the energy of the bob when it goes from one side at maximum height to the other side at maximum height?
- Why is it difficult to mark the exact height the pendulum bob rises to? How could you make your judgement of height more accurate?

Conservation of energy

Scientists have done lots of tests to find out if the total energy after a transfer is the same as the energy before the transfer. All the tests so far show it is the same.

This important result is known as the **conservation of energy**.

It tells us that **energy cannot be created or destroyed**.

Energy can be stored in various ways. For example:
- elastic energy is stored in a rubber band by stretching it
- gravitational potential energy is stored in an object when it is lifted.

Study tip

Never use the terms 'movement energy' or 'motion energy' in an exam – you will only gain marks by using 'kinetic energy'.

Bungee jumping

What energy transfers happen to a bungee jumper after jumping off the platform?

- When the rope is slack, some of the gravitational potential energy of the bungee jumper is transferred to kinetic energy as the jumper falls.
- Once the slack in the rope has been used up, the rope slows the bungee jumper's fall. Most of the gravitational potential energy and kinetic energy of the jumper is transferred into elastic potential energy of the rope. This elastic energy is stored in the rope.
- After reaching the bottom, the rope pulls the jumper back up. As the jumper rises, most of the elastic energy of the rope is transferred back to gravitational potential energy and kinetic energy of the jumper.

The bungee jumper doesn't return to the same height as at the start. This is because some of the initial gravitational potential energy has been transferred to the surroundings by heating as the rope stretched then shortened again.

Figure 3 *Bungee jumping*

Practical

Bungee jumping

You can try out the ideas about bungee jumping using the experiment shown in Figure 4.

Safety: Make sure the stand is secure. Protect the bench and your feet from falling objects.

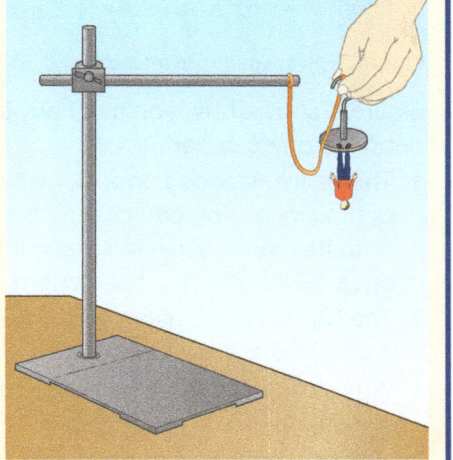

Figure 4 *Testing a bungee jump*

Summary questions

1 When a roller coaster gets to the bottom of a descent, what energy transfers happen if:
   a the brakes are applied to stop it?
   b it goes up and over a second 'hill'?

2 a A ball dropped onto a trampoline returns to almost the same height after it bounces. Describe the energy transfer of the ball from the point of release to the top of its bounce.
   b What can you say about the energy of the ball at the point of release compared with at the top of its bounce?
   c Describe how would you use the test in a above to see which of three trampolines is the bounciest.

3 One exciting fairground ride acts like a giant catapult. The capsule, in which you are strapped, is fired high into the sky by the rubber bands of the catapult. Explain the energy transfers taking place in the ride.

Key points

- Energy cannot be created or destroyed.
- Conservation of energy applies to all energy changes.
- Energy can be stored in various ways.

5.5 Kinetic energy

Learning objectives

After this topic, you should know:

- what the kinetic energy of an object depends on
- how to calculate kinetic energy
- what elastic potential energy is
- how to calculate the elastic potential energy stored in a stretched spring.

Did you know …?

Sports scientists design running shoes:

- to reduce the force of each impact when the runner's foot hits the ground
- to return as much kinetic energy as possible to the foot in each impact.

Figure 2 *A sports shoe*

Practical

Investigating a catapult

Use rubber bands to catapult a trolley along a horizontal runway. Find out how the speed of the trolley depends on how much the catapult is pulled back before the trolley is released. For example, see if the distance needs to be doubled to double the speed. Figure 1 shows how the speed of the trolley can be measured.

Practical

Investigating kinetic energy

- The **kinetic energy** of an object is the energy it has because of its motion. Its kinetic energy depends on its mass and its speed.

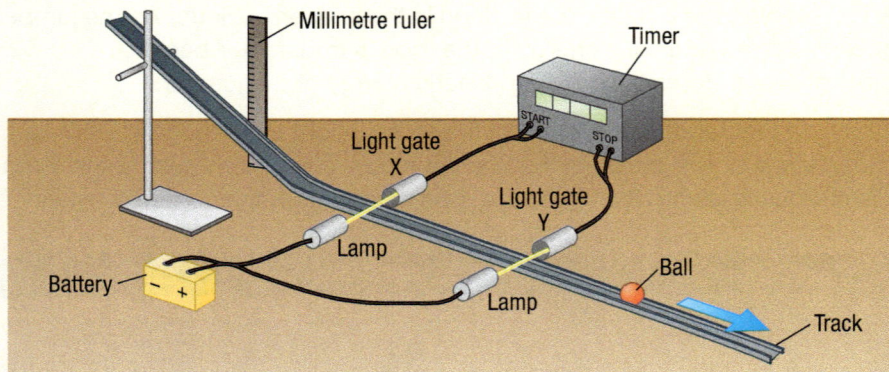

Figure 1 *Investigating kinetic energy*

Figure 1 shows how you can investigate how the kinetic energy of a ball depends on its speed.

1. The ball is released on a slope from a measured height above the foot of the slope. You can calculate the gravitational potential energy it loses from its mass × gravitational field strength × its fall in height. This is equal to its gain of kinetic energy.

2. The ball is timed, using light gates, over a measured distance between X and Y after the slope.

- Why do light gates improve the quality of the data you can collect in this investigation?

Some sample measurements for a ball of mass 0.5 kg are shown in Table 1.

Table 1 *Results of investigating kinetic energy*

Fall in height to foot of slope (metres, m)	0.05	0.10	0.16	0.20
Initial kinetic energy of ball (joules, J)	0.25	0.50	0.80	1.00
Time to travel 1.0 m from X to Y (seconds, s)	0.98	0.72	0.57	0.50
Speed (metres/second, m/s)	1.02			2.00

Work out the speed in each case. The first and last values have been worked out for you. Can you see a link between speed and the fall in height? The results show that the greater the fall in height, the faster the speed is. So the kinetic energy of the ball increases if the speed increases.

The kinetic energy formula

Table 1 shows that when the fall in height is increased by four times from 0.05 m to 0.20 m, the speed doubles. The height drop is directly proportional to the (speed)^2. Since the fall in height is proportional to the ball's kinetic energy, the ball's kinetic energy is directly proportional to the square of its speed.

The exact link between the kinetic energy of an object and its speed is given by the equation:

kinetic energy = ½ × mass × speed²
(joules, J)   (kilograms, kg)   (metres/second, m/s)²

You can write this word equation using symbols:

$$E_K = \tfrac{1}{2} \times m \times v^2$$

where:
E_K = kinetic energy in J
m = mass in kg
v = speed in m/s.

Using elastic potential energy

When you stretch a rubber band or a bowstring, the work you do is stored in it as **elastic potential energy**. As explained in Topic 1.5 'Forces and elasticity', the force F needed to stretch a spring varies with its extension e and is given by the equation for Hooke's Law: $F = k \times e$, where k is the spring constant.

For a spring stretched to an extension e,

its elastic potential energy = ½ × the spring constant × (its extension)²

You can write this word equation using symbols:

$$E_e = \tfrac{1}{2} \times k \times e^2$$

where:
E_e = elastic potential energy in J
e = extension in m
k = spring constant in N/m.

Elastic potential energy is the energy stored in an elastic object when work is done on it to change its shape.

Figure 3 *Using elastic potential energy*

Summary questions

1 **a** Calculate the kinetic energy of:
    i a vehicle of mass 500 kg moving at a speed of 12 m/s
    ii a football of mass 0.44 kg moving at a speed of 20 m/s.
  b Calculate the velocity of a vehicle of mass 500 kg with twice as much kinetic energy as you calculated in **a i**.

2 **a** A catapult is used to fire an object into the air. Describe the energy transfers when the catapult is:
    i stretched
    ii released.
  b An object of weight 2.0 N fired vertically upwards from a catapult reaches a maximum height of 5.0 m. Calculate:
    i the gain of gravitational potential energy of the object
    ii the speed of the object when it left the catapult.

3 A car moving at a constant speed has 360 000 J of kinetic energy. When the driver applies the brakes, the car stops in a distance of 100 m.
  a Calculate the force that stops the vehicle.
  b The speed of the car was 30 m/s when its kinetic energy was 360 000 J. Calculate its mass.

4 A mobility aid to assist walking uses a steel spring to store energy when the walker's foot goes down, and it returns energy as the foot is lifted. The spring has a spring constant of 250 N/m. Calculate the elastic potential energy stored in the spring when its extension is 0.21 m.

5.6 Useful energy

Learning objectives

After this topic, you should know:

- what 'useful' energy is
- what is meant by 'wasted' energy
- what eventually happens to wasted energy
- whether energy is still as useful after it is used.

Energy for a purpose

Can you imagine a world without **machines**? People use washing machines at home. Manufacturers use machines in factories to make products. Peope use machines in the gym to keep fit and for transport from place to place.

A machine transfers energy for a purpose. Friction between the moving parts of a machine causes the parts to warm up. So, not all of the energy supplied to a machine is usefully transferred. Some energy is wasted.

- **Useful energy** is energy transferred to where it is wanted, in the form that is wanted.
- **Wasted energy** is energy that is not usefully transferred.

Practical

Investigating friction

Friction in machines causes energy to be wasted. Figure 2 shows two examples of friction in action. Try one of them out.

In **a**, friction acts between the drill bit and the wood. The bit becomes hot as it bores into the wood. Some of the electrical energy supplied to the bit heats up the drill bit (and the wood).

In **b**, when the brakes are applied, friction acts between the brake blocks and the wheel. This slows the bicycle and the cyclist down. Some of the kinetic energy of the bicycle and the cyclist is transferred by heating to the brake blocks (and the bicycle wheel).

a

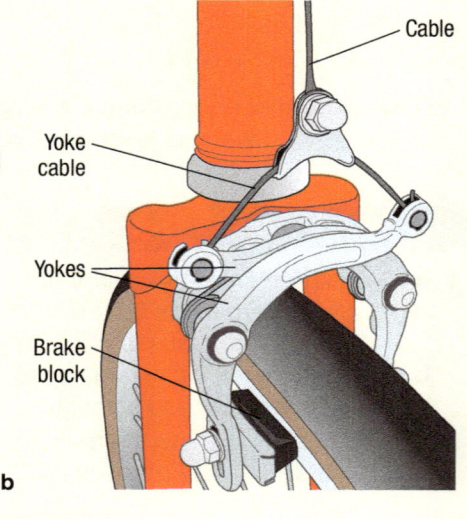
b

Figure 2 *Friction in action*
a Using a drill b Braking on a bicycle

Safety: Use a battery drill. Do not touch the drill bit or wheel until they have stopped moving.

Figure 1 *Using energy*

Study tip

Try to decide what type of energy you want from an appliance – for example, from a lamp you want to light.

??? Did you know ...?

Lots of energy is transferred in a car crash. The faster the car travels, the more kinetic energy it has and the more it has to transfer before stopping. In a crash, kinetic energy is quickly transferred to elastic strain energy, distorting the car's shape, and energy is transferred by heating the metal. There is usually quite a lot of sound too!

Disc brakes at work

The next time you are in a car slowing down at traffic lights, think about what is making the car stop. Figure 3 shows how the disc brakes of a car work. When the brakes are applied, the pads are pushed on to the disc in each wheel. Friction between the pads and each disc slows the wheel down. Some of the kinetic energy of the car is transferred by heating to the disc pads and the discs. In Formula One racing cars you can sometimes see the discs glow red hot.

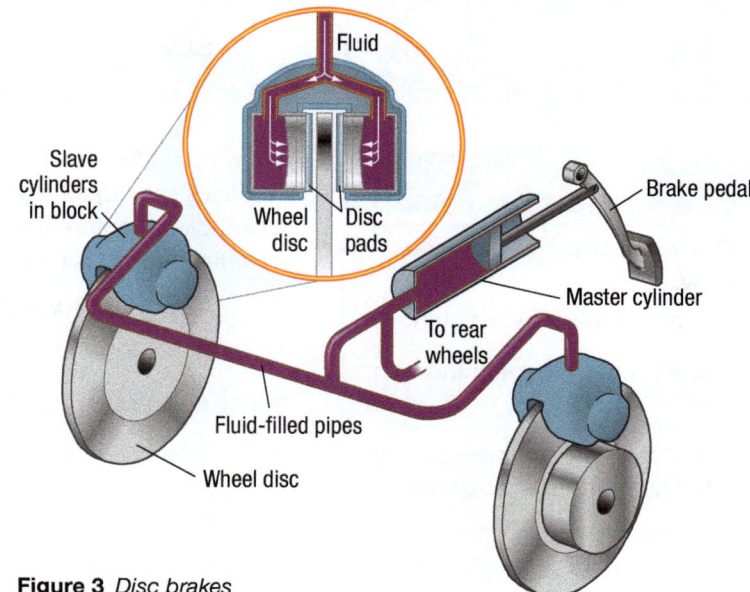

Figure 3 *Disc brakes*

Spreading out

- **Wasted energy is dissipated (spreads out) to the surroundings.**
  For example, the gears of a car get hot because of friction when the car is running. So energy transfers from the gear box to the surrounding air.
- **Useful energy eventually transfers to the surroundings too.**
  For example, the useful energy supplied to the wheels of a car is transferred by heating to the tyres. This energy is then transferred to the road and the surrounding air.
- **Energy becomes less useful the more it spreads out.**
  For example, the hot water from the cooling system of a combined heat and power (CHP) power station is used to heat nearby buildings. The energy supplied to heat the buildings will eventually be transferred to the surroundings.

Summary questions

1 Copy and complete the table below.

Energy transfer by	Useful energy output	Wasted energy output
a An electric fan heater	warms the air and surrounding objects	
b A television		
c An electric kettle		
d Headphones		

2 What will happen, in terms of energy transfer, to:
  a a gear box that is insulated so it cannot transfer energy by heating to the surroundings?
  b the running shoes of a jogger if the shoes are well insulated?
  c a blunt electric drill if you use it to drill into hard wood?
  d the metal wheel discs of the disc brakes of a car when the brakes are applied?

3 **a** Describe the energy transfers of the pendulum in Topic 5.5, as it swings from one side to the middle, then to the opposite side.
  b Explain why a swinging pendulum eventually stops.

Key points

- Useful energy is energy in a useful form.
- Wasted energy is energy that is not useful energy.
- Wasted energy is eventually transferred to the surroundings, which become warmer.
- As energy spreads out (dissipates), it gets less and less useful.

5.7 Energy and efficiency

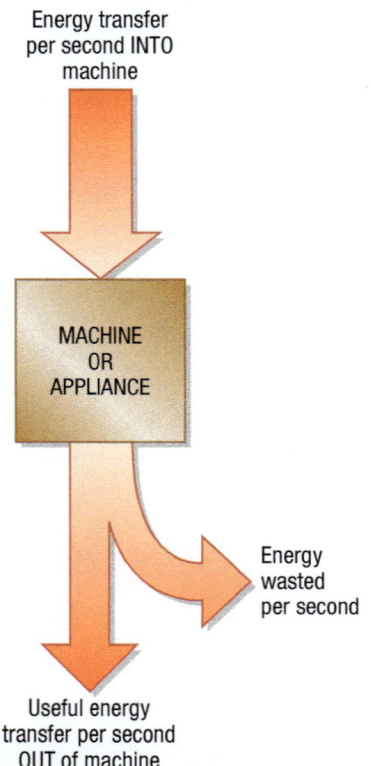

Energy transfer per second INTO machine

MACHINE OR APPLIANCE

Energy wasted per second

Useful energy transfer per second OUT of machine

Figure 1 *A Sankey diagram. The width of each arrow can be drawn to scale to represent the amount of energy transferred*

When you lift an object, the energy from your muscles that is usefully transferred increases the object's gravitational potential energy. As you saw in Topic 5.3 'Gravitational potential energy', you can calculate the gain in gravitational potential energy using the formula:

change in gravitational potential energy = weight of object × gain in height
        (joules, J)   (newtons, N)   (metres, m)

Sankey diagrams

Figure 1 represents the energy flow through a system. It shows how you can represent any energy transfer in which energy is wasted. This type of diagram is called a **Sankey diagram**.

Because energy cannot be created or destroyed:

input energy (energy supplied) = useful energy transferred + energy wasted

For any device that transfers energy:

$$\text{efficiency} = \frac{\text{useful energy transferred out of the device}}{\text{total energy supplied into the device}} \times 100\%$$

Note: Because power is energy transferred per second, you can also calculate efficiency in terms of power, using the equation:

$$\text{efficiency} = \frac{\text{useful power out}}{\text{total power in}} \times 100\%$$

Maths skills

Efficiency can be written as a number (which is never more than 1), or as a percentage (which is never more than 100%).

For example, a light bulb with an efficiency of 0.15 would radiate 15 J of energy as light for every 100 J of electrical energy supplied to it.

- Its efficiency (as a number) $= \frac{15}{100} = 0.15$

- Its percentage efficiency $= 0.15 \times 100\% = 15\%$

Worked example

An electric motor is used to raise an object. The object gains 60 J of gravitational potential energy when the motor is supplied with 200 J of electrical energy. Calculate the percentage efficiency of the motor.

Solution

Total energy supplied to the device = 200 J
Useful energy transferred by the device = 60 J
Percentage efficiency of the motor

$$= \frac{\text{useful energy transferred out of the motor}}{\text{total energy supplied into the motor}} \times 100\%$$

$$= \frac{60\,\text{J}}{200\,\text{J}} \times 100\% = 0.30 \times 100\% = \textbf{30\%}$$

Study tip

Remember that in a calculation, if you obtain an answer for efficiency of over 100% or greater than 1, you have made a mistake.

Efficiency limits

No machine can be more than 100% efficient, because you can never get more energy from a machine than you put into it.

Practical

Investigating efficiency

Figure 2 shows how you can use an electric winch to raise a weight. You can use the joulemeter to measure the electrical energy supplied.

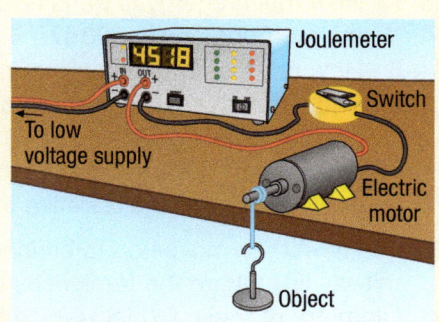

Figure 2 *An electric winch*

- If you double the weight for the same increase in height, do you need to supply twice as much electrical energy to do this task?

gravitational potential energy (J) gained by the weight
= weight (N) × height increase (m)

- Use this equation and the joulemeter measurements to work out the percentage efficiency of the winch.

Safety: Protect the floor and your feet from falling objects. Stop the winch before the masses wrap around the pulley.

Improving efficiency

	Why machines waste energy	How to reduce the problem
1	Friction between the moving parts causes heating.	Lubricate the moving parts to reduce friction.
2	The resistance of a wire causes the wire to get hot when a current passes through it.	In circuits, use wires with as little electrical resistance as possible.
3	Air resistance causes energy transfer by heating the surroundings.	Streamline the shapes of moving objects to reduce air resistance.
4	Sound created by machinery causes energy transfer to the surroundings.	Cut out noise (e.g., tighten loose parts to reduce vibration).

Summary questions

1. A certain light bulb has an efficiency of 15%.
   a How much energy is wasted for every 100 J of electrical energy supplied?
   b What happens to the wasted energy?
   c Draw a Sankey diagram for the light bulb showing the useful and wasted energies.

2. An electric motor is used to raise a weight. When you supply 60 J of electrical energy to the motor, the weight gains 24 J of gravitational potential energy. Work out:
   a the energy wasted by the motor
   b the efficiency of the motor.

3. A machine is 25% efficient. If the total energy supplied to the machine is 3200 J:
   a how much useful energy can be transferred?
   b what is the output power of the machine if the energy is supplied in 16 seconds?

Key points

- The efficiency of a device = useful energy transferred out of the device ÷ total energy supplied into the device (×100%).

- Devices can be made more efficient, but never more than 100% efficient.

- Sankey diagrams are used to show energy flow in a system.

Chapter summary questions

1 A train on a straight level track is pulled at a constant speed of 23 m/s by an engine with an output power of 700 kW.

a i How much energy is transferred from the train to the surroundings in 300 s?

ii How far does the train travel in 300 s?

iii Show that the resistive force on the engine is approximately equal to 30 000 N.

iv Explain why the driving force of the engine is equal and opposite to the resistive force on the train.

b The train then moves on to an inclined section of the railway line where the track rises by 1 m for every kilometre of track. Explain why the output power of the engine needs to be increased to maintain the same speed of 23 m/s.

2 A student pushes a trolley of weight 150 N up a slope of length 20 m. The slope is 1.2 m high.

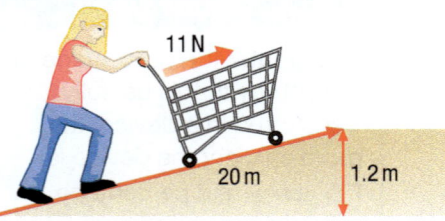

a Calculate the gravitational potential energy gained by the trolley.

b The student pushes the trolley up the slope with a force of 11 N. Show that the work done by the student is 220 J.

c If the student pushes the trolley up the slope at constant speed, explain why all the work done by the student is not transferred to the trolley as gravitational potential energy.

3 a A low-energy light bulb has an efficiency of 80%. Using an energy meter, a student found the light bulb used 1200 J of electrical energy in 100 s.

i How much useful energy did the light bulb transfer in this time?

ii How much energy was wasted by the light bulb?

iii Draw a Sankey diagram for the light bulb.

b A filament light bulb that gives the same light intensity as the low-energy light bulb has an efficiency of 16%. Calculate the energy wasted by this filament light bulb in 100 s.

4 On a building site, an electric winch and a pulley are used to lift bricks from the ground.

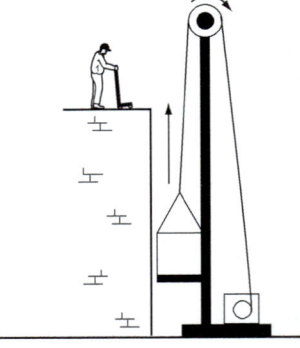

The winch raises a load of 500 N through a height of 3.0 m in 25 s. During this time, the average power supplied to the winch is 600 W.

a i How much useful energy is transferred by the motor?

ii Calculate the energy wasted.

iii Calculate the percentage efficiency of the system.

b i How could the efficiency of the winch be improved?

ii Explain why the efficiency of a winch can never be as much as 100%.

5 a A stone is fired into the air from a catapult and falls to the ground some distance away. Describe the energy transfers that take place after the catapult is released.

b A stone of mass 0.015 kg is catapulted into the air at a speed of 25 m/s. It gains a height of 20 m before it descends and hits the ground some distance away.

i Calculate its initial kinetic energy.

ii Calculate the increase of gravitational potential energy of the stone when it reaches its maximum height (g = 10 N/kg).

iii Estimate its speed when it is at its maximum height.

6 A parachutist of total mass 75 kg jumps from an aeroplane moving at a speed of 60 m/s at a height of 900 m above the ground.

a Calculate her kinetic energy when she leaves the aeroplane.

b Her parachute reduces her speed of descent to 5 m/s.

i Calculate her kinetic energy at this speed.

ii Calculate her loss of gravitational potential energy as a result of the descent.

c Calculate the work done by air resistance during her descent.

Practice questions

1 Energy is measured in various units.

a Which of the following are units of energy?

joule   kilojoule   kilowatt   newton   watt   (2)

b The picture shows a woman using a step machine in a gym.

The display panel shows the readings at the end of the exercise.

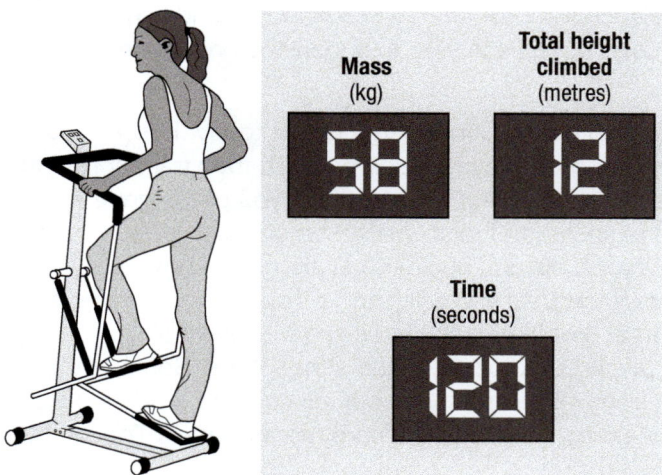

i Calculate the woman's weight.   (2)

ii Calculate the total amount of work done during the exercise.   (2)

iii Calculate the average power developed.   (3)

2 A high jumper runs along a short track before taking off in the jump.

The high jumper has a mass of 65 kg.

To jump over the bar, the high jumper must raise his mass 1.25 m.

a i Discuss the energy transfers that take place.   (3)

ii Calculate the gain in the high jumper's gravitational potential when he just clears the bar.   (2)

b i Calculate the speed the high jumper must reach along the track in order to just clear the bar.   (4)

ii Explain why the high jumper's speed will need to be more than this.   (3)

3 a Two students carry out an experiment to measure their own power by running up some stairs.

Describe how the students should carry out the experiment to obtain the data needed.

Your description should include the equipment they will need.   (6)

b One of the students has a mass of 45 kg. The stairs are 400 cm high and she takes 2.5 s to climb them.

Calculate her power.   (3)

4 a A ball of mass 200 g is dropped from a height of 2 m. It hits the floor and bounces back to a height of 1.5 m.

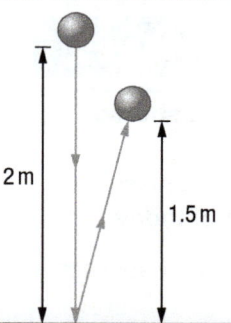

i What type of energy does the ball have just before it is released?

Choose from the list.

gravitational potential   kinetic   strain potential   thermal   (1)

ii What percentage of its energy does the ball lose when it hits the floor?   (3)

iii Explain what happens to the 'lost' energy and why it is difficult to re-use.   (3)

b i Calculate the kinetic energy of the ball just before it hits the floor.   (3)

ii Calculate the speed of the ball just before it hits the floor.   (4)

iii Explain why the ball accelerates when it hits the floor, even though its speed is less when it leaves the floor.   (4)

5 A battery-powered mobility scooter and its driver have a total mass of 220 kg and can travel at 1.8 m/s on a flat pavement. Its batteries supply power to its electric motor.

a On a level surface, the driver takes 5.0 seconds to accelerate the scooter to a speed of 1.8 m/s from rest. Calculate the resultant force acting on the scooter.   (2)

b When the scooter is moving at a speed of 1.8 m/s on a level surface, the driving force is the same as the value calculated in **a**.

i Calculate the work done by the motor in 1 second when its speed is 1.8 m/s.   (2)

ii The battery supplies 500 J of electrical energy to the motor each second when the scooter is moving at a speed of 1.8 m/s. Calculate the percentage efficiency of the electric motor.   (2)

Energy demands

Learning objectives

After this topic, you should know:

- how most of your energy demands are met today

- what other energy resources are used

- how nuclear fuels are used in power stations

- what other fuels are used to generate electricity.

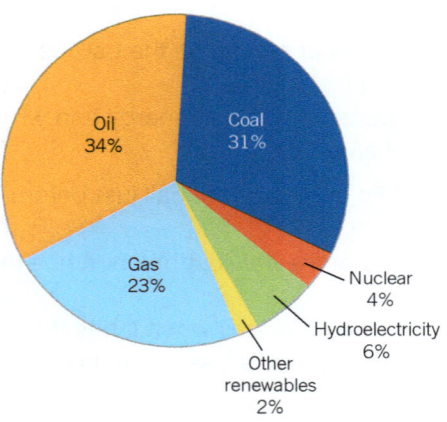

Figure 1 *World energy demand and sources of energy in 2010*

Most of the energy you use comes from burning fossil fuels, mostly gas or oil or coal. The energy in homes, offices, and factories is mostly supplied by gas or by electricity generated in coal or gas-fired power stations. Oil is needed to keep road vehicles, ships, and aeroplanes moving. Burning one kilogram of fossil fuel releases about 30 million joules of energy. You use about 5000 joules of energy each second, which is about 150 thousand million joules each year. But because of the inefficiencies in how energy is distributed and used, a staggering 10 000 kg of fuel is used each year to supply the energy needed just for you!

Figure 1 shows how the global demand for energy is met today. Fossil fuels are extracted from underground or under the sea bed and then transported to oil refineries and power stations. Much of the electricity you use is generated in fossil-fuel power stations. Instead of fossil fuels, some power stations use biofuels or nuclear fuel. Fossil fuels and nuclear fuel are non-renewable because they cannot be replaced. As you will learn in Topic 6.4, their use is causing major environmental problems and increasing the levels of greenhouse gases, such as carbon dioxide, in the atmosphere. Some of the electricity you use is from renewable energy resources such as wind energy, hydroelectricity, and solar energy, which you'll learn more about in Topics 6.2 and 6.3.

Inside a power station

In coal- or oil-fired power stations, and in most gas-fired power stations, the burning fuel heats water in a boiler. This produces steam. The steam drives a **turbine** that turns an electricity **generator**. Coal, oil, and gas are fossil fuels, which are fuels that come from long-dead animals and plants.

Biofuels

Methane gas can be collected from cows or animal manure, from sewage works, decaying rubbish, and other resources. It can be used in small gas-fired power stations. Methane is an example of a biofuel.

A **biofuel** is any fuel taken from living or recently living organisms. Animal waste is an example of a biofuel. Biofuels can be used instead of fossil fuel in modified engines for transport and in generators at power stations. Biodiesel uses waste vegetable oil and plants such as rapeseed. Other examples of biofuels are ethanol (from fermented sugar cane), straw, nutshells, and woodchip.

A biofuel is:

- **renewable** because its biological source either regrows (vegetation) or is continually produced (sewage and rubbish). This means it is used at the same rate that it is replaced.

- **carbon-neutral** because, in theory, the carbon that the living organism takes in from the atmosphere as carbon dioxide can balance the amount that is released when the biofuel is burnt.

Figure 2 *Using biofuel to generate electricity*

Nuclear power

Nuclear fuel takes energy from atoms. Figure 3 shows that every atom contains a positively charged **nucleus** surrounded by electrons.

The fuel in a nuclear power station is uranium (or plutonium). The uranium fuel is in sealed cans in the core of the reactor. The nucleus of a uranium atom is unstable and can split in two. Energy is released when this happens. Because there are lots of uranium atoms in the core, it becomes very hot.

The energy of the core is transferred by a fluid (called the coolant) that is pumped through the core.

- The coolant is very hot when it leaves the core. It flows through a pipe to a heat exchanger, then back to the reactor core.

- The energy transferred by the coolant is used to turn water into steam in the heat exchanger. The steam drives turbines that turn electricity generators.

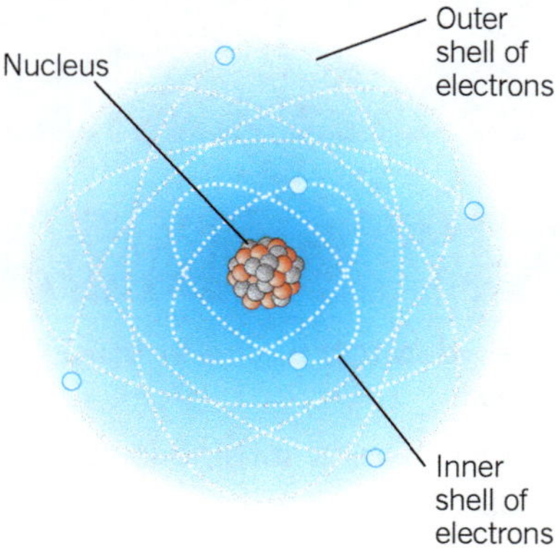

Figure 3 *The structure of the atom*

Table 1 *Comparing nuclear power and fossil fuel power*

	Nuclear power station	Fossil fuel power station
Fuel	Uranium or plutonium	Coal, oil, or gas
Energy released per kg of fuel	$\approx 300\,000$ MJ (= about $10\,000 \times$ energy released per kg of fossil fuel)	≈ 30 MJ
Waste	Radioactive waste that needs to be stored for many years	Non-radioactive waste
Greenhouse gases (e.g., carbon dioxide)	No – because uranium releases energy without burning	Yes – because fossil fuels produce gases such as carbon dioxide when they burn

Summary questions

1 a i Name the types of power stations that release carbon dioxide into the atmosphere.
   ii Name the type of power station that does not release carbon dioxide into the atmosphere.
  b Nuclear fuel releases about 10 000 times as much energy as the same mass of fossil fuel. Give one disadvantage of nuclear fuel compared with other types of fuel.

2 a Give one advantage and one disadvantage of:
   i an oil-fired power station compared with a nuclear power station
   ii a gas-fired power station compared with a coal-fired power station.
  b Look at Table 1.
   Calculate how many kilograms of fossil fuel would give the same amount of energy as 1 kilogram of uranium fuel.

3 a Explain why ethanol is described as a biofuel.
  b Ethanol is also described as carbon-neutral. Explain what a carbon-neutral fuel is.

4 Global energy usage is currently about 500×10^{20} joules per year. The global population is about 6×10^9 people. Estimate how much energy per second each person uses on average.

Key points

- Your energy demands are met mostly by burning oil, coal, and gas.

- Nuclear power, biofuels, and renewable resources provide energy to generate some of the energy you use.

- Uranium or plutonium is used as the fuel in a nuclear power station. Much more energy is released per kilogram from uranium or plutonium than from fossil fuels.

- Biofuels are renewable sources of energy. Biofuels such as methane and ethanol can be used to generate electricity.

6.2

Energy from wind and water

Learning objectives

After this topic, you should know:

- what a wind turbine is made up of

- how waves can be used to generate electricity

- the type of power station that uses water running downhill to generate electricity

- how the tides can be used to generate electricity.

Figure 1 *A wind farm is a group of wind turbines*

Go further!

The mass *m* of wind passing through a wind turbine each second is proportional to the wind speed *v*. If the wind speed doubles, the mass of wind passing through the wind turbine each second also doubles. As kinetic energy $= \frac{1}{2}mv^2$, the kinetic energy of the wind passing through each second is therefore 2^3 times greater, because *m* increases by 2 and v^2 increases by 4. In other words, the power of the wind is proportional to v^3.

Strong winds can cause lots of damage on a very stormy day. Even when the wind is much weaker, it can still turn a wind turbine. Energy from the wind and other sources such as **waves** and **tides** is called **renewable energy**. That's because such natural sources of energy can never be used up because they are always being replenished (i.e., replaced) by natural processes.

As well as this, no fuel is needed to produce electricity from these natural sources, so they are carbon-free to run.

Wind power

A wind turbine is an electricity generator at the top of a narrow tower. The force of the wind drives the turbine's blades around. This turns a generator. The power generated increases as the wind speed increases. Wind turbines are unreliable because when there is little or no wind they do not generate any electricity.

Wave power

A wave generator uses the waves to make a floating generator move up and down. This motion turns the generator so it generates electricity. A cable between the generator and the shoreline delivers electricity to the grid system.

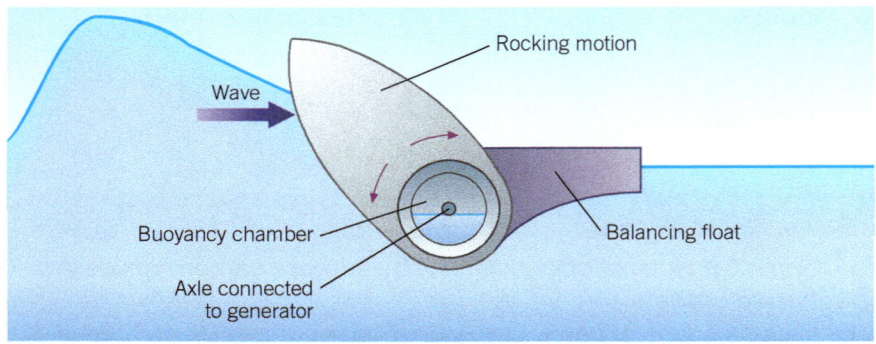

Figure 2 *Energy from waves*

Wave generators need to withstand storms, and they don't produce a constant supply of electricity. Also, lots of cables (and buildings) are needed along the coast to connect the wave generators to the electricity grid. This can spoil areas of coastline. Tidal flow patterns might also change, affecting the habitats of marine life and birds.

Hydroelectric power

Hydroelectricity can be generated when rainwater that's collected in a reservoir (or water in a pumped storage scheme) flows downhill. The flowing water drives turbines that turn electricity generators at the bottom of the hill.

Tidal power

A tidal power station traps water from each high tide behind a barrage. The high tide can then be released into the sea through turbines. The turbines drive generators in the barrage.

One of the most promising sites for a tidal power station in Britain is the Severn estuary. This is because the estuary rapidly becomes narrower as you move up-river away from the open sea. So it funnels the incoming tide and makes it higher.

Figure 3 *A hydroelectric power station. Some hydroelectric power stations are designed as pumped storage schemes. When electricity demand is low, electricity can be supplied from other power stations and electricity generators to pumped storage schemes to pump water uphill into a reservoir. When demand is high, the water can be allowed to run downhill to generate electricity*

Figure 4 *A tidal power station*

Summary questions

1 Hydroelectricity, tidal power, wave power, and wind power are all renewable energy resources.
   a Explain what a renewable energy resource is.
   b Name the renewable energy resource listed above that:
      i does not need energy from the Sun
      ii does not need water and is unreliable.

2 a Use the table below for this question. The output of each source is given in millions of watts (MW).
      i Calculate how many wind turbines would give the same total power output as a tidal power station.
      ii Calculate how many kilometres of wave generators would give the same total output as a hydroelectric power station.
   b Use the words below to complete the location column in the table.
   coastline   estuaries   hilly or coastal areas   mountain areas

	Output	Location	Total cost in £ per MW
Hydroelectric power station	500 MW per station		50
Tidal power station	2000 MW per station		300
Wave power generators	20 MW per kilometre of coastline		100
Wind turbines	2 MW per wind turbine		90

3 The last column of the table above shows an estimate of the total cost per MW of generating electricity using different renewable energy resources. The total cost for each resource includes its running costs and the capital costs to set it up.
   a The capital cost per MW of a tidal power station is much higher than that of a hydroelectric power station. Give one reason for this difference.
   b i Name the energy resource that has the lowest total cost per MW.
      ii Give two reasons why this resource might be unsuitable in many locations.

4 a Explain what a pumped storage scheme is.
   b Describe the main benefit to electricity users of a pumped storage scheme.

Key points

- A wind turbine is an electricity generator on top of a tall tower.
- Waves generate electricity by turning a floating generator.
- Hydroelectricity generators are turned by water running downhill.
- A tidal power station traps each high tide and uses it to turn generators.

6.3 Power from the Sun and the Earth

Learning objectives

After this topic, you should know:

- what solar cells are and how they are used
- the difference between a panel of solar cells and a solar heating panel
- what geothermal energy is
- how geothermal energy can be used to generate electricity.

Solar radiation transfers energy to you from the Sun. That can sometimes be more energy than is healthy if you get sunburnt. But the Sun's energy can be used to generate electricity using solar cells. The Sun's energy can also be used to heat water directly in solar heating panels.

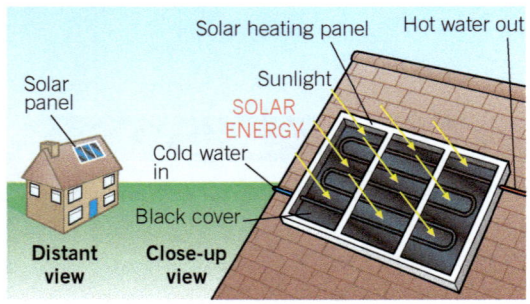

Figure 2 *Solar water heating*

Today's solar cells convert less than 10% of the solar energy they absorb into electrical energy (Figure 1). They can be connected together to make solar cell panels.

- They are useful where only small amounts of electricity are needed (e.g., in watches and calculators) or in remote places (e.g., on small islands in the middle of an ocean).
- They are very expensive to buy but they cost nothing to run.
- Lots of them are needed – and plenty of sunshine – to generate enough power to be useful. Solar panels can be unreliable in areas where the Sun is often covered by clouds.

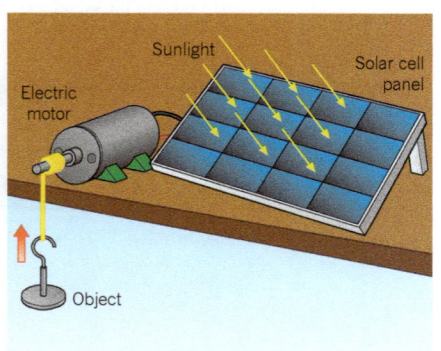

Figure 1 *Solar cells at work*

A solar heating panel heats water that flows through it and on a sunny day in Britain, a solar heating panel on a house roof can supply plenty of hot water for domestic use (Figure 2).

A solar power tower uses thousands of flat mirrors to reflect sunlight on to a big water tank at the top of a tower (Figure 3). The mirrors on the ground surround the base of the tower.

- The water in the tank is turned to steam by the heating effect of the solar radiation directed at the water tank.
- The steam is piped down to ground level, where it turns electricity generators.
- The mirrors are controlled by a computer so that they track the Sun.

A solar power tower in a hot dry climate can generate more than 20 MW of electrical power, which is enough to power a few thousand homes.

Figure 3 *A solar power tower*

Geothermal energy

Geothermal energy comes from energy released by radioactive substances deep within the Earth.

- The energy released by these radioactive substances heats the surrounding rock.
- So energy is transferred by heating towards the Earth's surface.

Geothermal power stations can be built in volcanic areas or where there are hot rocks deep below the surface. Water gets pumped down to these rocks to produce steam. Then the steam that is produced drives electricity turbines at ground level (Figure 4).

In some areas, buildings can be heated using geothermal energy directly. Heat flow from underground is sometimes called ground source heat. It can be used to heat water in long underground pipes. The hot water is then pumped around the buildings. In some big eco-buildings, this geothermal heat flow is used as under-floor heating.

⚭ links

Solar heating panels use solar energy to heat water that flows through the panel. You can use the equation $\Delta E = mc\,\Delta\theta$ from Topic 12.2 to estimate the temperature increase of mass m of water that flows through the panel, where c is the specific heat capacity of water and E is the solar energy absorbed by the panel.

Study tip

Make sure you know the difference between a solar cell panel (in which sunlight is used to make electricity) and a solar heating panel (in which sunlight is used to heat water).

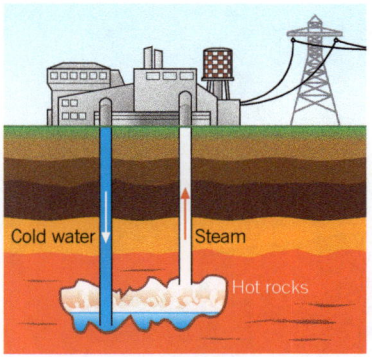

Figure 4 *A geothermal power station*

Key points

- Solar cells are flat solid cells and use the Sun's energy to generate electricity directly.

- Solar heating panels use the Sun's energy to heat water directly.

- Geothermal energy comes from the energy released by radioactive substances deep inside the Earth.

- Water pumped into hot rocks underground produces steam to drive turbines at the Earth's surface that generate electricity.

Summary questions

1. **a** What is the source of geothermal energy?
   b State why geothermal energy is more reliable than solar energy for heating water.

2. A satellite in space uses a solar cell panel for electricity. The panel generates 300 W of electrical power and has an area of 10 m².
   a Each cell generates 0.2 W. Calculate how many cells are in the panel.
   b The satellite carries batteries that are charged by electricity from the solar cell panels. State why batteries are carried as well as solar cell panels.

3. A certain geothermal power station has a power output of 200 000 W.
   a Calculate how many kilowatt-hours of electrical energy the power station generates in 24 hours.
   b Give one advantage and one disadvantage of a geothermal power station compared with a wind turbine.

4. A solar water panel heats the water flowing through from 14 °C to 35 °C when water flows through it at a rate of 0.010 kilograms per second.
   a Calculate the thermal energy per second transferred by the hot water from the solar panel.
   b Estimate the output temperature of the hot water if the flow rate is reduced to 0.007 kg/s. The specific heat capacity of water is 4200 J/kg °C.

6.4 Energy and the environment

Learning objectives

After this topic, you should know:

- what fossil fuels do to your environment
- why people are concerned about nuclear power
- the advantages and disadvantages of renewable energy resources
- how to evaluate the use of different energy resources.

Can you get energy without creating any problems? Figure 1 shows the energy sources people use today to generate electricity. What effect does each one have on your environment?

Fossil fuel problems

When coal, oil, or gas is burnt, **greenhouse gases** such as carbon dioxide are released. The amount of these gases in the atmosphere is increasing, and most scientists believe that this is causing more global warming and climate change. Some electricity comes from oil-fired power stations. People use much more oil to produce fuels for transport.

Burning fossil fuels can also produce sulfur dioxide. This gas causes **acid rain**. The sulfur can be removed from a fuel before burning it, to stop acid rain. For example, natural gas has its sulfur impurities removed before it is used.

Fossil fuels are non-renewable. Sooner or later, people will have used up the Earth's reserves of fossil fuels. Alternative sources of energy will then have to be found. But how soon? Oil and gas reserves could be used up within the next 50 years. Coal reserves will last much longer.

Carbon capture and storage (CCS) technology could be used to stop carbon dioxide emissions into the atmosphere from fossil fuel power stations. Old oil and gas fields could be used for storage.

Nuclear versus renewable

People need to use less fossil fuels in order to stop global warming. Should people rely on nuclear power or on renewable energy in the future?

Nuclear power

Advantages

- No greenhouse gases (unlike fossil fuel).
- Much more energy is released from each kilogram of uranium (or plutonium) fuel than from fossil fuel.

Disadvantages

- Used fuel rods contain radioactive waste, which has to be stored safely for centuries.
- Nuclear reactors are safe in normal operation. However, an explosion in a reactor could release radioactive material over a wide area. This would affect this area, and the people living there, for many years.

Renewable energy sources and the environment

Advantages

- They will never run out because they are always being replenished by natural processes.
- They do not produce greenhouse gases or acid rain.
- They do not create radioactive waste products.

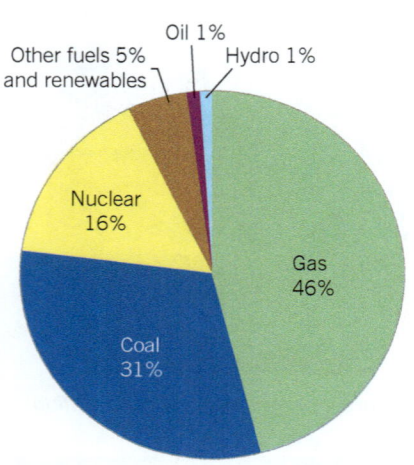

Figure 1 *Energy sources for electricity*

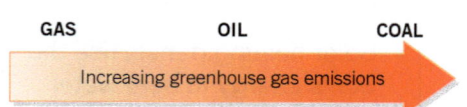

Figure 2 *Greenhouse gases from fossil fuels*

○○ links

See Topic 13.4 for more about how greenhouse gases cause global warming.

- They can be used where connection to the National Grid is uneconomic. For example, solar cells can be used for road signs and to provide people with electricity in remote areas.

Disadvantages

- Renewable energy resources are not currently able to meet the world demand. So fossil fuels are still needed to provide some of the energy demand.
- Wind turbines create a whining noise that can upset people nearby, and some people consider them unsightly.
- Tidal barrages affect river estuaries and the habitats of creatures and plants there.
- Hydroelectric schemes need large reservoirs of water, which can affect nearby plant and animal life. Habitats are often flooded to create dams.
- Solar cells need to cover large areas to generate large amounts of power.
- Some renewable energy resources are not available all the time or can be unreliable. For example, solar power is not produced at night and is affected by cloudy weather. Wind power is reduced when there is little or no wind, and hydroelectricity is affected by droughts if reservoirs dry up.

Figure 3 *The effects of acid rain*

Summary questions

1 **a** Name the type of fuel that is used to generate most of Britain's electricity.
   b Name two problems caused by burning fossil fuel.
   c State two advantages and two disadvantages of using renewable energy sources instead of fossil fuels.

2 Match each energy source with a problem it causes:

Energy source	Problem
a Coal	**A** Noise
b Hydroelectricity	**B** Acid rain
c Uranium	**C** Radioactive waste
d Wind power	**D** Takes up land

3 **a** Name three possible renewable energy resources that could be used to generate electricity for people on a remote flat island in a hot climate.
   b Name three types of power stations that do not release greenhouse gases into the atmosphere.

4 A tidal power station, a nuclear power station, or 1000 wind turbines can each supply enough power to meet the electricity needs of a large city on an estuary. Describe the advantages and disadvantages of each type of power station for this purpose.

Key points

- Fossil fuels produce increased levels of greenhouse gases, which could cause global warming.

- Nuclear fuels produce radioactive waste.

- Renewable energy resources will never run out, they do not produce harmful waste products (e.g., greenhouse gases or radioactive waste), and they can be used in remote places. But they cover large areas, and they can disturb natural habitats.

- Different energy resources can be evaluated in terms of reliability, environmental effects, pollution, and waste.

Chapter summary questions

1 a i Explain what is meant by a renewable energy source.

ii State with the aid of a suitable example what is meant by renewable fuel.

b Discuss whether or not renewable fuels contribute to greenhouse gases in the atmosphere.

2 a Give two similarities and two differences of a tidal power station compared with a hydroelectric power station.

b Name the renewable energy resource that transfers:
i the kinetic energy of moving air to electrical energy
ii the gravitational potential energy of water running downhill into electrical energy
iii the kinetic energy of water moving up and down to electrical energy.

3 a i Name the energy resource that does not produce greenhouse gases and uses energy which is from inside the Earth.

ii Name the energy resource that uses running water and does not produce greenhouse gases.

iii Name the energy resource that releases greenhouse gases and causes acid rain.

iv Name the energy resource that does not release greenhouse gases but does produce waste products that need to be stored for many years.

b Wood can be used as a fuel. State whether it is:
i renewable or non-renewable
ii a fossil fuel or a non-fossil fuel.

4 a Use the data in the table below to discuss whether or not wind turbines are less expensive to build than nuclear power stations and are also cheaper to run.

b Discuss the reliability and environmental effects of the non-fossil fuel resources listed in the table below.

	Output	Total cost in £ per MW
Hydroelectric power station	500 MW per station	50
Tidal power station	2000 MW per station	300
Wave power generators	20 MW per kilometre of coastline	100
Wind turbines	2 MW per wind turbine	90

5 a Figure 1 shows a landscape with three different renewable energy resources, numbered 1 to 3.

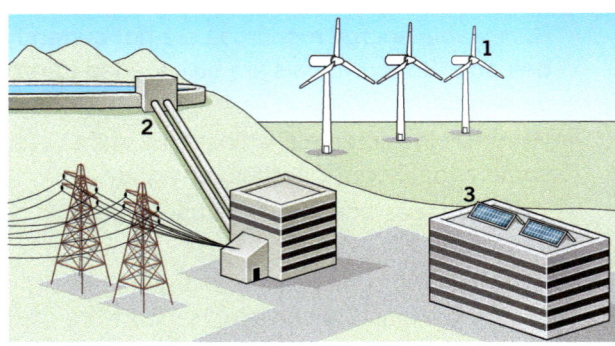

Figure 1 *Renewable energy*

Match each type of energy resource with one of the labels below.

hydroelectricity   solar energy   wind energy

b Determine which of the three resources shown is not likely to produce as much energy as the others if the area is:
i hot, dry, and windy
ii wet and windy.

6 A hydroelectric power station has an upland reservoir that is 400 m above the power station.
The power station is designed to produce 96 MW of electrical power with an efficiency of 60%.

a Estimate the loss of gravitational potential energy per second when the hydroelectric power station generates 96 MW of power.

b Use your estimate to calculate the volume of water per second that flows from the reservoir through the power station generators when 96 MW of power is generated. The density of water is 1000 kg/m^3.

Practice questions

1 Which two of the following statements are **not** good reasons for using renewable energy resources?

A supplies of renewable energy are unlimited

B renewable energy cannot generate electricity all the time

C renewable energy can be replenished

D renewable energy cannot supply electricity to millions of homes

E renewable energy does not produce carbon dioxide (2)

2 a Figure 1 shows a lagoon tidal barrage system used to generate electricity.

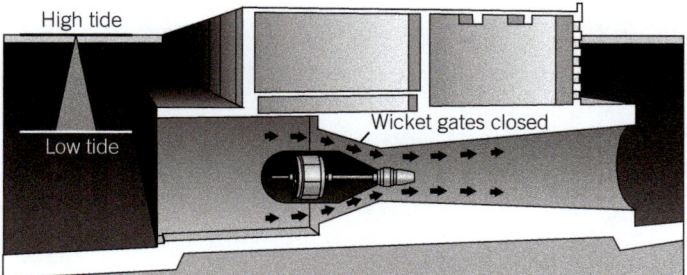

Figure 1

The sentences are not in the correct order. Copy the correct letters into the boxes. The first one is done for you.

A *When the tide is high the gates open and water rushes through the turbines into the lagoon.*

B The water flowing out again turns the turbines and generates electricity.

C The water coming in turns the turbines and generates electricity.

D When the tide is low the gates are opened and the water rushes out of the lagoon.

E The lagoon fills up with water and the gates are closed. (3)

A ☐ ☐ ☐ ☐

b The lagoon tidal barrage system is claimed to be able to generate electricity for 14 hours a day. Give two reasons why it is necessary to connect houses to the National Grid rather than to the lagoon tidal barrage system alone. (2)

c Suggest a reason, apart from cost, why people may object to the planned system. (1)

3 A group of students investigated whether changing the direction of a solar panel affected the amount of electricity generated. The tests were performed on a sunny day at midday. The students method is given below.

1 Fix the solar panel to a board and incline the panel 45° to the horizontal.

2 Attach a voltmeter to the solar panel.

3 Point the solar panel to the north and record the voltage reading.

4 Repeat the test pointing the solar panel in different directions.

The students recorded their results in **Table 1**.

Table 1

Direction	N	NE	E	SE	S	SW	W	NW
Volts	1.2	1.9	2.5	3.9	4.3	3.7	2.4	1.8

a Name the independent and dependent variable. (2)

b Suggest what advice you would give to a school about to install solar panels. (1)

c One student stated that the voltage readings would be higher if the investigation was carried out on the roof. Is the student correct? Give a reason for your decision. (2)

d The students want to check if their results are repeatable. They intend to carry out the investigation using the same apparatus at the same location. State one other factor they must keep the same. (1)

4 a The UK Government has set a target to use renewable energy for 20% of the country's energy use by 2020. Explain why it is important to increase the use of renewable energy sources to generate electricity. (3)

b The UK Government has agreed to build a nuclear power plant by 2023. It is estimated that the power plant will provide at least 7% of the electricity needed in the UK.

Give two advantages and one disadvantage of using nuclear energy to produce electricity. (3)

5 Electricity can be generated using ethanol as an energy source. Ethanol can be produced from sugar cane. Brazil is the biggest grower of sugar cane, and the biggest exporter of ethanol made from sugar cane in the world.

Evaluate the benefits of growing and using ethanol made from sugar cane. Your answer should include environmental as well as economic reasons. (3)

7.1

The nature of waves

Learning objectives

After this topic, you should know:

- what waves can be used for
- what transverse waves are
- what longitudinal waves are
- which types of wave are transverse and which are longitudinal.

Figure 1 *Waves in water are examples of mechanical waves*

Waves transfer energy without transferring matter. Waves are also used to transfer information – for example, when you use a mobile phone or listen to the radio.

There are different types of wave. These include:

- sound waves, water waves, waves on springs and ropes, and seismic waves produced by earthquakes. These are examples of **mechanical waves**, which are vibrations that travel through a **medium** (a substance).
- light waves, radio waves, and microwaves. These are examples of **electromagnetic waves**, which can all travel through a vacuum at the same speed of 300 000 kilometres per second. No medium is needed.

Practical

Observing mechanical waves

Figure 2 shows how you can make waves on a rope by moving one end up and down.

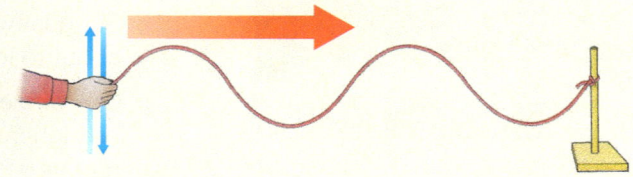

Figure 2 *Transverse waves*

Tie a ribbon to the middle of the rope. Move one end of the rope up and down. You will see that the waves move along the rope but the ribbon doesn't move along the rope – it just moves up and down. This type of wave is known as a **transverse wave**. The ribbon **vibrates** or **oscillates**. This means it moves repeatedly between two positions. When the ribbon is at the top of a wave, it is at the **peak** (or crest) of the wave.

Repeat the test with the slinky. You should observe the same effects if you move one end of the slinky up and down.

However, if you push and pull the end of the slinky as shown in Figure 3, you will see a different type of wave, known as a **longitudinal wave**. Notice that there are areas of **compression** (coils squashed together) and areas of **rarefaction** (coils spread further apart) moving along the slinky.

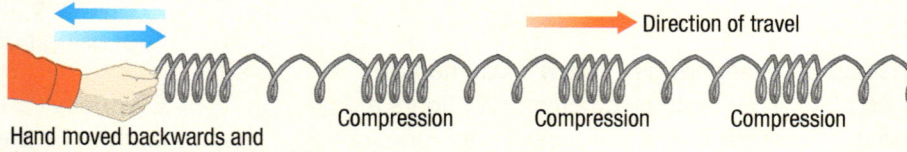

Direction of travel

Compression   Compression   Compression

Hand moved backwards and forwards along the line of the slinky

Figure 3 *Making longitudinal waves on a slinky*

- How does the ribbon move when you send **longitudinal** waves along the slinky?

Safety: Handle the slinky spring carefully.

∞ **links**

For more information on electromagnetic waves, see Topic 8.1 'The electromagnetic spectrum'.

Transverse waves

Imagine you send waves along a rope which has a white spot painted on it. You would see the spot move up and down without moving along the rope. In other words, the spot would oscillate **perpendicular** (at right angles) to the direction in which the waves are moving. The waves on a rope are called **transverse waves** because the vibrations (called oscillations) are for example up and down or from side to side. All electromagnetic waves are transverse waves.

For a transverse wave the oscillations are perpendicular to the direction of energy transfer.

Longitudinal waves

The slinky spring in Figure 3 is useful to demonstrate how sound waves travel. When one end of the slinky is pushed in and out repeatedly, vibrations travel along the spring. These oscillations are parallel to the direction in which the waves transfer energy, along the spring. Waves that travel in this way are called **longitudinal waves**.

Sound waves are longitudinal waves. When an object vibrates in air, it makes the air around it vibrate as it pushes and pulls on the air. The oscillations (**compressions** and **rarefactions**) that travel through the air are sound waves. The oscillations are along the direction in which the wave travels.

For a longitudinal wave the oscillations are parallel to the direction of energy transfer.

Mechanical waves may be either transverse or longitudinal.

links

For more information on sound, see Topics 7.4 'Diffraction' and 9.1 'Sound'.

Study tip

- Make sure that you understand the difference between transverse waves and longitudinal waves.
- Remember that electromagnetic waves are transverse and sound waves are longitudinal.

??? Did you know ... ?

If you pluck a guitar string, it oscillates because you send transverse waves along the string. The oscillating string sends sound waves into the surrounding air. The sound waves are longitudinal.

Summary questions

1 a What is the difference between a longitudinal wave and a transverse wave?
 b State **one** example of:
  i a transverse wave
  ii a longitudinal wave.
 c When a sound wave passes through air, what happens to the air particles at a compression?

2 A long rope with a knot tied in the middle lies straight along a smooth floor. A student picks up one end of the rope. This sends waves along the rope.
 a Are the waves on the rope transverse or longitudinal waves?
 b What can you say about:
  i the direction of energy transfer along the rope?
  ii the movement of the knot?

3 a Describe how to use a slinky spring to demonstrate to a friend the difference between longitudinal waves and transverse waves.
 b A blue slinky spring has one of its coils painted red. Describe the motion of the red coil when longitudinal waves travel along the slinky.

Key points

- Waves transfer energy and they can be used to transfer information.

- Transverse waves oscillate perpendicular to the direction of energy transfer of the waves. All electromagnetic waves are transverse waves.

- Longitudinal waves oscillate parallel to the direction of energy transfer of the waves. Sound waves are longitudinal waves.

- Mechanical waves, which need a medium (a substance) to travel through, may be transverse or longitudinal waves.

7.2

Measuring waves

You can make measurements on waves, to find out how much energy or information they carry. Figure 1 shows a snapshot of waves on a rope. The **crests**, or peaks, are at the tops of the waves. The **troughs** are at the bottom. They are equally spaced.

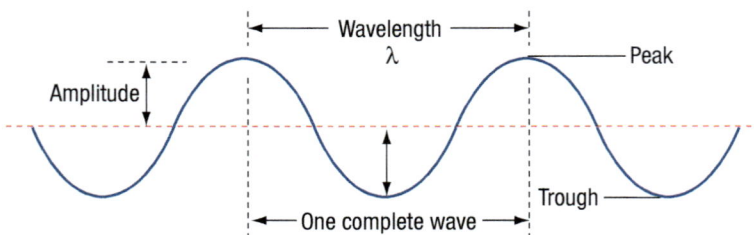

Figure 1 *Waves on a rope*

- The **amplitude** of the waves is the height of the wave crest (or the depth of the wave trough) from the middle, which is the position of the rope at rest.
- **The bigger the amplitude of the waves, the more energy the waves carry.**
- The **wavelength** of the waves is the distance from one wave crest to the next crest.

Frequency

If you made a video of the waves on the rope, you would see the waves moving steadily across the screen. The number of wavecrests passing a fixed point every second is the **frequency** of the waves.

The unit of frequency is the **hertz** (Hz). One wave crest passing each second is a frequency of 1 Hz.

Wave speed

Figure 2 shows a ripple tank, which is used to study water waves in controlled conditions. You can make straight waves by moving a ruler up and down on the water surface in a ripple tank. Straight waves are called **plane** waves. The waves all move at the same speed and stay the same distance apart.

The **speed** of the waves is the distance travelled by a wave crest (or a wave trough) every second.

For example, sound waves in air travel at a speed of 340 m/s. In 5 seconds, sound waves travel a distance of 340 m/s × 5 s = **1700 m**.

For waves of constant frequency, the speed of the waves depends on the frequency and the wavelength as follows:

$$\text{wave speed} = \text{frequency} \times \text{wavelength}$$
$$\text{(metres/second, m/s)} \quad \text{(hertz, Hz)} \quad \text{(metres, m)}$$

Learning objectives

After this topic, you should know:

- what is meant by the amplitude, frequency, and wavelength of a wave
- the relationship between the speed, wavelength, and frequency of a wave
- how to use the wave speed equation in calculations.

Study tip

A common error is to think that the amplitude is the distance from the top of the crest to the bottom of the trough (but that is twice the amplitude).

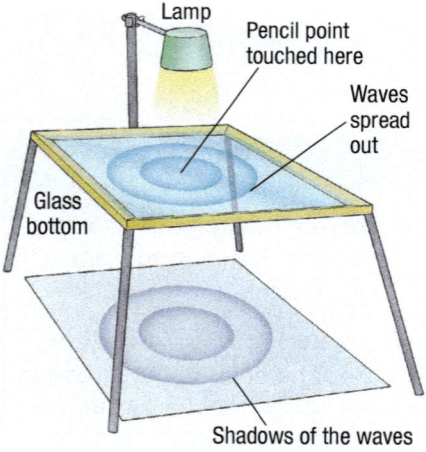

Figure 2 *Circular waves on a ripple tank*

Maths skills

You can write the wave speed equation as $v = f \times \lambda$, where:

v = speed in m/s
f = frequency in Hz
λ = wavelength in m.
Note: The Greek letter λ is pronounced 'lambda'.

Study tip

- Be careful with powers of ten when using $v = f \times \lambda$.
- Frequencies can have high positive powers of ten – for example, light waves might have a frequency of 5×10^{14} Hz.
- Wavelengths can have high negative powers of ten – for example, X-rays might have a wavelength of 1×10^{-12} m.

Practical

Making straight (plane) waves

To measure the speed of the waves:

Use a stopwatch to measure the time it takes for a wave to travel from the ruler to the side of the ripple tank.

Measure the distance the waves travel in this time.

Use the equation: speed = distance ÷ time to calculate the speed of the waves.

Observe the effect on the waves of moving the ruler up and down more frequently. More waves are produced every second and they are closer together.

- Find out if the speed of the waves has changed.

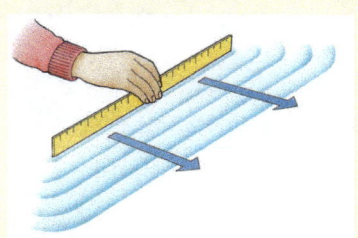

Figure 3 *Making plane water waves*

To understand what the wave speed equation means, look at Figure 4. The surfer is riding on the crest of some unusually fast waves.

Suppose the frequency of the waves is 3 Hz and the wavelength of the waves is 4 m.

- At this frequency, 3 wave crests pass a fixed point once every second (because the frequency is 3 Hz).
- The surfer therefore moves forward a distance of 3 wavelengths every second, which is 3 × 4 m = 12 m.

The speed of the surfer is therefore **12 m/s**.

This speed is equal to the frequency × the wavelength of the waves: $v = f \times \lambda$.

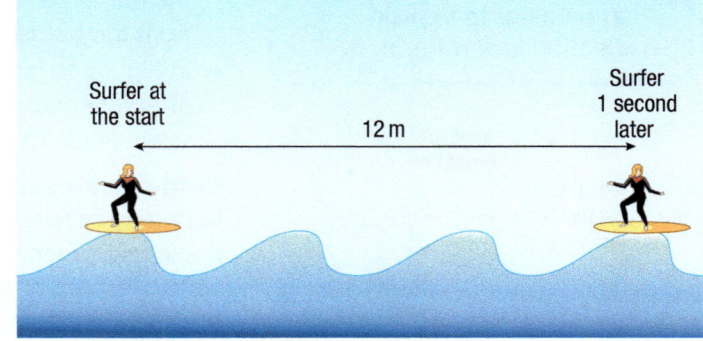

Figure 4 *Surfing*

Summary questions

1. **a** Use a millimetre ruler to measure the amplitude and the wavelength of the waves in Figure 1.
   b What is meant by the frequency of a wave?

2. The figure shows a snapshot of a wave travelling from left to right along a rope.

   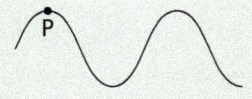

   a Copy the figure and mark on your diagram:
     i one wavelength
     ii the amplitude of the waves.
   b Describe the motion of point P on the rope when the wave crest at P moves along by a distance of one wavelength.

3. **a** A speedboat on a lake sends waves travelling across a lake at a frequency of 2.0 Hz and a wavelength of 3.0 m. Calculate the speed of the waves.
   b If the waves had been produced at a frequency of 1.0 Hz and travelled at the speed calculated in **a**:
     i what would be their wavelength?
     ii calculate the distance travelled by a wave crest in 60 seconds.

Key points

- For any wave, its amplitude is the height of the wave crest (or the depth of the wave trough) from the position at rest.

- For any wave, its frequency is the number of wave crests passing a point in one second.

- For any wave, its wavelength is the distance from one wave crest to the next wave crest. This is the same as the distance from one wave trough to the next wave trough.

- Wave speed = frequency × wavelength:

  $$v = f \times \lambda$$

7.3 Reflection and refraction

After this topic, you should know:

- the patterns of reflection and refraction of plane waves in a ripple tank

- what causes refraction

- how the behaviour of waves can be used to explain reflection and refraction.

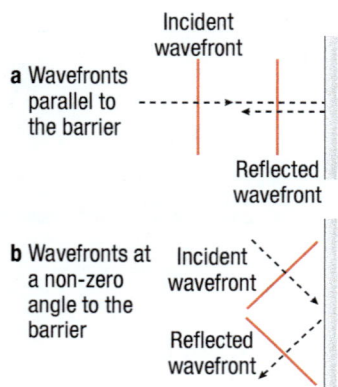

a Wavefronts parallel to the barrier

Incident wavefront

Reflected wavefront

b Wavefronts at a non-zero angle to the barrier

Incident wavefront

Reflected wavefront

Figure 1 *Reflection of plane waves*

Practical

A reflection test

Use a ruler to create and direct plane waves at a straight barrier, as shown in Figure 1. Find out if the reflected waves are always at the same angle to the barrier as the incident waves. You could align a second ruler with the reflected waves and measure the angle of each ruler to the barrier. Repeat the test for different angles.

Investigating waves using a ripple tank

Reflection of plane waves can be investigated using the ripple tank shown in Topic 7.2. The waves travelling across the water surface are called wavefronts. Plane (i.e., straight) waves, produced by repeatedly dipping a ruler in water, are directed at a metal barrier in the water. These waves are referred to as the incident waves to distinguish them from the reflected waves. The incident waves are reflected by the barrier.

Figures 1a and b each show a wavefront before and after hitting the barrier.
- In Figure 1a, the incident wavefront is parallel to the barrier as it approaches the barrier. It is still parallel to the barrier after reflection as it travels away from the barrier.
- In Figure 1b, the incident wavefront is not parallel to the barrier before or after reflection. The reflected wavefront moves away from the barrier at the same angle to the barrier as the incident wavefront.

Refraction is a change of the direction of travel of waves when they cross a boundary between one medium and another medium. This can be seen in a ripple tank when water waves cross a boundary between deep and shallow water. An area of shallow water can be created by placing a glass or transparent plastic plate flat in the water. The water above the plate is shallower than the rest of the water. Plane waves are directed at a non-zero angle to a boundary. The wavefronts change direction as they cross the boundary, as shown in Figure 2.

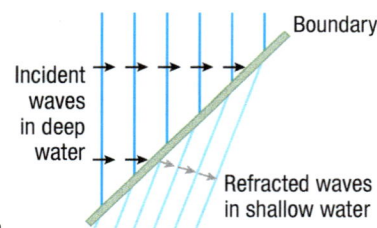

Incident waves in deep water

Boundary

Refracted waves in shallow water

Figure 2 *Refraction*

Practical

Refraction tests

Use a vibrating beam to create plane waves continuously in a ripple tank containing a transparent plastic plate.

Arrange the plate so the waves cross a boundary between the deep and shallow water. The water over the plate needs to be very shallow.

- **At a non-zero angle to a boundary.** The waves change direction when they cross the boundary. Find out if plane waves change direction towards or away from the boundary when they cross from deep to shallow water.
- **Parallel to a boundary.** The waves cross the boundary without changing direction. However, their speed changes.

Find out if the waves travel slower or faster when they cross the boundary. You will see later in this that this change of speed explains why the waves are refracted when they cross a boundary at a non-zero angle.

Explaining reflection and refraction

To explain how a wavefront moves forward, imagine that each tiny section creates a wavelet that travels forward, as shown in Figure 3. The wavelets move forward together to recreate the wavefront that created them.

Refraction: When plane waves cross a boundary at a non-zero angle to the boundary they slow down, and each wavefront changes its direction. When the wavelets reach the boundary they line up with the previous wavelets that have crossed the boundary. These wavelets form a refracted wavefront.

In Figure 3, the wavefronts move more slowly after they have crossed the boundary. They are not as far from the boundary as they would have been if their speed had not changed. So the refracted wavefronts are at a smaller angle to the boundary than the incident wavefronts.

Reflection: When plane waves reflect from a flat barrier, the reflected waves are at the same angle to the barrier as the incident waves. When each point on the wavefront reaches the barrier, it creates a wavelet moving away from the barrier. This wavelet lines up with the previous reflected wavelets to form a reflected wavefront moving away from the barrier. All parts of a wavefront move at the same speed. This means that the reflected wavefront is at the same angle to the barrier as the incident wavefront, as shown in Figure 4.

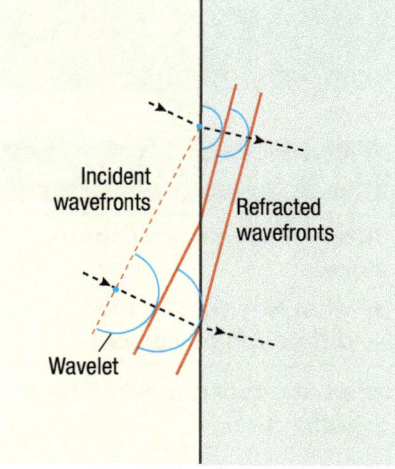

Figure 3 *Explaining refraction*

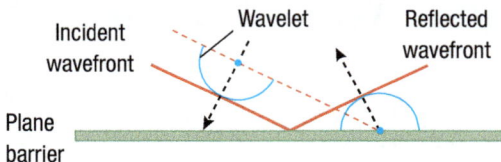

Figure 4 *Explaining reflection*

Summary questions

1 **a** When plane waves reflect from a straight barrier, what can be said about the angle of each reflected wavefront to the barrier and the angle of each incident wavefront to the barrier?
   b When waves speed up on crossing a boundary, what can be said about the angle of each refracted wavefront to the boundary and the angle of each incident wavefront to the boundary?

2 Copy the figure, which shows plane waves passing from deep to shallow water. They travel more slowly in shallow water than in deep water. Draw some refracted wavefronts, indicating their direction.

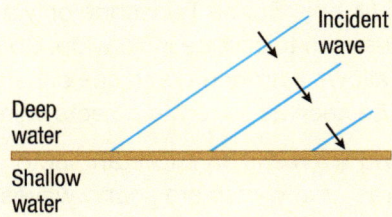

3 Sea waves rolling up a sandy beach are not reflected.
   a Why are the sides of a ripple tank sloped?
   b What would happen if the sides of a ripple tank were vertical instead of sloped?

Materials and waves

When waves travel through a material, the wave is said to be **transmitted** by the material.

The amplitude of the waves gradually decreases if the substance **absorbs** some energy from the waves.

When waves cross a boundary between two materials, **partial reflection** occurs at the boundary, as well as refraction. This is why you might see a faint mirror image of yourself when you look at a window.

The waves that cross the boundary lose energy at the boundary and so have a smaller amplitude than that of the incident waves.

Key points

- Plane waves reflected from a straight barrier are at the same angle to the barrier as the incident waves.

- Refraction is the change in direction of waves caused by a change in their speed when they cross a boundary between one medium and another.

7.4 Diffraction

Diffraction is the spreading of waves when they pass through a gap or move past an obstacle. The waves passing through the gap or past the edges of the obstacle spread out without changing their speed, their wavelength, or their frequency. Figure 1 shows waves in a ripple tank spreading out after they pass through two gaps. You can see from Figure 1 that for the same wavelength:

- the narrower the gap, the more the waves spread out
- the wider the gap, the less the waves spread out.

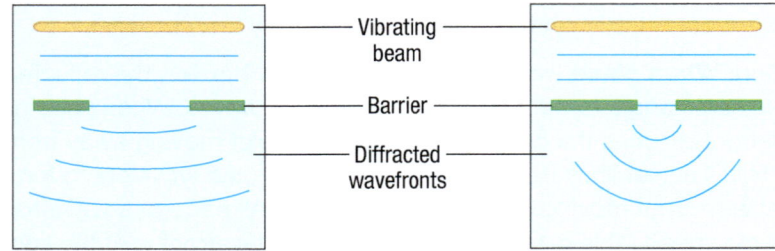

Figure 1 *Diffraction of waves by a gap* **a** *A wide gap* **b** *A narrow gap*

Practical

Investigating diffraction

Use a ripple tank as in Figure 1 to direct plane waves continuously at a gap between two metal barriers. Notice that the waves spread out after they pass through the gap. In other words, they are diffracted by the gap.

Change the gap spacing or the wavelength and observe that the diffraction of the waves increases as the wavelength is increased or the gap is made narrower, as shown in Figure 1.

Remove one of the metal barriers and observe the waves spreading behind the **edge** of the remaining barrier. See Figure 2. If the wavelength of the waves is increased, you should find they spread out more behind the edge of the barrier.

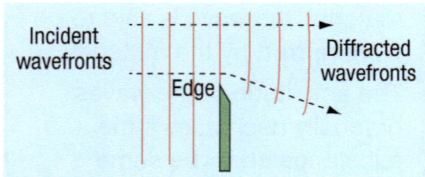

Figure 2 *Diffraction at an edge*

Figure 3 *Image of two colliding galaxies taken by the Hubble Space Telescope*

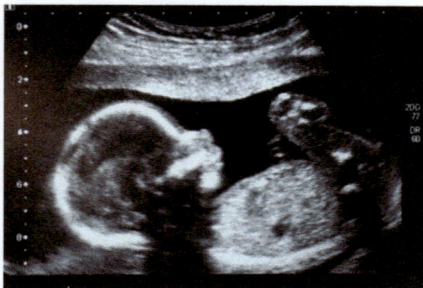

Figure 4 *An ultrasound scan of a baby in the womb*

links

To learn more about ultrasound, see Topic 9.3 'Ultrasound'.

Diffraction details

Diffraction of light is important in any optical instrument. The Hubble Space Telescope in its orbit above the Earth has provided amazing images of objects far away in space. Its focusing mirror has a diameter of 2.4 m. When light passes through the Hubble Space Telescope only a small amount of diffraction occurs, because the telescope is so wide. So its images are very clear and detailed. This allows astronomers to see separate images of objects that are far too close to be seen as separate objects using a narrower telescope.

Diffraction of ultrasound waves is an important factor in the design of an ultrasound scanner. Ultrasound waves are sound waves at frequencies above the range that can be heard by the human ear. An ultrasound scan can be taken of a baby in the womb. The ultrasound waves spread out from a hand-held transmitter and then reflect from the tissue boundaries inside the womb. If the transmitter is too narrow, the waves spread out too much and the image becomes too faint.

Demonstration

Tests using microwaves

A microwave transmitter and a detector can be used to demonstrate the diffraction of microwaves. The transmitter produces microwaves of wavelength 3.0 cm.

1 Place a metal plate between the transmitter and the detector across the path of the microwaves. Microwaves can still be detected behind the metal plate. This is because some microwaves diffract around the edge of the plate.

● Why do the microwaves not go through the metal plate?

2 Place two metal plates separated by a gap across the path of the microwaves, as shown in Figure 5. The microwaves pass through the gap but not through the plates. When the detector is moved along an arc centred on the gap, it detects microwaves that have spread out from the gap.

When the gap is made wider, the microwaves passing through the gap spread out less. The detector needs to be nearer the centre of the arc to detect the microwaves.

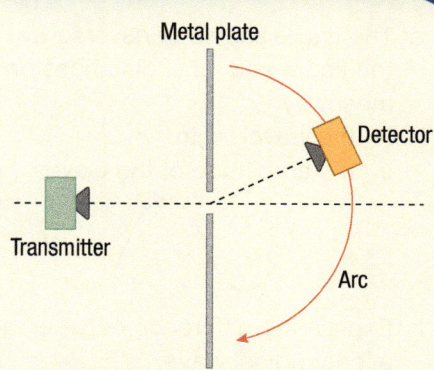

Figure 5 *Diffraction in microwaves (top view)*

Signal problems

People in hilly areas often have poor TV reception. The signal from a TV transmitter mast is carried by radio waves. If there are hills between a TV receiver and the transmitter mast, the signal may not reach the receiver. The radio waves passing the top of a hill are diffracted by the hill but they do not spread enough behind the hill.

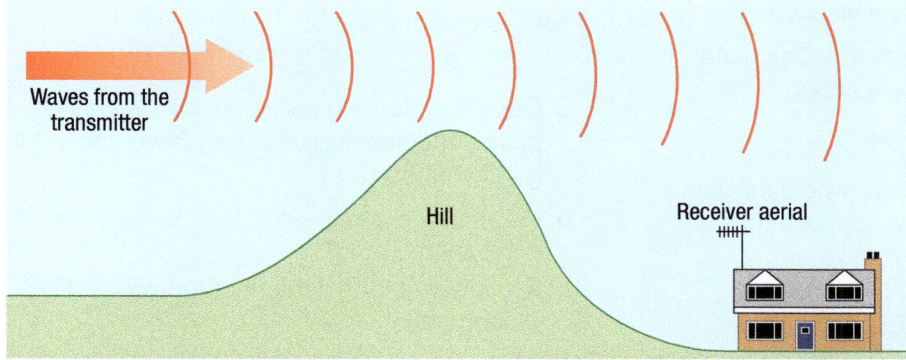

Figure 6
Poor reception

Summary questions

1 Waves spread out when they pass through a gap.
  a What can be said about the wavelength of the waves before and after passing through the gap?
  b What happens to the waves if the gap is made:
    i narrower?   ii wider?

2 a State what is meant by diffraction.
  b A transmitter mast transmits TV signals.
    i Explain why the TV reception can be poor in hilly areas.
    ii The transmitter also transmits electromagnetic waves of a much longer wavelength that carry the signal from a radio station. Explain why this signal can be detected at some locations where the TV signal cannot be detected.

3 A small portable radio inside a room can be heard all along a corridor that runs past the room when its door is open. Explain why it can be heard by someone in the corridor who is not near the door.

Chapter summary questions

1 a The figure shows transverse waves on a string. Copy the figure and label distances on it to show what is meant by:
  i the wavelength
  ii the amplitude of the waves.

b Explain the difference between a transverse wave and a longitudinal wave.

c Give **one** example of:
  i a transverse wave
  ii a longitudinal wave.

2 A speedboat on a lake creates waves that make a buoy bob up and down as shown in the figure.

a The buoy bobs up and down three times in one minute. Calculate the frequency of the waves.

b The waves travel 24 metres in one minute. Calculate the speed of the waves in metres per second.

c Calculate the wavelength of the waves.

3 a When a wave is refracted at a boundary where its speed is reduced, state what change, if any, happens to:
  i its wavelength
  ii its frequency.

b When a wave is reflected, what change, if any, happens to:
  i its wavelength?
  ii its frequency?
  iii its speed?

4 a Copy and complete this diagram to show the reflection of a straight wavefront at a straight reflector.

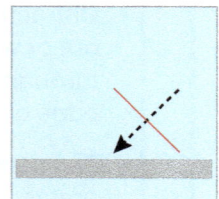

b Copy and complete this diagram to show the refraction of a straight wavefront at a straight boundary as the wavefront moves from deep to shallow water.

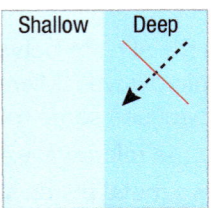

5 a Calculate the frequency in air of electromagnetic waves of wavelength 3.0 m. The speed of electromagnetic waves in a vacuum is 300 000 km/s.

b The signal from a radio station is carried by radio waves of wavelength 3.0 m. Explain why the signal may be difficult to receive in hilly locations.

6 The diagram shows straight waves directed at a gap in a barrier.

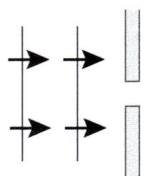

a Copy the diagram and draw **two** waves showing their shape and spacing after they have passed through the gap in the barrier.

b i What is the name for the change in the shape of the waves?
  ii How would the shape of the waves differ if the gap had been wider?

7 a When a wave is diffracted, which of the following, if any, will change:
  i its wavelength?
  ii its frequency?
  iii its speed?

b i Draw a diagram to show the diffraction of plane waves when they move past the edge of an obstacle.
  ii Describe the effect on the diffraction of these waves if their wavelength is reduced.

Practice questions

1 The diagrams **A**, **B**, and **C** show three processes that can happen to waves.

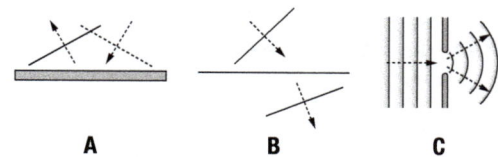

    A   **B**   **C**

a Which diagram, **A**, **B**, or **C**, shows:

  i diffraction?

  ii refraction? (2)

b i State what the diagram you did not choose in **a** shows. (1)

  ii In which diagram does the speed of the waves change? (1)

2 The diagrams **X** and **Y** show two types of wave travelling along a slinky spring.

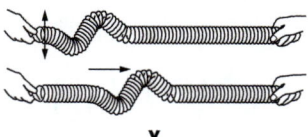 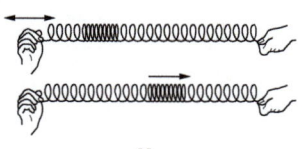

    X   **Y**

a i What type of wave is shown in diagram **X**? (1)

  ii What type of wave is shown in diagram **Y**? (1)

  iii Describe the oscillations of the slinky spring that cause the wave shown in diagram **X**. (1)

  iv Describe the oscillations of the slinky spring that cause the wave shown in diagram **Y**. (1)

b Light, sound, and radio waves can all transfer information.

What are the similarities and differences in the way that light, sound, and radio waves travel? (6)

3 a A student showed three waves, **P**, **Q**, and **R**, on an oscilloscope screen.

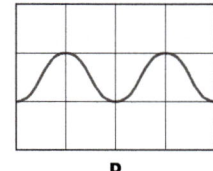 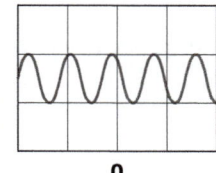 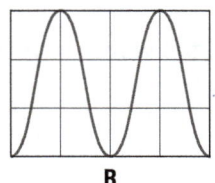

    P   **Q**   **R**

Which wave, **P**, **Q**, or **R**, has:

  i the largest amplitude? (1)

  ii the highest frequency? (1)

b The behaviour of waves is often demonstrated by teachers using microwaves with a wavelength of 3 cm. Microwaves have a wave speed of 3×10^8 m/s. Calculate the frequency of these microwaves. (4)

c A teacher demonstrated the reflection of microwaves using the apparatus below.

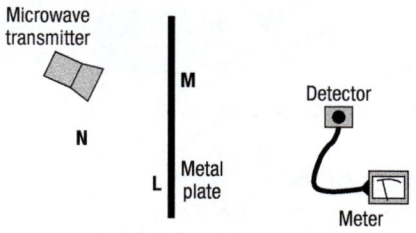

The meter read zero when the detector was at **M** but had a high reading when the detector was at **N**.

  i Explain why the meter read zero when the detector was at **M**. (3)

  ii Explain why the meter had a high reading when the detector was at **N**. (2)

  iii What would the meter read if the detector was placed at **L**?

    Choose an answer from the list.

    higher than N   lower than N   zero (1)

4 Waves can spread through gaps.

a Looking at the diagrams above, what conclusion can be reached about how waves spread through gaps? (2)

b The front door of a house is open at night. The lights are on in the house and shine down the front path. The television is also on. A woman stands in the garden to the side of the front path. She is in shadow.

Explain why she can hear the television even though the light from the house does not reach her. (3)

c A communication mast transmits both radio and television programmes.

The carrier wave for Radio 4 has a frequency of 200 kHz. The carrier wave for BBC 1 television has a frequency of 600 MHz.

One householder can receive Radio 4 but cannot receive BBC 1 television.

There is a large hill between the householder and the communication mast.

  i How does the wavelength of the Radio 4 carrier wave compare with that of the BBC 1 television carrier wave? (2)

  ii Explain how the householder can receive Radio 4 even though the mast is on the other side of the hill. (2)

  iii Explain why he cannot receive BBC 1 television. (2)

The electromagnetic spectrum

Electromagnetic waves are electric and magnetic disturbances that transfer energy from one place to another. There is a range of electromagnetic waves that makes up the **electromagnetic spectrum**.

Electromagnetic waves do not transfer matter. The energy they transfer depends on the **wavelength** of the waves. This is why waves of different wavelengths have different effects. Figure 1 shows some of the uses of each part of the electromagnetic spectrum.

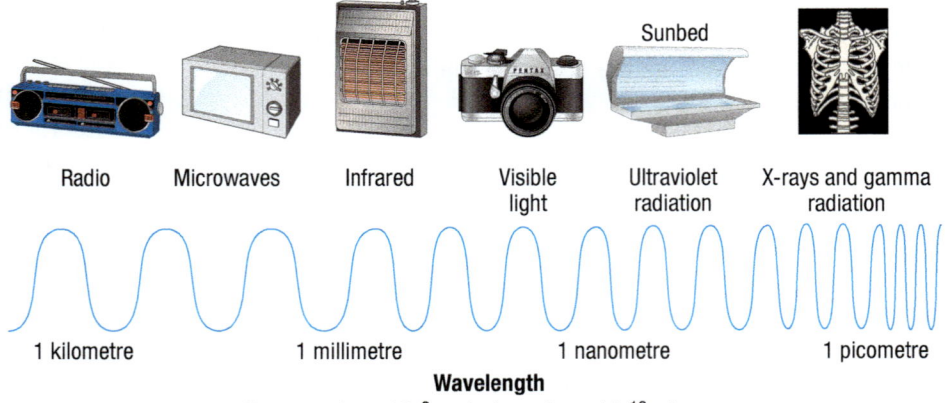

Radio   Microwaves   Infrared   Visible light   Ultraviolet radiation   X-rays and gamma radiation

Wavelength
(1 nanometre = 10^{-9} m, 1 picometre = 10^{-12} m)

Figure 1 *The electromagnetic spectrum is continuous. The boundaries of frequency and wavelength between the types of wave are not defined – one type of wave merges into the next*

Waves from different parts of the electromagnetic spectrum have different wavelengths.

- Long-wave radio waves have wavelengths as long as 10 km (10^4 m).
- X-rays and gamma rays have wavelengths as short as 0.001 picometre.

The speed of electromagnetic waves

All electromagnetic waves travel at a speed of 300 million m/s through space or in a vacuum.

You can link the speed of the waves to their frequency and wavelength using the **wave speed** equation:

$$v = f \times \lambda$$

where:
v = wave speed in m/s
f = frequency in Hz
λ = wavelength in m.

⦾ links

For more information on the wave speed equation, look back at Topic 7.2 'Measuring waves'.

You can work out the wavelength if you know the frequency and the wave speed. To do this, you rearrange the wave speed equation into:

$$\lambda = \frac{v}{f}$$

You can work out the frequency if you know the wavelength and the wave speed. To do this, you rearrange the equation into:

$$f = \frac{v}{\lambda}$$

where:

v = speed in m/s
f = frequency in Hz
λ = wavelength in m.

A mobile phone gives out electromagnetic waves of frequency 900 million Hz. Calculate the wavelength of these waves.

The speed of electromagnetic waves in air = 300 million m/s.

Solution

$$\text{wavelength } \lambda \text{ (m)} = \frac{\text{wave speed } v \text{ (m/s)}}{\text{frequency } f \text{ (Hz)}}$$

$$= \frac{300\,000\,000\,\text{m/s}}{900\,000\,000\,\text{Hz}} = 0.33\,\text{m}$$

Remember that the wavelength decreases from radio waves to gamma rays.

Energy and frequency

The wave speed equation shows that the shorter the wavelength of the waves, the higher their frequency. The energy of the waves increases as the frequency increases. Therefore, the energy and frequency of the waves increases from radio waves to gamma rays as the wavelength decreases.

Summary questions

1 a Which is greater, the wavelength of radio waves or the wavelength of visible light waves?
   b What can you say about the speed in a vacuum of different electromagnetic waves?
   c Which is greater, the frequency of X-rays or the frequency of infrared radiation?
   d Where in the electromagnetic spectrum would you find waves of wavelength 10 millimetres?

2 Fill in the missing parts of the electromagnetic spectrum in the list below.

   radio _ _ _ _ _ _ **infrared** **visible light** _ _ _ _ _ _ **X-rays** _ _ _ _ _ _

3 Electromagnetic waves travel through space at a speed of 300 million m/s. Calculate:
   a the wavelength of radio waves of frequency 600 million Hz
   b the frequency of microwaves of wavelength 0.30 m.

4 A distant star explodes and emits visible light and gamma rays simultaneously.
   a Which of these two types of wave has the greater frequency?
   b Explain why the gamma rays and the visible light waves reach the Earth at the same time.

- The electromagnetic spectrum (in order of decreasing wavelength, and increasing frequency and energy) is made up of:
  - radio waves
  - microwaves
  - infrared radiation
  - visible light
  - ultraviolet radiation
  - X-rays
  - gamma radiation.

- The wave speed equation is used to calculate the frequency or wavelength of electromagnetic waves.

Light, infrared, microwaves, and radio waves

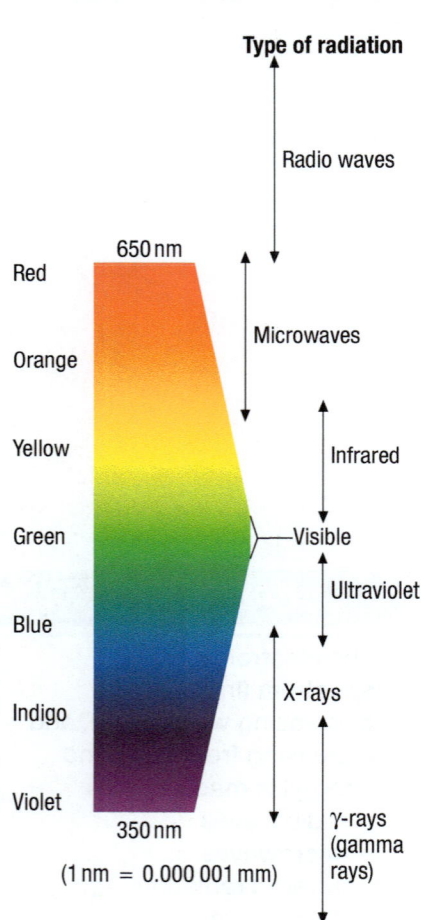

Figure 1 *The electromagnetic spectrum with an expanded view of the visible range*

Type of radiation

Radio waves

650 nm
Red
Microwaves
Orange

Yellow
Infrared
Green — Visible

Ultraviolet
Blue

X-rays
Indigo

Violet
γ-rays (gamma rays)
350 nm

(1 nm = 0.000 001 mm)

links

For more information on infrared radiation, look at Topic 13.4 'Infrared radiation'.

Light and colour

Light from ordinary lamps and from the Sun is called **white light**. It has all the colours of the visible spectrum in it. The wavelength decreases across the spectrum from red to violet (Figure 1).

You see the colours of the spectrum when you look at a rainbow. You can also see them if you use a glass prism to split a beam of white light.

Photographers need to know how shades and colours of light affect the photographs they take.

1 In a film camera, the light is focused by the camera lens on to a light-sensitive film. The film then needs to be developed to see the image of the objects that were photographed.

2 In a digital camera, the light is focused by the lens on to a sensor. This consists of thousands of tiny light-sensitive cells called pixels. Each pixel gives a dot of the image. You can see the image on a small screen at the back of the camera. When you take the photograph, the image is stored electronically on a memory card.

Infrared radiation

All objects emit infrared radiation.

- The hotter an object is, the more infrared radiation it emits.
- Infrared radiation is absorbed by the skin. It can burn or kill skin cells because it heats up the cells.

Infrared devices

- **Optical fibres** in communications systems use infrared radiation instead of visible light. This is because infrared radiation is absorbed less than visible light in the glass fibres.
- **Remote control handsets** for TV and video equipment transmit signals carried by infrared radiation. When you press a button on the handset, it sends out a sequence of infrared pulses.
- **Infrared scanners** are used in medicine to detect 'hot spots' on the body surface. These hot areas can mean the underlying tissue is unhealthy.
- **Infrared cameras** provide night vision, to see people and animals in darkness.

Microwaves

Microwaves lie between radio waves and infrared radiation in the electromagnetic spectrum. They are called **microwaves** because they are shorter in wavelength than radio waves.

Microwaves are used for communications, for example, **satellite TV**, because they can pass through the atmosphere and reach satellites above the Earth. They are also used to beam signals from one place to another, because microwaves don't spread out as much as radio waves. Microwaves (as well as radio waves) are used to carry **mobile phone** signals.

Radio waves

Radio wave frequencies range from about 300 000 Hz to 3000 million Hz (where microwave frequencies start). Radio waves are longer in wavelength and lower in frequency than microwaves.

As you will see in Topic 8.3 'Communications', radio waves are used to carry **radio**, **TV**, and **mobile phone** signals.

Radio waves can also be used instead of cables to connect a computer to other devices such as a printer or a mouse. For example, Bluetooth-enabled devices can communicate with each other over a range of about 10 metres. No cables are needed – just a Bluetooth radio in each device and the necessary software. These wireless connections work at frequencies of about 2400 million Hz, and they operate at low power.

Bluetooth was set up by electronics manufacturers. They realised that they all needed to use the same radio frequencies for their software.

Microwaves and radio waves can be hazardous because they penetrate the human body, and can heat the internal parts.

Practical

Testing infrared radiation

Can infrared radiation pass through paper? Use a remote control handset to find out.

Summary questions

1 a When you watch a TV programme, what type of electromagnetic wave is:
      i   detected by the aerial?
      ii emitted by the screen?
   b What type of electromagnetic wave is used:
      i   to carry signals to and from a satellite?
      ii to send signals to a printer from a computer without using a cable?

2 Mobile phones use electromagnetic waves in a wavelength range that includes short-wave radio waves and microwaves.
   a What would be the effect on mobile phone users if remote control handsets operated in this range as well?
   b Why do our emergency services use radio waves in a wavelength range that no else is allowed to use?

3 a The three devices listed below each emit a different type of electromagnetic radiation. State the type of radiation each one emits.
      i   a TV transmitter mast
      ii   a TV satellite
      iii a TV remote handset
   b The speed of electromagnetic waves in air is 300 000 km/s. Calculate the wavelength in air of electromagnetic waves of frequency 2400 MHz.

⊂⊃ links
Is mobile phone radiation dangerous? See Topic 8.3 'Communications'.

Demonstration

Demonstrating microwaves

Look at the demonstration.

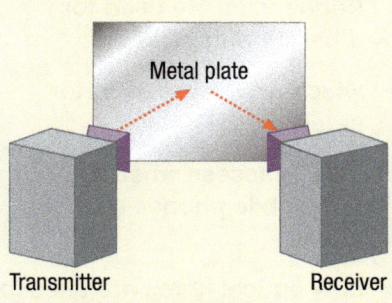

Metal plate

Transmitter   Receiver

Figure 2 *Demonstrating microwaves*

• What does this show?

Key points

• White light contains all the colours of the visible spectrum.

• Infrared radiation is used to carry signals from remote handsets and inside optical fibres. Microwaves are used to carry signals for satellite TV programmes and mobile phone calls. Radio waves are used for radio and TV broadcasting, radio communications, and mobile phone calls.

• Different types of electromagnetic radiation are hazardous in different ways. Microwaves and radio waves can heat the internal parts of the human body. Infrared radiation can cause skin burns.

8.3 Communications

Figure 1 *Sending microwave signals to a satellite*

Figure 2 *A mobile phone mast*

Radio communications

Radio waves are emitted from an aerial when an alternating voltage is applied to the aerial. An alternating voltage is an electrical signal that regularly changes direction – see Topic 16.1. The frequency of the radio waves produced is the same as the frequency of the alternating voltage.

When the radio waves pass across a receiver aerial, they cause a tiny alternating voltage in the aerial. The frequency of the alternating voltage is the same as the frequency of the radio waves received. The aerial is connected to a loudspeaker (see Topic 9.1 'Sound'). The alternating voltage from the aerial is used to make the loudspeaker send out sound waves.

The radio and microwave spectrum is divided into **bands** of different wavelength ranges. This is because the shorter the wavelength of the waves:

- the more information they can carry
- the shorter their range (due to increasing absorption by the atmosphere)
- the less they spread out (because they diffract less).

Radio wavelengths

Microwaves and radio waves of different wavelengths are used for different communications purposes. Examples are given below.

- **Microwaves** are used for satellite phone and TV links and satellite TV broadcasting. This is because microwaves can travel between satellites in space and the ground. Also, they spread out less than radio waves do, so the signal doesn't weaken as much.
- **Radio waves of wavelengths less than about 1 metre** are used for TV broadcasting from TV masts, because they can carry more information than longer radio waves.
- **Radio waves of wavelengths from about 1 metre up to about 100 m** are used by local radio stations (and for the emergency services) because their range is limited to the area around the transmitter.
- **Radio waves of wavelengths greater than 100 m** are used by national and international radio stations, because they have a much longer range than shorter-wavelength radio waves.

Mobile phone radiation

When you phone a friend on a mobile phone, it sends out a radio signal. The signal is picked up by a local mobile phone mast and is sent through the phone network to you friend's phone. The signal from their phone is sent through the network back to the mast near you and then on to your phone. The signals to and from your local mast are carried by radio waves of different frequencies.

The radio waves to and from a mobile phone have a wavelength of about 30 cm. Radio waves at this wavelength are not quite in the microwave range, but they do have a similar heating effect to microwaves. So they are usually referred to as microwaves.

Study tip

Remember that in communications, electromagnetic waves carry the information.

Optical fibre communications

Optical fibres are very thin glass fibres. They are used to transmit signals carried by light or infrared radiation. The light rays can't escape from the fibre. When they reach the surface of the fibre, they are reflected back into the fibre, as shown in Figure 3.

In comparison with radio wave and microwave communications:

- optical fibres can carry much more information – this is because light has a much smaller wavelength than radio waves so can carry more pulses of waves
- optical fibres are more secure, because the signals stay in the fibre.

Is mobile phone radiation dangerous?

The radiation from a mobile phone is much weaker than the microwave radiation in an oven. However, when you use a mobile phone, it is very close to your brain. Some scientists think the radiation might affect the brain. As children have thinner skulls than adults, their brains might be more affected by mobile phone radiation. A UK government report published in May 2000 recommended that the use of mobile phones by children should be limited.

Activity

Mobile phone hazards

Here are some findings by different groups of scientists:

- The short-term memory of volunteers using a mobile phone was found to be unaffected by whether or not they had used their phone before the memory test.
- The brains of rats exposed to microwaves were found to respond less to electrical impulses than the brains of unexposed rats.
- Some scientists found that mice exposed to microwaves developed more cancers than unexposed mice. Other scientists were unable to confirm this effect.
- A survey of mobile phone users in Norway and Sweden found they experienced headaches and fatigue. No control group of people who did not use a mobile phone was surveyed.
- What conclusions can you make from the evidence above?
- Suggest how researchers could improve the validity of any conclusions.

Summary questions

1 a What types of electromagnetic wave are used to carry:
   - **i** mobile phone signals?
   - **ii** signals along a thin transparent fibre?
   - **b i** Why should signals to and from a mobile phone be at different frequencies?
   - **ii** Why are signals in an optical fibre more secure than radio signals?

2 a Why could children be more affected by mobile phone radiation than adults?
   - **b** Why can light waves carry more information than radio waves?

3 a Explain why microwaves are used for satellite TV, and radio waves for terrestrial TV.
   - **b** Why can microwave signals be sent a long distance between two transmitter dishes provided that the two transmitter masts are within sight of each other?

∞ links

For more information on optical fibres, see Topic 10.4 'Total internal reflection'.

??? Did you know … ?

Satellite TV signals are carried by microwaves. These signals can be detected on the ground because they pass straight through a layer of ionised gas in the upper atmosphere. This layer reflects lower-frequency radio waves.

Demonstration

Demonstrating an optical fibre

Observe light shone into an optical fibre. You should see the reflection of light inside an optical fibre. This is known as total internal reflection.

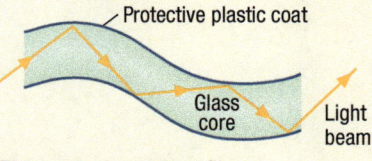

Figure 3 *An optical fibre*

Key points

- Radio waves of different frequencies are used for different purposes because the wavelength (and therefore the frequency) of waves affects:
  - how far they can travel
  - how much they spread
  - how much information they can carry.

- Microwaves are used for satellite TV signals.

- Further research is needed to evaluate whether or not mobile phones are safe to use.

- Optical fibres are very thin transparent fibres that are used to transmit signals by light and infrared radiation.

8.4

Ultraviolet rays, X-rays, and gamma rays

Ultraviolet radiation

Ultraviolet (often written as UV) radiation lies between violet light and X-rays in the electromagnetic spectrum. Some chemicals emit light in UV radiation. Posters and ink that glow in ultraviolet light contain these chemicals. Security marker pens containing this kind of ink are used to mark valuable objects. The chemicals absorb ultraviolet rays and then emit visible light as a result.

Ultraviolet radiation is harmful to humans because:

1 it damages the eyes and can cause blindness

2 too much UV radiation causes sunburn and can cause skin cancer.

UV wavelengths are smaller than visible light wavelengths. UV rays carry more energy than light rays.

- If you use a sunbed to get a suntan, don't exceed the recommended time. You should also wear special goggles to protect your eyes.

- If you stay outdoors in summer, use special skin creams to block UV radiation and prevent it reaching your skin.

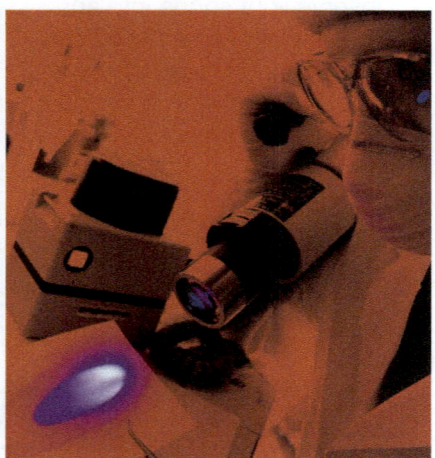

Figure 1 *Using an ultraviolet lamp to detect biological stains*

Demonstration

Ultraviolet radiation

Watch your teacher place different-coloured clothes under an ultraviolet lamp. The lamp must point downwards so you can't look directly at the glow from it. Observe what happens.

- What do white clothes look like under a UV lamp?

X-rays and gamma rays

X-rays and gamma rays both travel straight into substances, and may pass through them if the substances are not too dense and not too thick. A thick plate made of lead will stop them.

X-rays and gamma rays have similar properties because they both:

- are at the short-wavelength end of the electromagnetic spectrum

- carry much more energy per second than longer-wavelength electromagnetic waves.

They differ from each other because:

- X-rays are produced in an X-ray tube when electrons or other particles moving at high speeds are stopped

- gamma rays are produced by radioactive substances when unstable nuclei release energy (you will learn about radioactive substances in Chapter 17)

- gamma rays can have shorter wavelengths than X-rays and can therefore penetrate substances more than X-rays can.

As you will see in Topic 8.5 'X-rays in medicine', X-rays are used in hospitals to make X-ray images of broken limbs. The X-rays are stopped by bone, but travel though the surrounding soft tissues. So the bone shows up as a white area on the X-ray image. A crack in a bone is revealed as a break in the white area. X-rays are also used to detect internal cracks in metal objects. These kinds of application are usually possible because the more dense a substance is, the more X-rays it absorbs from an X-ray beam passing through it.

Did you know …?

The ozone layer in the atmosphere absorbs some of the Sun's ultraviolet radiation. Unfortunately, gases such as CFCs from old refrigerators react with the ozone layer and make it thinner. Many countries have strict regulations about how to dispose of unwanted refrigerators.

Using gamma radiation

High-energy gamma radiation has several important uses:

- **Killing harmful bacteria in food**
  About 20% of the world's food is lost through spoilage. One of the major causes is bacteria. The bacteria produce waste products that cause food poisoning. Exposing food to gamma radiation kills 99% of disease-carrying organisms, including *Salmonella* (found in poultry) and *Clostridium* (which causes botulism).

- **Sterilising surgical instruments**
  Exposing surgical instruments in sealed plastic wrappers to gamma radiation kills any bacteria on the instruments. This helps to stop infection spreading in hospitals.

- **Killing cancer cells**
  Doctors and medical physicists use gamma-ray therapy to destroy cancerous tumours. A narrow beam of gamma radiation from a radioactive source (cobalt-60) is directed at the tumour. The beam is aimed at it from different directions so as to kill the tumour but not the surrounding tissue. The cobalt-60 source is in a thick lead container. When it is not in use, it is rotated away from the exit channel (see Figure 2).

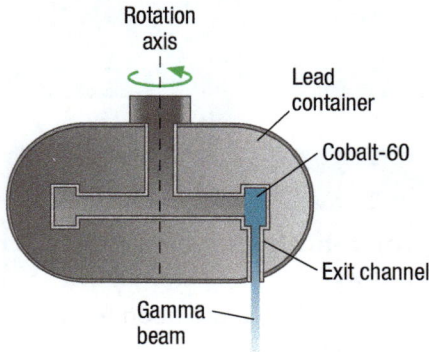

Safety matters

X-rays and gamma rays passing through substances can knock electrons out of atoms in the substance. The atoms become charged because they have lost electrons. This process is called **ionisation**. Charged atoms are called **ions**.

If ionisation happens to a living cell, it can damage or kill the cell. For this reason, exposure to too much X-ray or gamma radiation is dangerous and can cause cancer. High doses of the radiation kill living cells, and low doses cause genetic mutation and cancerous growth.

People who use equipment or substances that produce any form of ionising radiation (e.g., X-ray or gamma radiation) must wear a film badge. If the badge shows that it is over-exposed to ionising radiation, its wearer is not allowed to continue working with the equipment for a period of time.

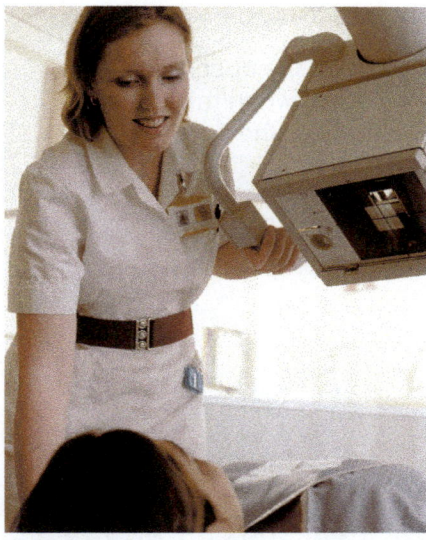

Figure 2 *Gamma treatment*

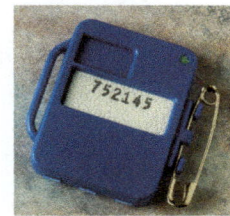

Figure 3 *A film badge shows how much ionising radiation the wearer has received. Who might wear these?*

Summary questions

1. **a** Why does a crack inside a metal object show up on an X-ray image?
   b Will gamma radiation pass through thin plastic wrappers?
   c Why does a film badge used for monitoring radiation need to have a plastic case, not a metal case?

2. **a** Why is ultraviolet radiation harmful?
   b **i** How does the Earth's ozone layer help to protect us from ultraviolet radiation from the Sun?
      ii Why do people need suncream outdoors in summer?

3. **a** Which types of electromagnetic radiation can penetrate thin metal sheets?
   b Which metal can be used most effectively to absorb X-rays and gamma rays?
   c Which types of electromagnetic radiation:
      i ionise substances they pass through?
      ii damage the human eye?

Key points

- Ultraviolet radiation has a shorter wavelength than visible light and harms the skin and the eyes.

- X-rays are used in hospitals to make X-ray images.

- Gamma rays are used to kill harmful bacteria in food, to sterilise surgical equipment, and to kill cancer cells.

- X-rays and gamma rays damage living tissue when they pass through it.

8.5 X-rays in medicine

Learning objectives

After this topic, you should know:

- what X-rays are used for in hospitals
- why X-rays are dangerous
- about the absorption of X-rays when they pass through the body
- what a CT scan is.

Have you ever broken one of your bones? If you have, you will have gone to your local hospital for an X-ray photograph. X-rays are also used by dentists to diagnose dental problems. X-rays are electromagnetic waves at the short-wavelength end of the electromagnetic spectrum. They are produced in an X-ray tube when fast-moving electrons hit a target. Their wavelengths are about the same as the diameter of an atom.

To make a **radiograph** or X-ray image, X-rays from an X-ray tube are directed at the patient. In the past, a lightproof cassette containing a photographic film was placed on the other side of the patient. In modern X-ray machines, a **flat-panel detector** is used instead of the film.

- When the X-ray tube is switched on, X-rays from the tube pass through the part of the patient's body under investigation.
- X-rays pass through soft tissue, but they are absorbed by bones, teeth, and metal objects that are not too thin. X-rays blacken a photographic film in the same way that light does. The parts of the film or the detector that the X-rays reach become darker than the other parts. So the bones appear lighter than the surrounding tissue, which appears dark. The radiograph shows a negative image of the bones. A hole or a cavity in a tooth shows up as a dark area in the bright image of the tooth.
- An organ that consists of soft tissue can be filled with a substance called a **contrast medium** which absorbs X-rays easily. This enables the internal surfaces in the organ to be seen on the radiograph. For example, to obtain a radiograph of the stomach, the patient first has a food containing a barium compound. The barium compound is a good absorber of X-rays.
- Lead plates between the X-ray tube and the patient stop X-rays reaching other parts of the body. Lead is used because it is a good absorber of X-rays. The X-rays reaching the patient pass through a gap between the plates.
- A flat-panel detector is a small screen that contains a **charge-coupled device (CCD)**. The sensors in the CCD are covered by a layer of a substance that converts X-rays to light. The light rays then create electronic signals in the sensors that are sent to a computer, which displays a digital X-ray image.

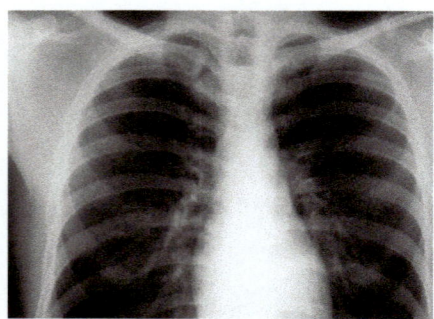

Figure 1 *A chest X-ray*

Safety matters

X-radiation, as well as gamma radiation, is dangerous because it ionises substances it passes through. High doses kill living cells. Low doses can cause cell mutation and cancerous growth. There is no evidence of a safe limit below which living cells would not be damaged.

Workers who use equipment or substances that produce X-radiation (or alpha, beta, or gamma radiation) must wear a film badge.

X-ray therapy

Doctors use X-ray therapy to kill cancer cells and destroy cancerous tumours in the body. Thick plates between the X-ray tube and the body stop X-rays from reaching healthy body tissues. A gap between the plates allows X-rays through to reach the tumour. X-rays for therapy are shorter in wavelength than X-rays used for imaging.

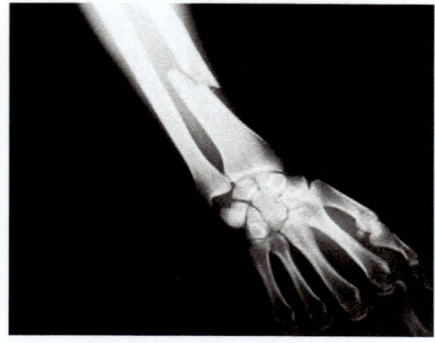

Figure 2 *Spot the break*

The CT scanner

A computerised tomography scanner (**CT scanner**) produces a digital image of any cross-section through the body. It can also be used to construct a three-dimensional (3-D) image of an organ.

Figure 3 shows an end-view of a CT scanner. The patient lies without moving on a bed inside a ring of detectors.

- The X-ray tube automatically moves round the inside of the ring in small steps.
- At each position, X-rays from the tube pass through the patient and reach the detector ring.
- Electronic signals from the detector are recorded by a computer until the tube has moved round the ring.
- The computer displays a digital image of the scanned area.

Each detector receives X-rays that have travelled through different types of tissue. The detector signal depends on:

- the different types of tissue along the X-ray path
- how far the X-rays pass through each type of tissue.

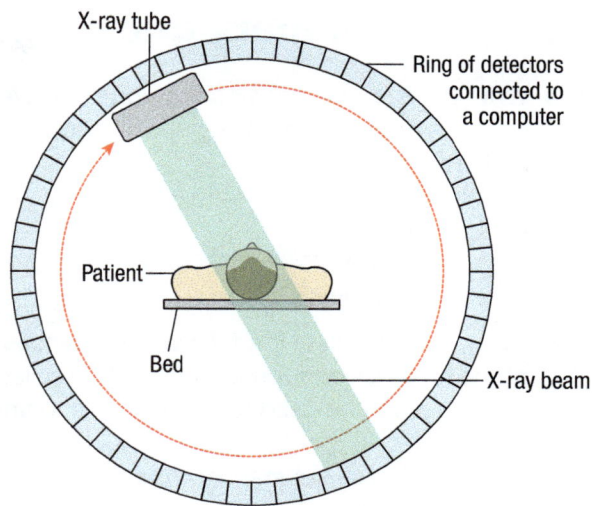

Figure 3 *The CT scanner can distinguish between different types of soft tissue, as well as bone*

Table 1 *Comparison of a CT scanner with an ordinary X-ray machine*

	CT scanner	Ordinary X-ray machine
Image distinguishes between bone and soft tissue	Yes	Yes
Image distinguishes between different types of soft tissue	Yes	No
Three-dimensional image	Yes	No
Radiation dose	CT scanner gives a much higher dose than an ordinary X-ray machine	
Cost	CT equipment cost is much greater than an ordinary X-ray machine	

Study tip

Make sure that you know the advantages and disadvantages of CT scanners.

Summary questions

1 a What is a contrast medium used for when an X-ray photograph of the stomach is taken?
  b What can X-ray therapy be used for?

2 When an X-ray photograph is taken, why is it necessary:
  a to place the patient between the X-ray tube and the film cassette?
  b to have the film in a lightproof cassette?
  c to shield the parts of the patient not under investigation from X-rays? Explain what would happen to healthy cells.

3 a What type of electromagnetic radiation is used in a CT scanner?
  b State one advantage and one disadvantage of a CT scanner in comparison with an ordinary X-ray machine.

Key points

- X-rays are used in hospitals:
  – to make images and CT scans
  – to destroy tumours at or near the body surface.
- X-rays can damage living tissue when they pass through it.
- X-rays are absorbed more by bones and teeth than by soft tissues.
- CT scans distinguish between different types of soft tissue as well as between bone (or teeth) and soft tissue.

Chapter summary questions

1 a Write the five types of electromagnetic wave listed below in order of increasing wavelength.
  A infrared waves
  B microwaves
  C radio waves
  D gamma rays
  E ultraviolet rays

b Copy and complete the following sentences using one of the types of electromagnetic radiation listed in **a**:
  i .... can be used to send signals to and from a satellite.
  ii .... ionise substances when they pass through them.
  iii .... are used to carry signals in thin transparent fibres.

2 a The radio waves from a local radio station have a wavelength of 2.9 metres in air. The speed of electromagnetic waves in air is 300 000 km/s.
  i Write down the equation that links frequency, wavelength, and wave speed.
  ii Calculate the frequency of the radio waves.

b A certain local radio station transmitter has a range of 30 km. State and explain the effect on the range if the power supplied to the transmitter is reduced.

3 Mobile phones send and receive signals using electromagnetic waves near or in the microwave part of the electromagnetic spectrum. New mobile phones are tested for radiation safety and given a specific absorption rate (SAR) value before being sold. The SAR is a measure of the energy per second absorbed by your head when using the phone. For use in the UK, SAR values must be less than 2.0 W/kg. SAR values for two different mobile phones are given below.

  • phone **A**   0.2 W/kg
  • phone **B**   1.0 W/kg

a What is the main reason that mobile phones are tested for radiation safety?

b Which phone, **A** or **B**, is safer? Give a reason for your answer.

c The UK government recommends caution in the use of mobile phones, particularly by children and young people, until scientists and doctors find out more. Explain why children and young people may be more at risk than adults.

4 The figure shows an X-ray source which is used to direct X-rays at a broken leg. A photographic film in a lightproof wrapper is placed under the leg. When the film is developed, an image of the broken bone is observed.

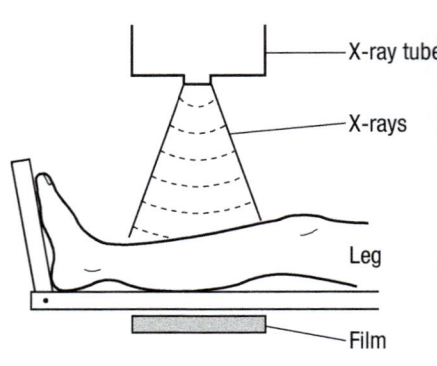

a i Explain why an image of the bone is seen on the film.
  ii Why is it possible to see the fracture on the image?

b When an X-ray photograph of the stomach is taken, the patient is given food containing barium before the photograph is taken.
  i Why is it necessary for the patient to be given this food before the photograph is taken?
  ii The exposure time for a stomach X-ray must be shorter than the X-ray time for a limb. Why?
  iii Low-energy X-rays from the X-ray tube can be absorbed by placing a metal plate between the patient and the X-ray tube. Such X-rays would otherwise be absorbed by the body. What is the benefit of removing such low-energy X-rays in this way?

c An ultrasound scanner is used to observe an unborn baby. Why is ultrasound instead of X-rays used to observe an unborn baby?

5 a Explain what is meant by ionisation.

b Name the **two** types of electromagnetic radiation that can ionise substances.

c Give **two** reasons why ionising radiation is harmful.

6 a Describe and explain how invisible ink works. In your explanation, state the type of radiation that is absorbed and the type of radiation that is emitted.

b Explain why a beam of infrared radiation cannot be used to carry signals to a detector that is more than a few metres from a transmitter.

c Explain why a local radio channel can broadcast only to a limited area, whereas a satellite TV channel can broadcast to a much larger area.

Practice questions

1 There are several types of wave in the electromagnetic spectrum.

Some are listed below.

infrared   light   radio waves   ultraviolet   X-rays

a Which of the waves in the list:
  i   has the highest frequency?
  ii has the longest wavelength?
  iii carries the most energy?
  iv causes skin cancer? (4)

b A newspaper headline states:

> ## AIRPORT BODY SCANNERS MAY POSE HEALTH RISK

The X-ray scanners were introduced in 2009 to detect hidden weapons and explosives.
  i   What properties of X-rays make them suitable for use in body scanners? (4)
  ii Explain why X-rays may pose health risks. (4)

2 Electromagnetic waves have many uses.

a Match each electromagnetic wave in the left-hand column to its use in the right-hand column. (4)

Electromagnetic wave		Use
Gamma		Carrying TV programmes
Infrared		In a TV remote control
Radio		Prolonging the shelf-life of fruit
Ultraviolet		Security marking of TV sets

b Microwaves are used to send signals to and from satellites.
  i   What property of microwaves makes them suitable for sending signals to satellites? (1)
  ii Some people are worried about the dangers of very young children using mobile phones.

  Discuss the reasons for their concerns. (6)

3 Hospitals use different types of scanner.

a **i**   What are the advantages of a CT scan over a normal X-ray scan? (2)
  ii What are the disadvantages of a CT scan when compared with a normal X-ray scan? (2)

b Why do hospitals not use X-ray scans on pregnant women? (2)

c The charts below give information about exposure to radiation.

The average annual exposure to radiation is given in millisieverts (mSv).

1980

Average annual exposure 3.6 mSv

▨ Natural background
☐ Artificial (medical)
▧ Artificial (other)

2010

Average annual exposure 7.3 mSv

Outline the changes in average annual exposure to radiation that occurred between 1980 and 2010.

Suggest reasons for the changes. (6)

4 Electromagnetic waves are used for communication.

a Which types of electromagnetic wave can be transmitted by optical fibres? (2)

b Radio waves have a large range of wavelengths.

The table gives information about some radio waves.

Wavelength	less than 1 m	1 to 100 m	greater than 100 m
Information	can carry a lot of information	have a limited range	are reflected by a layer of the atmosphere

Suggest a wavelength of radio wave suitable for carrying:
  i   international radio
  ii local radio
  iii national TV. (3)

c Bluetooth is a short-range radio communication system which can link mobile phones to laptop computers.

A typical Bluetooth frequency is 2.4×10^3 MHz.

Electromagnetic waves travel at a speed of 3×10^8 m/s.

Calculate the wavelength of these Bluetooth waves. (3)

Learning objectives

After this topic, you should know:

- the range of frequencies that can be detected by the human ear
- what sound waves are
- what echoes are.

Investigating sound waves

Sound waves are easy to produce. Your vocal cords vibrate and produce sound waves every time you speak. Any object vibrating in air makes the layers of air near the object vibrate, which makes the layers of air next to them vibrate. The vibrating object pushes and pulls repeatedly on the air. This sends out the vibrations of the air in waves of compressions and rarefactions. When the waves reach your ears, they make your eardrums vibrate in and out so you hear sound as a result.

The vibrations travelling through the air are sound waves. The waves are longitudinal because the air particles vibrate (or oscillate) along the direction in which the waves transfer energy. Energy transfer by sound waves is sometimes called acoustic energy.

Figure 1 *Making sound waves*

Practical

Investigating sound waves

You can use a loudspeaker to produce sound waves by passing alternating current through it. Figure 2 shows how to do this using a signal generator. This is an alternating current supply unit with a variable frequency dial.

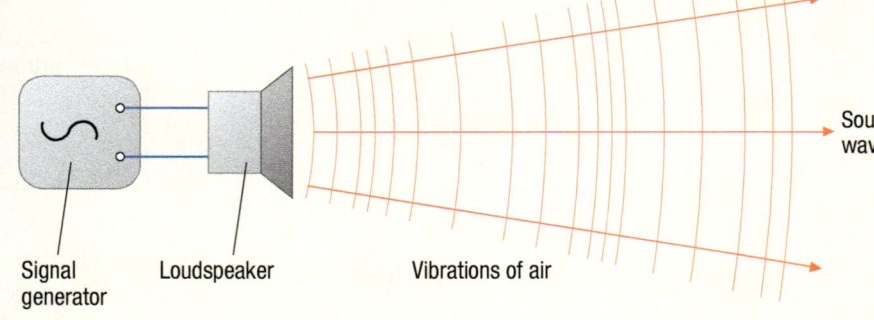

Signal generator   Loudspeaker   Vibrations of air   Sound waves

Figure 2 *Using a loudspeaker*

- If you observe the loudspeaker closely, you can see it vibrating. It produces sound waves as it pushes the surrounding air backwards and forwards.
- If you alter the frequency dial of the signal generator, you can change the frequency of the sound waves.

Find out the lowest and the highest frequency you can hear. Young people can usually hear sound frequencies from about 20 Hz to about 20 000 Hz. Older people, in general, can't hear frequencies at the higher end of this range.

⚭ **links**

You can find out more about alternating current in Topic 16.1 'Alternating current'.

Sound waves cannot travel through a vacuum. You can test this by listening to an electric bell in a bell jar – see Figure 3. As the air is pumped out of the bell jar, the ringing sound fades away.

Reflection of sound

Have you ever created an echo? An **echo** is an example of reflection of sound. Echoes can be heard in a large hall or gallery that has bare, smooth walls.

- If the walls are covered in soft fabric, the fabric will absorb sound instead of reflecting it. No echoes will be heard.
- If the wall surface is uneven (not smooth), echoes will not be heard because the reflected sound is broken up and scattered.

Diffraction of sound

Sound waves are diffracted when they pass through a suitable-sized opening such as a doorway. As waves pass through the opening, they are diffracted and so they spread out. This is partly why you can hear around corners. The width of the opening needs to be of the same order of magnitude as the wavelength of the sound waves for diffraction to be big enough to be noticeable. Sound waves of much smaller wavelengths would pass straight through with little diffraction.

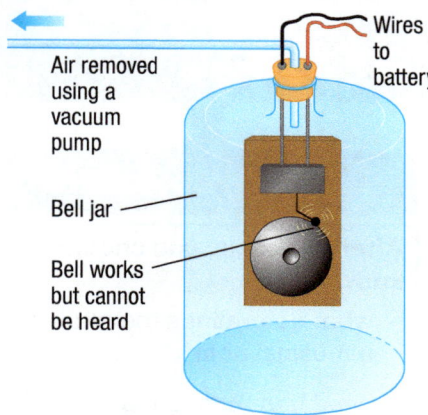

Figure 3 *A sound test – sound waves can't travel through a vacuum*

> ??? **Did you know … ?**
>
> When you blow a round whistle, you force a small ball inside the whistle to go round and round inside. Each time it goes round, its movement draws air in then pushes it out. Sound waves are produced as a result.

Study tip

Remember that sound waves are longitudinal, and need a medium to compress.

Summary questions

1 **a** The sound from a vehicle engine can be reduced in the seating compartment by placing a suitable material between the engine compartment and the seating compartment. What type of material would be most suitable for this purpose?

  b Traffic noise from an urban motorway can be reduced by planting tall bushes near the motorway and installing concrete fence panels behind the bushes. Explain why each of these measures reduces traffic noise for people in the area.

2 **a** What is the highest frequency of sound the human ear can hear?

  b Why does a round whistle produce sound at a constant frequency when you blow steadily into it?

3 **a** A boat is at sea in fog. The captain wants to know if the boat is near any cliffs, so he sounds the horn and listens for an echo.
    i Why would hearing an echo tell him he is near cliffs?
    ii The captain hears an echo 5.0 s after sounding the horn. Show that the distance from the ship to the cliff is 850 m. The speed of sound in air = 340 m/s.

  b Explain why a person in a large cavern may hear more than one echo of a sound.

Key points

- The frequency range of the normal human ear is from about 20 Hz to about 20 000 Hz.

- Sound waves are vibrations (longitudinal waves) that travel through a medium (a substance). They cannot travel through a vacuum (as in outer space).

- Echoes are caused by sound waves reflected from a smooth, hard surface.

9.2

More about sound

After this topic, you should know:

- what determines the pitch of a musical note

- what happens to the loudness of a note as the amplitude increases

- how sound waves are created by musical instruments.

Figure 1 *Making music*

What type of music do you like? Whatever your taste in music is, when you listen to it you usually hear sounds produced by specially designed instruments. Your voice is produced by a biological organ that has the function of producing sound.

- Music is usually rhythmic. The sound waves change smoothly and the wave pattern repeats itself regularly.

- Noise often consists of sound waves that vary in frequency without any pattern.

Practical

Investigating different sounds

Use a microphone connected to an oscilloscope to display the waveforms of different sounds. The waveform shows how the amplitude of the sound waves varies with time.

Figure 2 *Investigating different sound waves*

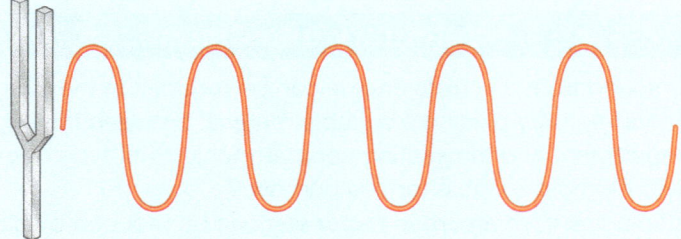

Figure 3 *Tuning-fork waves*

1 Test a tuning fork to see the waveform of a sound of constant frequency.

2 Compare the pure waveform of a tuning fork with the sound you produce when you talk, sing, or whistle. You may be able to produce a pure waveform when you whistle or sing but not when you talk.

3 Use a signal generator connected to a loudspeaker to produce sound waves. The waveform on the oscilloscope screen should be a pure waveform. See how the waveform changes when you change the loudness and frequency of the sound waves.

Study tip

Frequency controls the pitch of a note – how high it is – whilst amplitude controls how loud the note sounds.

Your investigations should show you that:

- **increasing the frequency of a sound** (the number of waves per second) increases its **pitch**. This makes more waves appear on the screen.

- **increasing the amplitude** of the waves increases the **loudness** of a sound. The waves on the screen become taller.

Figure 4 shows the waveforms for different sounds from the loudspeaker.

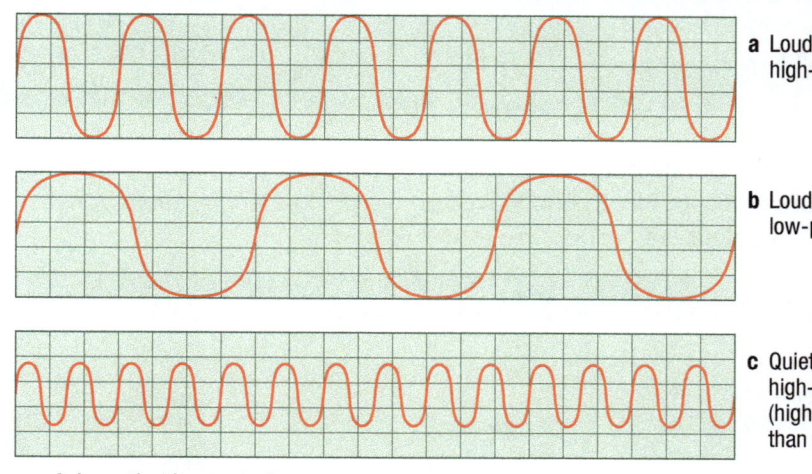

a Loud and high-pitched

b Loud and low-pitched

c Quiet and high-pitched (higher pitch than a)

Figure 4 *Investigating sounds*

Musical instruments

When a musician plays a musical instrument, they create sound waves by making the instrument and the air inside it vibrate. Each new cycle of vibrations makes the vibrations stronger at certain frequencies. The instrument **resonates** at these frequencies. Because the instrument and the air inside it vibrate strongly at these frequencies when it is played, you can hear recognisable notes from the instrument.

- A wind instrument such as a flute is designed so that the air inside resonates when it is played. You can make the air in an empty bottle resonate by blowing across the top gently.
- A string instrument such as a guitar produces sound when the strings vibrate. The vibrating strings make the surfaces of the instrument vibrate and produce sound waves in the air. In an acoustic guitar, the air inside the hollow body of the guitar (the sound box) vibrates too.
- A percussion instrument such as a drum vibrates and produces sound when it is struck.

Practical

Musical instruments

Investigate the waveform produced by a musical instrument, such as a flute.

You should find its waveform changes smoothly, like the one in Figure 5 – but only if you can play it correctly. The waveform is a mixture of frequencies instead of a single frequency waveform like Figure 3.

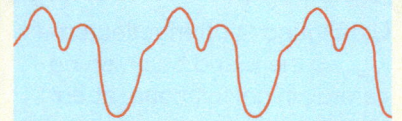

Figure 5 *Flute wave pattern*

Summary questions

1 a A tuning fork creates a note of sound when it is struck briefly. How does the waveform of the sound from a tuning fork change as the sound becomes quieter?

 b When a key of a piano is pressed, a hammer briefly strikes a stretched wire in the piano and a note of sound is produced. Describe and explain how the note changes after the key has been pressed.

2 A microphone and an oscilloscope are used to investigate sound from a loudspeaker connected to a signal generator. What change would you expect to see on the oscilloscope screen if the sound is:

 a made louder at the same frequency?

 b made lower in frequency at the same loudness?

3 a How does the note produced by a guitar string change if the string is:

  i shortened? ii tightened?

  iii loosened?

 b Compare the sound produced by a violin with the sound produced by a drum.

 c Explain why the sound from a vibrating tuning fork is much louder if the base of the tuning fork is held on a table.

Key points

- The pitch of a note increases if the frequency of the sound waves increases.

- The loudness of a note increases if the amplitude of the sound waves increases.

- Vibrations created in a musical instrument when it is played produce sound waves.

Ultrasound

Learning objectives

After this topic, you should know:

- the nature of ultrasound waves
- how ultrasound waves are used in medicine
- why ultrasound waves can be used to scan the human body
- why an ultrasound scan is safer than an X-ray.

⃝⃝ links

To learn more about what an oscilloscope is used for, look at Topic 16.1 'Alternating current'.

Study tip

Ultrasound is often reflected – it goes there and back, so be careful, in calculations, about distance.

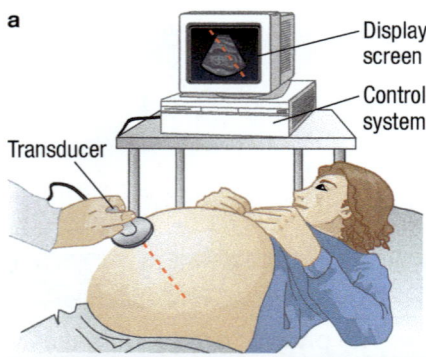

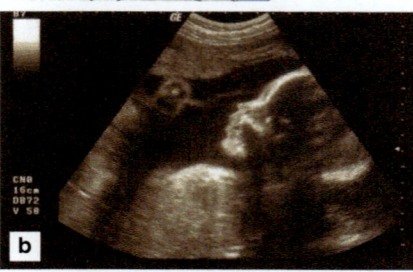

Figure 2 a *An ultrasound scanner system* **b** *An ultrasound image of a baby in the womb*

The human ear can detect sound waves in the frequency range from about 20 Hz to about 20 000 Hz. Sound waves above the highest frequency that humans can detect are called **ultrasound waves**.

Practical

Testing ultrasound

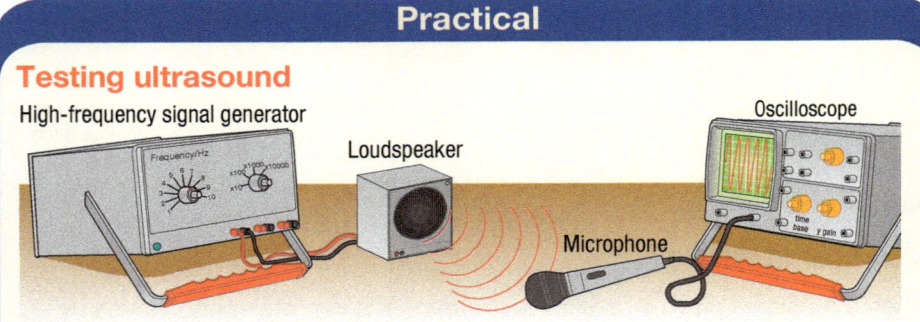

Figure 1 *Testing ultrasound*

Use a loudspeaker connected to a signal generator to produce ultrasound waves. Connect a microphone to an oscilloscope to detect the waves and display them. You can use the apparatus to:

- measure the frequency of the ultrasound waves
- test different materials to see if they absorb ultrasound waves
- show that ultrasound waves can be partly reflected.

Ultrasound scanners

Ultrasound waves are used for prenatal scans of a baby in the womb. They are also used to image organs in the body, such as a kidney or damaged ligaments and muscles. An ultrasound scanner consists of an electronic device called a **transducer** placed on the body surface, a control system, and a display screen. The transducer produces and detects pulses of ultrasound waves.

Each pulse from the transducer:

- is partially reflected from the different tissue boundaries in its path
- returns to the transducer as a sequence of reflected pulses from the boundaries, arriving back at different times.

The transducer is moved across the surface of part of the body. The pulses are then detected by the transducer. They are used to build up an image on a screen of the internal tissue boundaries in the body.

The advantages of using ultrasound waves instead of X-rays for medical imaging are that ultrasound waves (unlike X-rays) are:

- non-ionising and therefore harmless when used for scanning
- reflected at boundaries between different types of tissue (different media), so they can be used to scan organs and other soft tissues in the body.

Distance measurements

Sight can sometimes be restored to a blind person by replacing the eye lens with an artificial lens. Before this is done, the eye surgeon needs to know how long the eyeball is. This is to make sure the new lens gives clear vision. Figure 3 shows how ultrasound is used to measure the length of the eyeball. This type of scan is called an **A-scan**.

A transducer placed in front of the eye sends ultrasound pulses into the eye. The reflected pulses are detected by the transducer and displayed on an oscilloscope screen or on a computer monitor, as shown in Figure 3.

An oscilloscope can be used to measure the transit time of each pulse. This is the time taken by the pulse to travel from the transmitter at the surface to and from the boundary that reflected it. To calculate distance travelled:

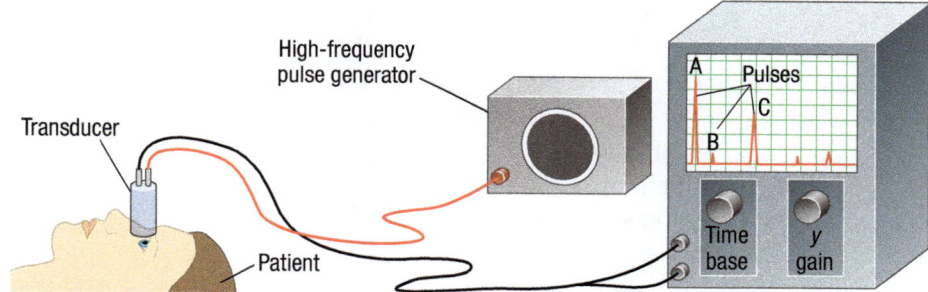

Figure 3 *Pulse A is due to partial reflection at the front surface of the eye. Pulse B is due to partial reflection at the surfaces of the eye lens. Pulse C is due to reflection at the back of the eye. Some further pulses are present due to partial reflection beyond the back of the eye*

$$\text{distance travelled by the pulse (m)} = \text{speed of ultrasound waves in body tissue (m/s)} \times \text{transit time (s)}$$

Because the pulse travels from the surface to the boundary then back to the surface, the depth of the boundary below the surface is half the distance travelled by each pulse to and from the boundary. So:

$$\text{depth of the boundary below the surface (m)} = \tfrac{1}{2} \times \text{speed of the ultrasound waves in body tissue (m/s)} \times \text{transit time (s)}$$

Ultrasound therapy

Kidney stones can be very painful. Powerful ultrasound waves can be used to break a kidney stone into tiny bits. The fragments are small enough to leave the kidney naturally. The transmitter is used in an A-scan system so that the waves are aimed exactly at the kidney stone.

Maths skills

You can write the distance equation in symbols:

$$s = v \times t$$

where:
s is the distance in m
v is the wave speed in m/s
t is the time taken in s.

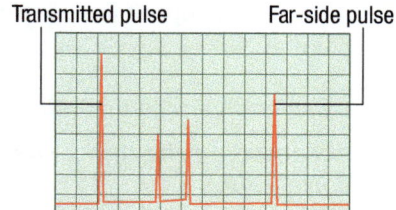

Figure 4 *The screen of an oscilloscope connected to an ultrasound detector on the surface of a patient's body*

Transmitted pulse   Far-side pulse

Summary questions

1 a Why are ultrasound waves partly reflected by body organs?

 b Why is an ultrasound scanner better than an X-ray scanner for imaging a body organ?

 c The wavelength of ultrasound waves used for scanning needs to be much smaller than the transducer width, otherwise the waves would spread out too much because of diffraction. How would too much diffraction affect the reflected pulses?

2 Look at the screen in Figure 4. It shows the reflected pulses that are detected for each transmitted pulse.

 a How many internal boundaries are present according to this display?

 b The oscilloscope beam takes 32 millionths of a second to travel across each grid square on the screen.

   i How long does each pulse take to travel from the body surface to the nearest internal boundary?

   ii The speed of ultrasound in the body is 1500 m/s. What is the distance from the body surface to the nearest tissue boundary?

3 a In an A-scan of a model eye, the distance from the front to the back of the eye was known to be 48 mm. If Figure 3 represented the oscilloscope display for the model eye, what would be the distance from the eye lens to the front of the model eye?

 b Estimate the accuracy of the distance you calculated in a.

Key points

- Ultrasound waves are sound waves of frequency above 20 000 Hz.

- Ultrasound waves are used in medicine for ultrasound scanning and for destroying kidney stones.

- Ultrasound waves are partly reflected at a boundary between two different types of body tissue.

- An ultrasound scan is non-ionising, so it is safer than an X-ray.

Chapter summary questions

1 a A loudspeaker is used to produce sound waves.
  i Describe how sound waves are created when an object in air vibrates.
  ii In terms of the amplitude of the sound waves, explain why the sound is fainter further away from the loudspeaker.

  b A microphone is connected to an oscilloscope. The figure shows the display on the screen of the oscilloscope when the microphone detects sound waves from a loudspeaker which is connected to a signal generator.

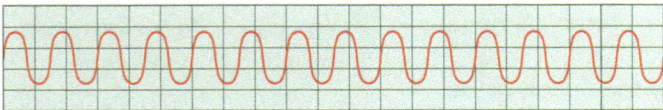

Describe how the waveform displayed on the oscilloscope screen changes if the sound from the loudspeaker is:
  i made louder
  ii reduced in pitch.

  c Describe how you would use this apparatus to measure the upper limit of frequency of a person's hearing.

2 Copy and complete **i** to **iii** using the words below.

absorbed   reflected   scattered   smooth
soft   rough

  i An echo is caused by sound waves that are .... from a .... wall.
  ii When sound waves are directed at a .... surface, they are broken up and ....
  iii When sound waves are directed at a wall covered with a .... material, they are .... and not reflected.

3 a What is the highest frequency the human ear can hear?

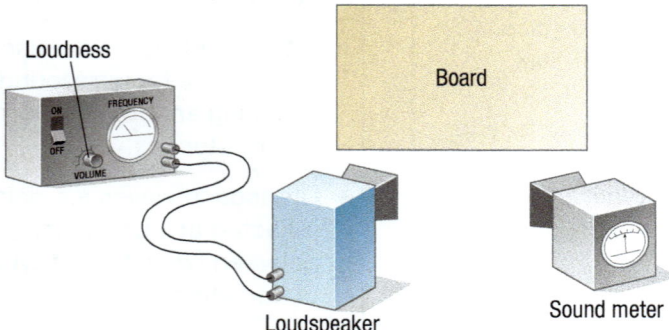

Loudness
Board
Loudspeaker
Sound meter

  b A sound meter is used to measure the loudness of the sound reflected from an object. Describe how you would use the meter and the arrangement shown in the figure to test if more sound is reflected from a board than from a cushion in place of the board. The control knob and a frequency dial can be used to change the loudness and the frequency of the sound from the loudspeaker. List the variables that you would need to keep constant in your test.

4 A person is standing a certain distance from a flat wall. She claps her hands and hears an echo.

  a Explain the cause of the echo.

  b She hears the sound 0.30 s after clapping her hands. Calculate how far she is from the nearest point of the wall. The speed of sound in air = 340 m/s.

5 In a test to measure the depth of the sea bed, ultrasound pulses took 0.40 s to travel from the surface to the sea bed and back. Given that the speed of sound in sea water is 1350 m/s, calculate the depth of the sea bed below the surface.

6 a Calculate the wavelength in air of ultrasound waves of frequency 40 kHz. The speed of sound in air = 340 m/s.

  b Bats navigate by detecting the echoes of ultrasound pulses they emit. Describe how such echoes enable bats to avoid flying into obstacles.

7 Ultrasound waves used for medical scanners have a frequency of 2000 kHz.

  a Use the equation: speed = frequency × wavelength to calculate the wavelength of these ultrasound waves in human tissue. The speed of ultrasound in human tissue is 1500 m/s.

  b Ultrasound waves of this frequency in human tissue are not absorbed much and they do not spread out.
    i Why is it important in a medical scanner that they are not absorbed?
    ii Why is it important in a medical scanner that they do not spread out?

  c State and explain which is better, ultrasound or X-rays, for producing images of babies in the womb.

Practice questions

1 a A teacher wants to display sound as a waveform so that the students can both hear and 'see' the sound.

    i What four pieces of equipment will she need? (4)

    ii The diagram shows two waveforms, **A** and **B**, that she displays.

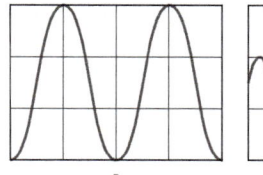 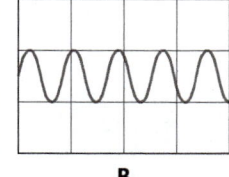

       A   **B**

      Describe how **A** sounds different from **B**. (2)

b The teacher takes the class outside to measure the speed of sound. All the students stand as far away from a wall of the sports hall as possible. The teacher bangs two sticks together. She asks them to listen to the sound reflected from the wall.

    i What name is given to the reflected sound? (1)

    ii Describe how the students would take the measurements needed and how they would calculate the speed of sound. (5)

    iii What should they do to minimise random errors in their measurements? (2)

2 a Both light and sound travel as waves.

    Choose words from the list to complete the following sentences.

    can   cannot   faster   longitudinal   slower   transverse

    i Sound waves are .... waves and light waves are .... waves.

    ii In air, sound waves travel .... than light waves.

    iii Sound waves .... travel through a vacuum, but light waves .... (3)

b The diagram shows a sound wave.

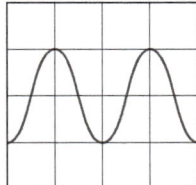

    Scale: X – each square equals 0.001 s

    Y – each square equals 1 cm.

    i What is the amplitude of the wave? (1)

    ii Calculate the frequency of the wave. (2)

    iii The speed of sound in air is 340 m/s. Calculate the wavelength of the sound wave. (2)

    iv Discuss whether this sound wave would be diffracted by an open doorway. (3)

3 a Sound and ultrasound waves travel through water at 1400 m/s.

    i Describe how sound waves travel through water. (3)

    ii Why can a sound wave not travel through a vacuum? (1)

    iii A sound wave has a frequency of 2000 Hz. Calculate the wavelength of the wave in water. (2)

b Echo sounders use ultrasound waves to detect underwater objects such as submarines.

The diagram shows how an echo sounder works.

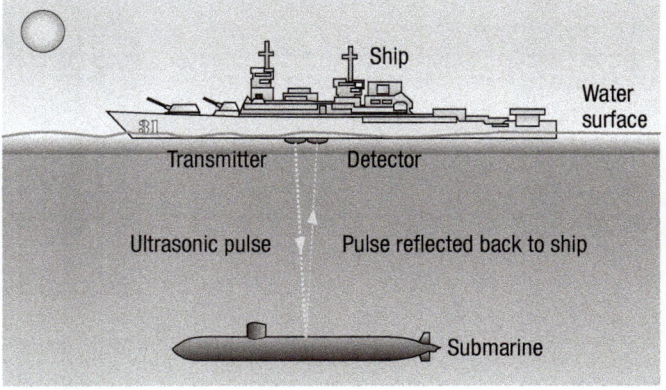

    i What are ultrasound waves? (2)

    ii An ultrasound pulse returns to the detector 0.1 s after leaving the transmitter. Calculate the distance of the submarine from the ship. (3)

    iii State **one** other use of ultrasound waves. (1)

4 a Dentists often use ultrasound waves to clean plaque from teeth.

Which of the following frequencies is an ultrasound frequency? (1)

3 Hz   30 Hz   300 Hz   3000 Hz   30 000 Hz

b The diagram shows an ultrasound scan being carried out on a pregnant woman.

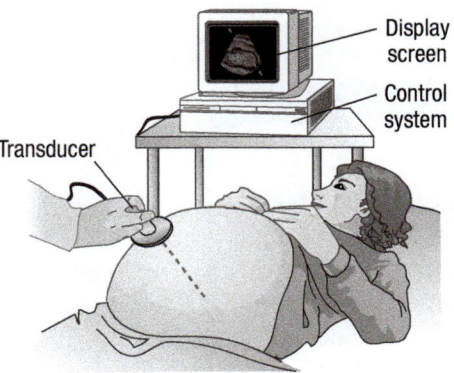

Explain how ultrasound waves are used to produce an image of the foetus. (6)

10.1 Reflection of light

Learning objectives

After this topic, you should know:

- what the normal is in a diagram showing light rays

- what is meant by the angle of incidence and the angle of reflection

- the law of reflection for a light ray at a plane mirror

- how an image is formed by a plane mirror.

If you visit a Hall of Mirrors at a funfair, you will see some strange images of yourself. A tall, thin image or a short, broad image of yourself means you are looking into a mirror that is curved. If you want to see an ordinary image of yourself, look in a **plane mirror**. A plane mirror is a perfectly flat mirror. You see an exact **mirror image** of yourself.

Figure 1 *A good image*

Light consists of waves. In Topic 7.3 'Reflection and refraction', you used a ripple tank to investigate the reflection of waves. The investigations showed that when plane (straight) waves reflect from a flat reflector, the reflected waves are at the same angle to the reflector as the incident waves.

The law of reflection

A light ray is the line which light waves move along. It tells the direction in which light waves are travelling. Figure 2 shows how you can investigate the reflection of a light ray from a ray box using a plane mirror.

Study tip

Remember that angles of reflection and angles of incidence are measured between the ray and the normal.

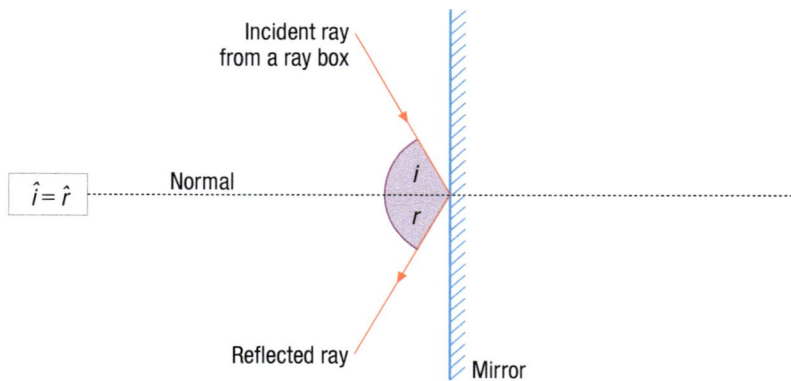

Figure 2 *The law of reflection, i = r*

- The line perpendicular to the mirror is called the **normal**.
- The **angle of incidence** *i* is the angle between the incident ray and the normal.
- The **angle of reflection** *r* is the angle between the reflected ray and the normal.

Measurements show that **for any light ray reflected by a plane mirror:**

the angle of incidence = the angle of reflection

Practical

A reflection test

Test the law of reflection by using a ray box to direct a light ray at a plane mirror as shown in Figure 2. Repeat the test for different angles of incidence.

Safety: If mirrors are glass, take care with sharp edges.

Image formation by a plane mirror

Figure 3 shows how an image is formed by a plane mirror. This ray diagram shows the path of two light rays from a point object that reflect off the mirror. To a person looking into the mirror, the light rays seem to come from the image behind the mirror. The image and the object in Figure 3 are at equal distances from the mirror.

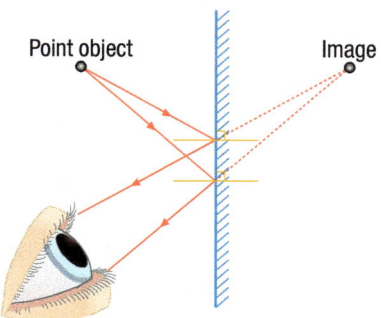

Point object   Image

Figure 3 *Image formation by a plane mirror*

Real and virtual images

The image formed by a plane mirror is virtual, upright (the same way up as the object), and laterally inverted (back to front but not upside down). Unlike movie images that you see at the cinema, a **virtual image** can't be projected onto a screen. An image on a screen is described as a **real image** because it is formed by focusing light rays onto the screen.

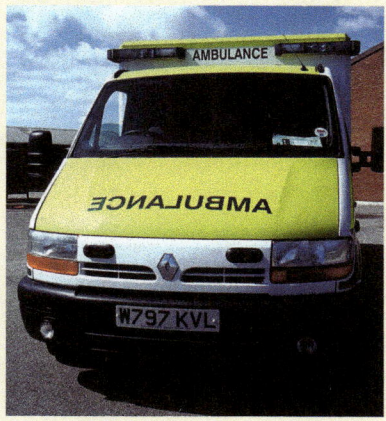

Summary questions

1 a In Figure 2, if the angle of reflection of a light ray from a plane mirror is 20°, what is:

   i the angle of incidence?

   ii the angle between the incident ray and the reflected ray?

  b If the mirror is turned so the angle of incidence is increased to 21°, what is the angle between the incident ray and the reflected ray?

2 A point object O is placed in front of a plane mirror, as shown.

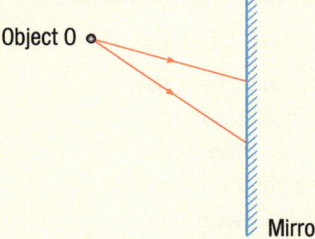

Object O

Mirror

  a Copy the diagram and complete the path of the two rays shown from O after they have reflected off the mirror.

  b i Use the reflected rays to locate the image of O.

   ii Show that the image and the object are the same distance from the mirror.

3 Two plane mirrors are placed perpendicular to each other.

  a Draw a ray diagram to show the path of a light ray at an angle of incidence of 60° that reflects off both mirrors.

  b i Measure the angle A between the final reflected ray and the incident ray.

   ii Show that angle A is always equal to 180° whatever the angle of incidence is.

Key points

- The normal at a point on a mirror is a line drawn perpendicular to the mirror.

- For a light ray reflected by a plane mirror:
  - The angle of incidence is the angle between the incident ray and the normal
  - The angle of reflection is the angle between the reflected ray and the normal.

- The law of reflection states that: the angle of incidence = the angle of reflection.

10.2 Refraction of light

Learning objectives

After this topic, you should know:

- where refraction of light can occur
- how a light ray refracts when it travels from air into glass
- how a light ray refracts when it travels from glass into air.

When you have your eyes tested, the optician might test different lenses in front of each of your eyes. Each lens changes the direction of light passing through it. This change of direction is known as **refraction**.

Refraction is a property of all forms of waves, including light and sound. In Topic 7.3 'Reflection and refraction', you investigated the refraction of water waves in a ripple tank. Water waves travel more slowly in shallow water than in deep water. Refraction occurs when the waves cross a boundary between the deep and the shallow water at a non-zero angle to the boundary. The change of speed at the **boundary**, or **interface**, causes them to change direction.

- Light waves are refracted as shown in Figure 1 when they travel across a boundary between air and a transparent medium or between two transparent media. This is because the speed of light changes at such a boundary.

Figure 1 shows light waves (in blue) entering then leaving a glass block. The direction in which the light waves are moving is represented by light rays (in red). The change of direction of each ray relative to the normal at each boundary is:

- towards the normal when light travels from air into glass
- away from the normal when light travels from glass to air.

Both changes occur because light travels more slowly in glass than in air. Because light travels more slowly in glass than in air, glass is described as optically more dense than air. In general:

- when light enters a more dense medium, it is refracted towards the normal
- when light enters a less dense medium, it is refracted away from the normal.

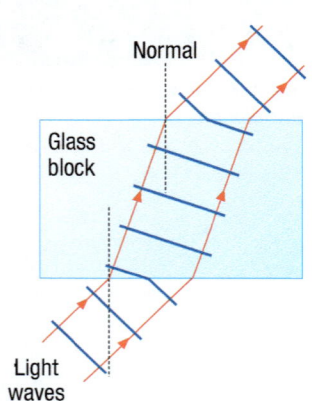

Figure 1 *Refraction of light waves*

Study tip

- Remember that angles of incidence, reflection, and refraction are always measured between the ray and the normal.
- Make sure you know that light bends towards the normal when entering a denser medium, and away from the normal when entering a less dense medium.

Required Practical

Investigating refraction of light

Figure 2 shows how you can use a ray box and a rectangular glass block to investigate the refraction of a light ray when it enters glass. The ray changes direction at the boundary between air and glass (unless it is along the normal).

At the point where the light ray enters the glass, compare the angle of refraction r (the angle between the refracted ray and the normal) with the angle of incidence i.

You should find that the angle of refraction at the point of entry is always smaller than the angle of incidence.

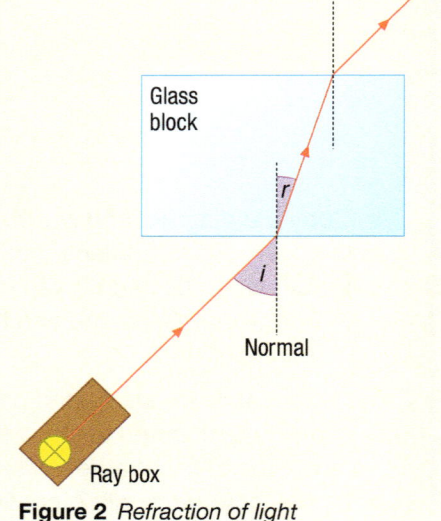

Figure 2 *Refraction of light*

Safety: Make sure the glass block doesn't have any sharp edges.

Refraction rules

Your investigation should show that a light ray:

- changes direction towards the normal when it travels from air into glass. The angle of refraction (*r*) is smaller than the angle of incidence (*i*).
- changes direction away from the normal when it travels from glass into air. The angle of refraction (*r*) is greater than the angle of incidence (*i*).

Refraction by a prism

Figure 3 shows what happens when a light ray from a white light source passes through a triangular glass prism. The ray comes out of the prism in a different directions to the incident ray and is split into the colours of the spectrum.

White light contains all the colours of the spectrum. Each colour of light is refracted slightly differently. So the prism splits the light into colours. The splitting of white light into the colours of the spectrum is called **dispersion**.

Figure 3 *Dispersion. In this picture the incident ray of white light enters the prism top left*

?? Did you know ... ?

The next time you jump into water, make sure you know how deep it is. Light from the bottom refracts at the surface. This makes the water appear shallower than it is.

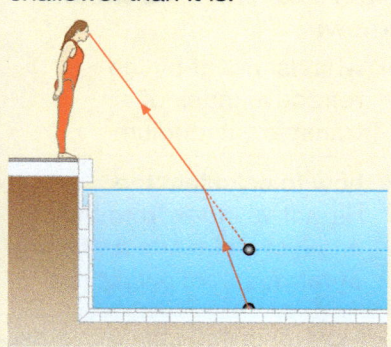
Figure 4 *Water looks shallower than it really is*

?? Did you know ... ?

A **rainbow** is caused by refraction of light when sunlight shines on rain droplets. The droplets refract sunlight and split it into the colours of the spectrum.

Figure 5 *A rainbow is caused by refraction*

Summary questions

1. When a light ray travels from air into glass, its speed changes at the boundary.
   a. State whether there is an increase or a decrease in the speed of the light waves when they cross the boundary.
   b. If the angle of incidence is zero, what is the angle of refraction?
   c. If the angle of incidence is non-zero, state whether the angle of refraction is greater than or smaller than the angle of incidence.

2. a. Copy the diagrams below, and complete the path of the light ray through each glass object.

   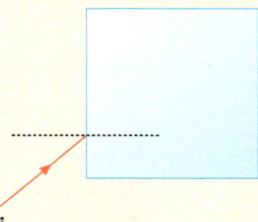
   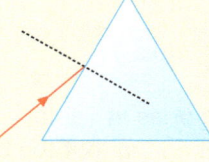

   i   ii

   b. i In Figure 3, which colour, blue or red, is refracted most?
      ii What does Figure 3 tell you about the speed of blue light in glass compared with the speed of red light in glass?

3. A light ray from the bottom of a swimming pool refracts at the water surface. Its angle of incidence is 40° and its angle of refraction is 75°.
   a. Draw a diagram to show the path of this light ray from the bottom of the swimming pool into the air above the pool.
   b. Use your diagram to explain why the swimming pool appears shallower than it really is when viewed from above.

Key points

- Refraction is the change of direction of waves when they travel across a boundary from one medium to another.

- When a light ray refracts as it travels from air into glass, the angle of refraction is smaller than the angle of incidence.

- When a light ray refracts as it travels from glass into air, the angle of refraction is greater than the angle of incidence.

10.3 Refractive index

Practical

Investigating how the angle of refraction varies with the angle of incidence

You can use a semicircular transparent glass block as shown in Figure 2.

- Measure the angle of refraction, r, for different angles of incidence, i.
- Record all your measurements in a table.

Safety: Make sure glass block does not have any sharp edges.

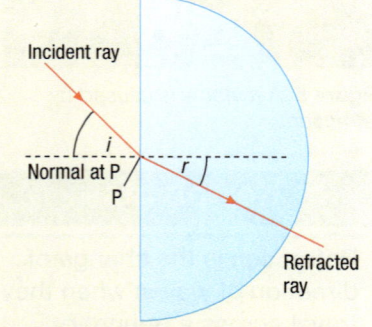

Figure 2 Using a semicircular block

Study tip

- Remember that angles i and r are measured between the ray and the normal.
- When calculating the refractive index, remember to use the sine of the angle, not the angle itself.

When a light ray travels from air into a transparent medium, the angle of refraction depends on the medium as well as on the angle of incidence. For example, for the same angle of incidence, glass refracts a light ray more than water does. Figure 1 shows the refraction of a light ray travelling from air into water. If glass had been used instead, the light ray would have been refracted much more.

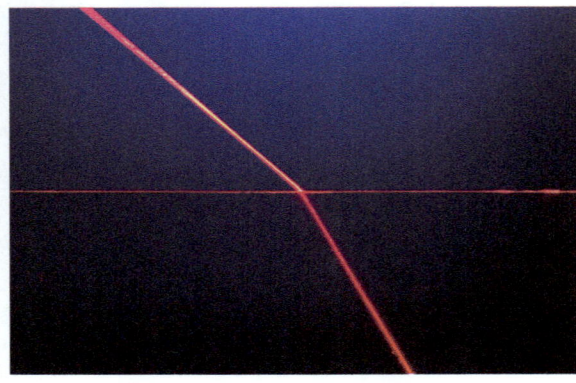

Figure 1 A laser beam is a narrow beam of light of a single colour

Look at the practical opposite. For a light ray travelling into glass from air as in Figure 2, your results should show that:

1 the angle of refraction is always smaller than the angle of incidence
2 the greater the angle of incidence, i, the greater the angle of refraction, r.

Snell's law

Some typical results of the investigation in the practical are shown in Table 1.

Table 1 Results of a refraction investigation

i (degrees)	r (degrees)	sin i	sin r	$\dfrac{sin\ i}{sin\ r}$
10.0	6.5	0.174	0.113	1.54
20.0	13.0	0.342	0.225	1.52
30.0	19.0			

Calculations like the ones in Table 1 show that sin i ÷ sin r always has the same value for the same medium that the block is made of, regardless of the angle of incidence. This relationship was discovered in 1618 and is known as **Snell's law** after its discoverer.

It can be shown that $\dfrac{\sin i}{\sin r}$ is equal to the ratio $\dfrac{\text{speed of light in vacuum (air)}}{\text{speed of light in the medium}}$

This ratio is defined as the **refractive index** of the medium. The symbol **n** is used for the refractive index.

$$\text{refractive index, } n = \frac{\text{speed of light in vacuum (air)}}{\text{speed of light in the medium}}$$

Calculate the mean value of sin i ÷ sin r for your own measurements. This is the refractive index of the block that you used in your investigations.

Maths skills

Using a calculator

To find the value of the sine of a given angle in degrees or the angle in degrees for a given sine value, make sure your calculator is in degree mode. Key the angle in degrees into your calculator, then press the button marked 'sin' (or on some calculators press 'sin' first). The calculator will then display the sine of the angle.

To find the angle for a given sine value, key the sine value into your calculator and press the button marked 'inv sin' (or 'sin⁻¹' on some calculators).

The law of refraction

For a light ray travelling from air into a transparent medium, the ratio $\sin i \div \sin r$ is always the same for the same medium.

This ratio is the refractive index of the medium. In other words:

the refractive index n of the medium $= \dfrac{\sin i}{\sin r}$

where i is the angle of incidence and r is the angle of refraction.

Rearranging the above equation to make $\sin i$ the subject gives: $\sin i = n \sin r$.

Rearranging the equation to make $\sin r$ the subject gives: $\sin r = \dfrac{\sin i}{n}$.

When a light ray travels from a transparent medium into air at a non-zero angle of incidence:

- the light ray is refracted **away** from the normal, as shown at point P in Figure 3
- the larger the angle of incidence, the larger the angle of refraction.

If the light ray in Figure 3 were reversed, the direction arrows would be reversed but the path would be the same. You can adapt the law of refraction to cover both situations by writing it as:

the sine of the angle in air $= n \times$ the sine of the angle in glass

Worked example

A light ray travels from glass into air across a straight boundary, as shown at P in Figure 3. The angle of incidence i of the light ray in the glass is 32.0°. The refractive index of the glass is 1.55. Calculate the angle of refraction r of the light ray in the air.

Solution

$\sin i = n \times \sin r$
Therefore:
$\sin r = 1.55 \times \sin 32.0° = 0.821$
Therefore:
  $r = \mathbf{55.2°}$

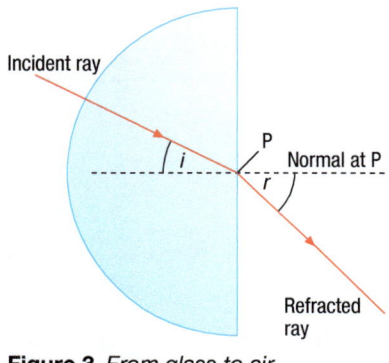

Figure 3 *From glass to air*

Summary questions

1 In Table 1, the angle of refraction is 19.0° for an angle of incidence of 30°.
   a Use this data to calculate the refractive index of the block.
   b Use your result in part **a** and the data in the table to calculate a mean value for the refractive index.

2 In an experiment like that shown in Figure 2 to measure the refractive index of a glass block, when $i = 40°$, $r = 26°$.
   a Calculate the refractive index of the glass.
   b Give **two** possible reasons why the value of refractive index in **a** differs so much from the values obtained in **1b**.

3 The refractive index of water is 1.33.
   a A light ray enters a flat water surface at an angle of incidence of 35.0°. Calculate the angle of refraction of the light ray in the water.
   b A light ray travels from water into air. The angle of incidence of the light ray in the water is 45.0°. Calculate the angle of refraction of the light ray in the air.
   c In Figure 3, when angle i is equal to 35.0°, angle r is equal to 60.5°. Calculate the refractive index of the glass.

Key points

- The refractive index, n, is a measure of how much a medium can refract a light ray.

- $$n = \frac{\text{speed of light in vacuum (air)}}{\text{speed of light in the medium}}$$

- The angle of incidence, i, and the angle of the refraction, r, are related to the refractive index, n, by the law of refraction:

$$n = \frac{\sin i}{\sin r}$$

10.4 Total internal reflection

Reflection and refraction of light

Optical fibres are very thin glass fibres that are designed to transmit light or infrared radiation. Doctors use them in an endoscope to see inside the body without cutting the body open. In telecommunications they are used to send signals securely (see Topic 8.3 'Communications'). The light rays can't escape from the fibre. Each light ray entering a fibre at one end leaves the fibre at the other end even if the fibre is bent. This is because a light ray in the fibre is **totally internally reflected** each time it reaches the boundary at the edge (Figure 1).

Investigating total internal reflection

In Topic 10.3 'Refractive index', you saw that a light ray travelling from glass into air at a non-zero angle of incidence is refracted away from the normal. As well as the refracted ray, you also see a partially reflected ray, as shown in Figure 2. The angle of reflection of this ray in the glass is the same as the angle of incidence.

- If the angle of incidence in the glass is gradually increased, the angle of refraction increases until the refracted ray emerges along the boundary, as shown in Figure 3. The angle of incidence at this position is referred to as the **critical angle**, labelled c in Figure 3.

- If the angle of incidence is increased beyond the critical angle, the light ray is **totally internally reflected** at P, as shown in Figure 4. When **total internal reflection** occurs, the angle of reflection r at P is equal to the angle of incidence i.

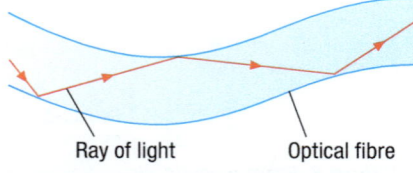

Figure 1 *Light rays in an optical fibre*

Critical angle and refractive index

If the light ray in Figure 2 was reversed, the direction arrows would be reversed but the path would be the same. You can adapt the law of refraction for this situation by writing it as:

the sine of the angle in air = n × the sine of the angle in glass

You can apply this equation to find the critical angle in Figure 3. The angle in air is 90° and the angle in glass is c:

$$\sin 90° = n \times \sin c$$

where n is the refractive index of the glass. Because $\sin 90° = 1$, the equation above becomes $1 = n \times \sin c$. Rearranging gives:

$$n = \frac{1}{\sin c} \quad \text{or} \quad \sin c = \frac{1}{n}$$

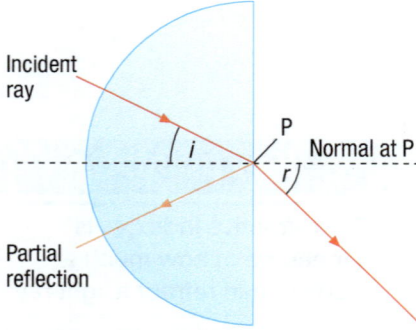

Figure 2 *Partial reflection and refraction*

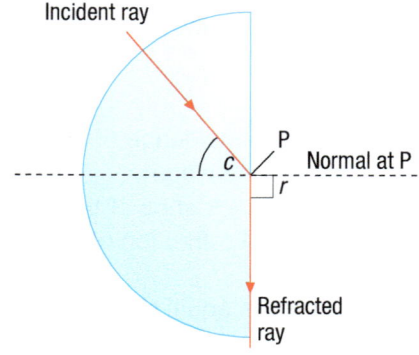

Figure 3 *At the critical angle*

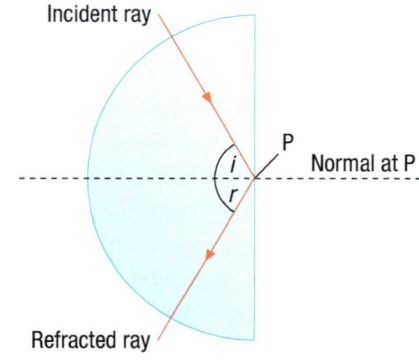

Figure 4 *Total internal reflection*

The endoscope

The **endoscope** is used by a surgeon to see inside a body cavity, such as the stomach, without cutting the body open. The endoscope is inserted into the stomach via the patient's throat.

The endoscope contains two bundles of optical fibres alongside each other. One of the bundles is used to shine light into the cavity, and the other to transmit light back so the surgeon can see the internal surfaces of the cavity. A tiny lens over the second bundle forms an image at the ends of the fibres. The image can be seen at the other end of the fibre, either directly or by using a digital camera.

For example, an endoscope can be used to observe a stomach ulcer or a bone fragment in the knee joint. The surgeon can then use tools in the endoscope to to remove them. This is called keyhole surgery.

Laser light may be used as a source of energy in an endoscope to carry out some surgical procedures. It can cut or burn away and destroy diseased tissue. It can also seal off (cauterise) leaking blood vessels. This is possible with laser light because the energy can be focused on to a very small area.

In addition, the colour of laser light can be matched to the type of tissue being treated. Choosing an appropriate laser source ensures the most effective absorption. Eye surgery on the retina can be carried out by applying laser treatment through the pupil of the eye for a very short time.

Safety note: Never look into or along a laser beam, even after reflection. It will damage the retina and may cause permanent blindness. Special safety goggles should always be worn in the presence of a laser beam.

Worked example

Calculate the critical angle for glass of refractive index 1.59.

Solution

$$\sin c = \frac{1}{n} = \frac{1}{1.59} = 0.629$$

Therefore, $c = \textbf{39.0°}$.

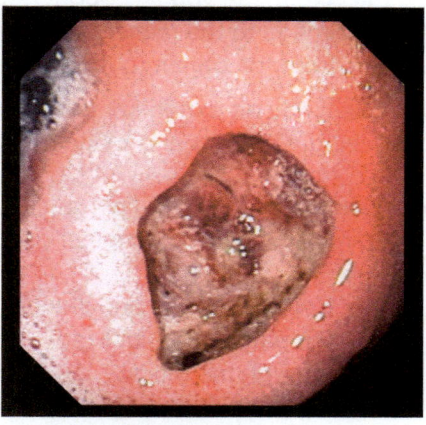

Figure 5 *A stomach ulcer viewed through an endoscope*

Summary questions

1 a When a light ray travelling in a refractive medium reaches a boundary with a less refractive medium, what is the condition under which total internal reflection can occur?

  b What is the angle between the normal and the refracted ray when the angle of incidence is equal to the critical angle?

  c The critical angle for a certain type of glass is 43.0°. Calculate the refractive index of this glass.

2 a The figure shows a light ray in an optical fibre. The angle of incidence of the light ray at P is greater than the critical angle of the optical fibre. Copy the diagram and complete the path of the light ray inside the optical fibre.

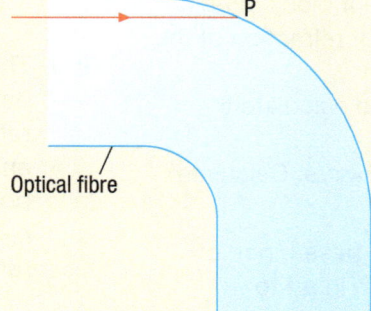

Optical fibre

  b State **two** advantages of using an endoscope instead of X-rays to observe fragments of bone in a knee joint.

3 a The refractive index of water is 1.33. Calculate the critical angle of water.

  b i The critical angle for a certain type of glass is 42.0°. Calculate the refractive index of the glass.

   ii The critical angle in **i** was measured to a precision of 0.5°. Show that this gives an uncertainty of 0.02 in the value of the refractive index.

Key points

- The critical angle, c, is the angle of incidence of a light ray in a transparent medium that produces refraction along the boundary.

- Refractive index $= \dfrac{1}{\sin c}$

- Total internal reflection occurs when the angle of incidence of a light ray in a transparent medium is greater than the critical angle.

- An endoscope uses total internal reflection to see inside the body directly.

Chapter summary questions

1 a The figure shows an incomplete ray diagram of image formation by a plane mirror.

Object O

Mirror

x
y

   i What can you say about the angles *x* and *y* in the diagram?

   ii Copy and complete the ray diagram to locate the image.

   iii What can you say about the distance from the image to the mirror compared with the distance from the object to the mirror?

b Describe an experiment to test the law of reflection using a ray box and plane mirror.

2 a The figure shows a light ray directed into a rectangular glass block.

   i Copy the figure and sketch the path of the light ray through the block.

   ii Explain why the light ray that emerges from the block is exactly parallel to the incident light ray.

b The figure shows a ray of red light directed into a triangular glass prism.

   i Copy the drawing and complete the path of the red light ray through the prism.

   ii The red light ray is replaced by a white light ray, and the light coming out from the prism is observed on a white screen. Describe and explain what is observed on the screen.

3 a A light ray is directed into a transparent block at an angle of incidence of 25°. The angle of refraction of the light ray is 16°.

   i Use the equation $n = \sin i \div \sin r$ to calculate the refractive index of the glass.

   ii The speed of light in air is 300 000 km/s. Calculate the speed of light in the glass.

b A light ray is directed from air into a glass block of refractive index 1.5. Use the equation in **a i** to calculate:

   i the angle of refraction when the angle of incidence is 30°

   ii the angle of incidence when the angle of refraction is 41°.

4 a The figure shows a light ray directed at a semicircular glass block. Copy the diagram and draw the path of the light ray through the glass block into the air.

Ray box   Semicircular glass block

b Describe how you would measure the refractive index of a rectangular glass block using the arrangement shown in the figure and any other necessary items of equipment.

5 a i Explain what is meant by total internal reflection.

   ii What is the condition for the angle of incidence of a light ray in a transparent medium for it to be totally internally reflected at a boundary with air?

   iii Use the equation $n = 1 \div \sin c$ to calculate the refractive index *n* of a transparent medium for which the critical angle *c* is 40°.

b An endoscope is used to see inside the body. The figure shows a light ray entering the end of the transparent core of an endoscope.

X

   i The core of the endoscope is surrounded by air. The refractive index of the core is 1.52. Use the equation $n = 1 \div \sin c$ to calculate the critical angle of a boundary between the core and air.

   ii The angle of incidence of the light ray in the figure when it reaches the core boundary at X is 70°. Copy and complete the figure to show the path of the light ray along the optical fibre.

   iii Explain why an endoscope needs to have two bundles of optical fibres.

6 a The figure shows a light ray entering a glass block of refractive index 1.59 at an angle of incidence of 40° at point P. Use the equation

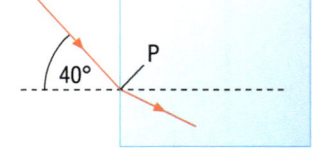

P
40°

$n = \sin i \div \sin r$ to show by calculation that the angle of refraction at P is 24°.

b i Use the equation $n = 1 \div \sin c$ to calculate the critical angle *c* of the glass.

   ii Copy the diagram and continue the path of the ray in the glass until it reaches a point Q at the bottom of the block. Explain why the angle of incidence at Q is 66°.

   iii Explain why the light ray does not enter the air at Q.

Practice questions

1 **a** A plane mirror produces a virtual image that is laterally inverted.

    i What is a *virtual* image? (2)

    ii What does *laterally inverted* mean? (1)

b A man stands in a lift. The walls of the lift have plane mirrors fixed to them.

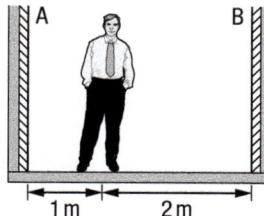

The man looks into mirror **A** and sees lots of images of himself, one behind the other.

When he looks into mirror **B**, he again sees lots of images of himself, one behind the other.

    i Calculate the distance between the nearest image that he sees in mirror **A** and the nearest image that he sees in mirror **B**. (3)

    ii Explain why he sees lots of images when he looks into mirror **A**. (2)

2 Light waves change direction when they travel from air into another transparent medium. The diagram shows a wavefront XY in air moving towards water.

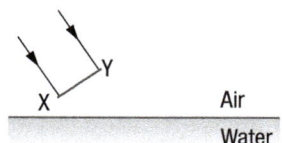

a **i** Explain how light waves change direction when they travel from air into water. (3)

    ii Under what conditions do light waves not change direction when they travel from air into water? (1)

b Some students use the apparatus opposite to find the refractive index of glass.

    i Copy and complete the diagram to show what is meant by *angle of incidence* and *angle of refraction*. (3)

    ii What would they use to measure the angles? (1)

    iii Describe what they need to do to obtain an accurate value for the refractive index of glass. (3)

    iv When the angle of incidence was 35°, the angle of refraction was 22°. Calculate the critical angle of glass. (4)

3 Below is part of an advert.

> **Brighten your darkest room with our light pipe**
>
> Use one of our light pipes to catch sunlight and send it into the darkest corner of your house.
>
> The pipes contain hi-tech optical fibres that can make light go round corners.

a The diagram shows how light passes through an optical fibre.

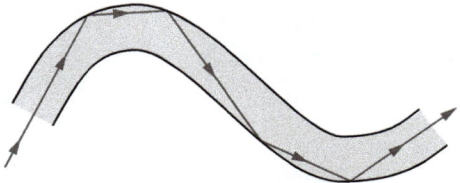

    i What is the process shown in the diagram? (2)

    ii Give two other uses for optical fibres. (2)

b A student investigated the process shown above. The diagrams give three stages of the investigation.

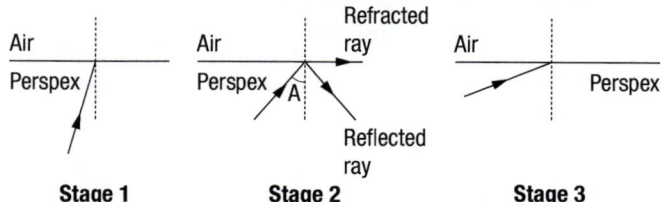

Stage 1   **Stage 2**   **Stage 3**

    i What is the name of the dotted line in the diagrams? (1)

    ii Copy and complete the Stage 1 diagram to show what happened to the ray of light after it hit the boundary between the Perspex and the air. (2)

    iii Name angle A in the Stage 2 diagram. (1)

    iv Copy and complete the Stage 3 diagram to show what happened to the ray of light after it hit the boundary between the Perspex and the air. (1)

4 A ray of red light is incident on one face of a glass prism.

a Copy and complete the ray diagram. (2)

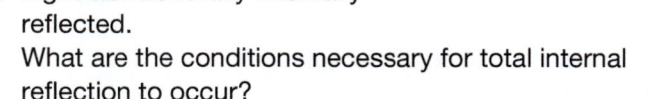

b Light can be totally internally reflected. What are the conditions necessary for total internal reflection to occur? (2)

c Doctors sometimes use endoscopes to see inside patients' stomachs. The diagram shows an endoscope.

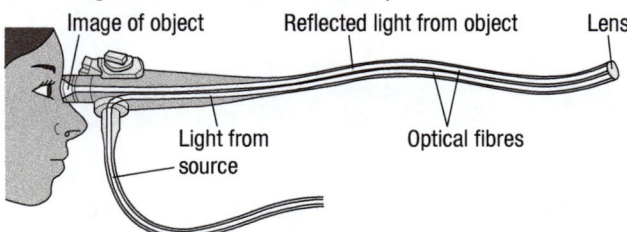

Explain how this endoscope works, and discuss the advantages of this method of investigation over other methods of investigation of the stomach. (6)

11.1 Lenses

Lenses are used in optical devices such as the camera. Although a digital camera is very different from the first cameras made over 160 years ago, they both contain a lens that is used to form an image.

Types of lenses

A lens works by changing the direction of light passing through it. Figure 1 shows the effect of a lens on the light rays from a ray box. The curved shape of the lens surfaces refracts the rays so they meet at a point.

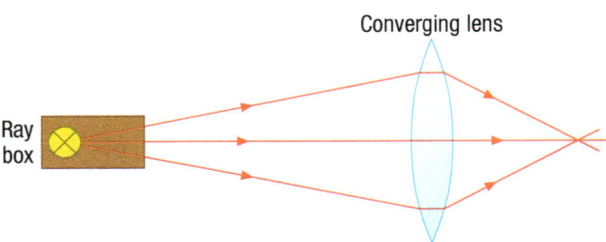

Figure 1 *Investigating lenses*

Each section of the lens acts like a tiny prism, refracting light as it enters the lens and again as it comes out. Because each section refracts light differently, the overall effect is to make the light rays converge, as shown in Figure 2.

Lenses of different shapes have different effects.

- A **converging (convex) lens** makes parallel rays converge to a focus. The point where parallel rays are focused is the **principal focus** (or focal point) of the lens. See Figure 3. A converging lens may be used as a **magnifying glass** and in a camera to form a clear image of a distant object.
- A **diverging (concave) lens** makes parallel rays diverge (spread out). The point where the rays appear to come from is the principal focus of the lens. See Figure 4. Diverging lees are used to correct short sight.

In both cases, the distance from the centre of the lens to the principal focus is the **focal length** of the lens. Notice that the principal focus is usually shown in ray diagrams on each side of the lens.

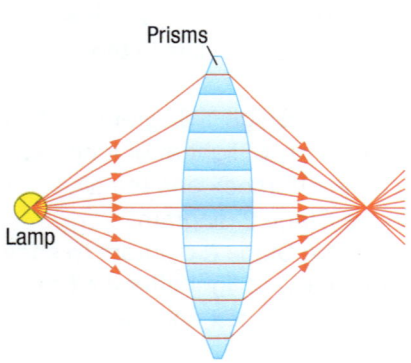

Figure 2 *How a lens works*

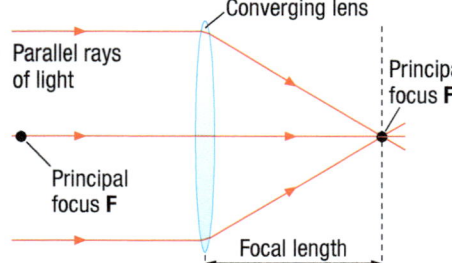

Figure 3 *The focal length of a converging (or convex) lens*

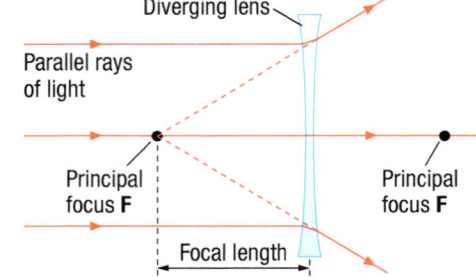

Figure 4 *The focal length of a diverging (or concave) lens*

Practical

Investigating the converging lens

Use the arrangement in Figure 5 to investigate the image formed by a converging lens.

Safety: Make sure glass lenses are not sharp.

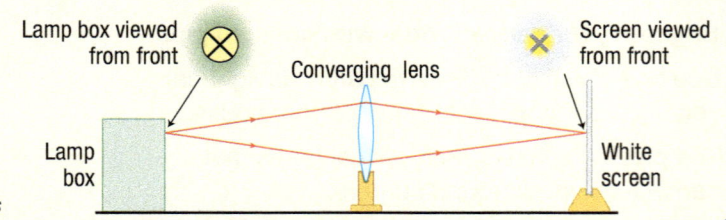

Figure 5 *Investigating images*

The converging lens

1 With the object at different distances beyond the principal focus of the lens, you can adjust the position of the screen until you see a clear image of the object on it. This is called a **real image** because it is formed on the screen where the light rays meet.

- When the object is a long distance away, the image is formed at the principal focus of the lens. This is because the rays from any point on a distant object are effectively parallel when they reach the lens.
- If the object is moved nearer the lens, towards its principal focus, the screen must be moved further from the lens to see a clear image. The nearer the object is to the lens, the larger the image is.

2 *With the object nearer to the lens than the principal focus*, a magnified image is formed. The image is called a **virtual image** because it is formed where the rays appear to come from. But you can see the image only when you look into the lens from the side opposite to the object. The lens acts as a magnifying glass in this situation.

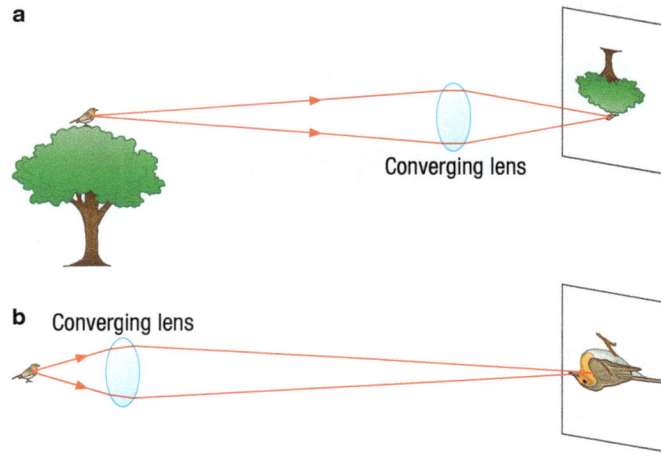

Figure 6 a *The image of a distant object* **b** *An enlarged image*

Magnification

The **magnification** produced by a lens = $\dfrac{\text{image height}}{\text{object height}}$

If the image is larger than the object, as in Figure 6b and Figure 7, the magnification is greater than 1.

If the image is smaller than the object, as in Figure 6a, the magnification is less than 1.

Figure 7 *A magnifying glass*

Summary questions

1 a What is the difference between a real image and a virtual image?
b State whether a real image or a virtual image is formed when:
i a converging lens is used to form an image of a distant object on a screen
ii a converging lens is used as a magnifying glass.

2 a A postage stamp is inspected using a converging lens as a magnifying glass. Describe the image.
b A converging lens is used to form a magnified image of a slide on a screen.
i Describe the image formed by the lens.
ii The screen is moved away from the lens. What adjustment must be made to the position of the slide to focus its image on the screen again?
c i Estimate the magnification of the flower in Figure 7.
ii How would the magnification change if the lens is moved away from the flower?

3 a Describe the image of the bird in Figure 6b and estimate the magnification of the lens.
b Describe how the image changes if the lens is moved further away from the bird and the card is moved to obtain a new clear image.

Key points

- A converging (convex) lens focuses parallel rays to a point called the principal focus.
- A diverging (concave) lens makes parallel rays spread out as if they had come from a point called the principal focus.
- A real image is formed by a converging lens if the object is further away than the principal focus.
- A virtual image is formed by a converging lens if the object is nearer to the lens than the principal focus.
- Magnification = $\dfrac{\text{image height}}{\text{object height}}$

11.2 Using lenses

The position and nature of the image formed by a lens depends on:

- the focal length, *f*, of the lens
- the distance from the object to the lens.

If you know the focal length and the object distance, you can find the position and nature of the image by drawing a ray diagram.

Formation of a real image by a converging lens

To form a real image using a converging (convex) lens, the object must be beyond the principal focus, F, of the lens. See Figure 1. The image is formed on the other side of the lens to the object.

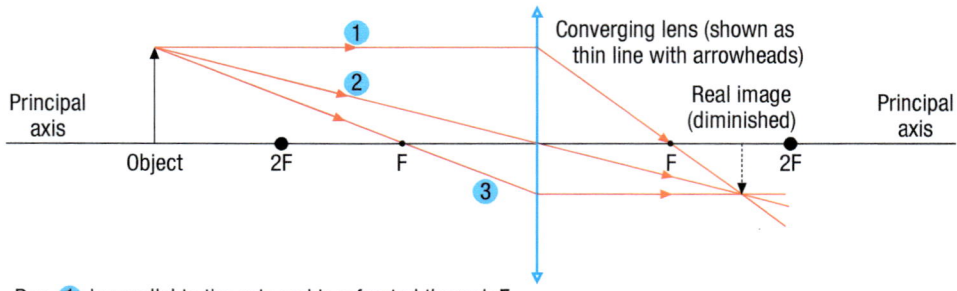

Ray **1** is parallel to the axis and is refracted through F
Ray **2** passes straight through the centre of the lens
Ray **3** passes through F and is refracted parallel to the axis

Figure 1 *Formation of a real image by a converging lens*

The diagram shows how you can use three key construction rays from a single point on the object to locate the image.

- The **principal axis** of the lens is the straight line that passes along the normal at the centre of each lens surface. Notice that the lens is drawn as a straight line with outward arrows to show it is a converging lens.
- The image is real, inverted, and smaller than the object (diminished).

Notice how light acts along the different ray paths:

- *ray 1* is refracted through F, the principal focus of the lens, because it is parallel to the principal axis of the lens before it passes through the lens
- *ray 2* passes through the centre of the lens (its pole) without changing direction, because the lens surfaces at the principal axis are parallel to each other
- *ray 3* passes through F, the principal focus of the lens, before the lens, so it is refracted by the lens parallel to the principal axis.

The image is smaller than the object because the object distance is greater than twice the focal length, *f*, of the lens. This is the type of image produced in a **camera** (Figure 2).

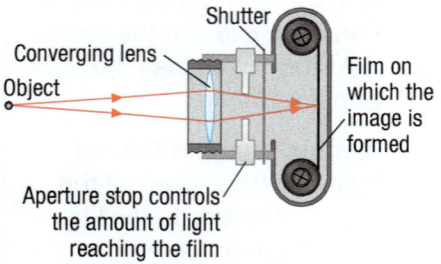

Figure 2 *A traditional camera*

The camera

In a traditional camera, a converging lens is used to produce a real image of an object on a film (Figure 2). (In a digital camera, the image is produced on an array of pixels on a charge-coupled device or CCD.) The position of the lens is adjusted to focus the image on the film.

- For a distant object, the distance from the lens to the film must be equal to the focal length of the lens.
- The nearer an object is to the lens, the greater the distance must be from the lens to the film.

Formation of a virtual image by a converging lens

To form a virtual image with a converging lens, the object must be between the lens and its principal focus, as shown in Figure 3. The image is formed on the same side of the lens as the object.

The image is virtual, upright, and larger than the object (magnified).

The image can be seen only by looking at it through the lens. This is how a **magnifying glass** works.

Formation of a virtual image by a diverging lens

The image formed by a diverging (concave) lens is always virtual, upright, and smaller than the object. Figure 4 shows why. A diverging lens is shown as a line with 'inward' arrows.

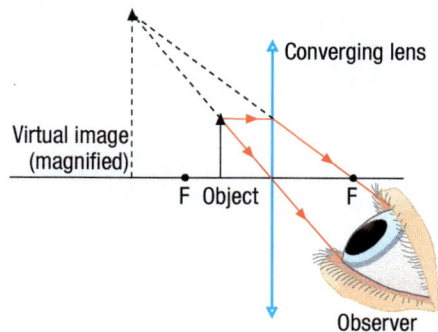

Figure 3 *Formation of a virtual image by a converging lens*

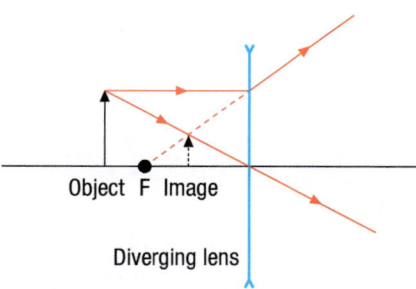

Figure 4 *Formation of an image by a diverging lens*

Summary questions

1 **a** Copy and complete the ray diagram in the figure to show how a converging lens forms an image of an object that is smaller than the object, as in a camera.

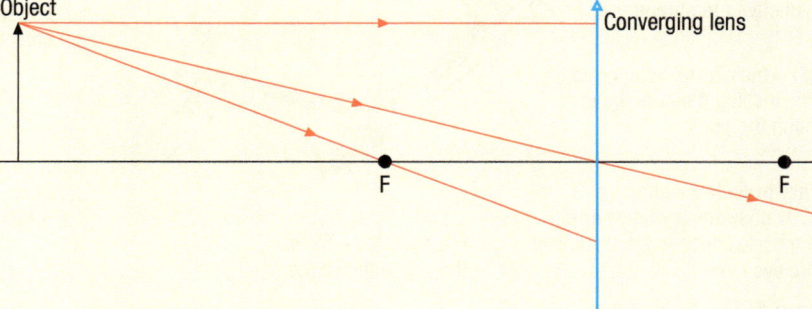

  b State whether the image is:
    i real or virtual   **ii** magnified or diminished
    iii upright or inverted.

2 **a** Draw a ray diagram to show how a converging lens is used as a magnifying glass.
  b State whether the image is:
    i real or virtual   **ii** magnified or diminished
    iii upright or inverted.
  c Why is a diverging lens no use as a magnifying glass?

3 A converging lens produces a magnification of ×2 when it is used to form a real image that is at a distance of 8.0 cm from the object.
  a Draw a scale ray diagram to show the formation of this image.
  b Use your diagram to find the focal length of the lens.
  c Describe how the position and height of the image would change if the object is moved gradually towards the focal point of the lens.

Key points

- A ray diagram can be drawn to find the position and nature of an image formed by a lens.

- When an object is placed between a converging lens and its principal focus F, the image formed is virtual, upright, magnified, and on the same side of the lens as the object.

- A camera contains a converging lens that is used to form a real image of an object.

- A magnifying glass is a converging lens that is used to form a virtual image of an object.

- A virtual image is formed by a diverging lens.

- A diverging lens forms a virtual image.

11.3 The eye

Learning objectives

After this topic, you should know:

- how the eye forms an image
- the range of vision of a normal human eye
- how the eye and the camera compare as optical devices.

?? Did you know …?

The conjuctiva membrane is a thin transparent membrane over the front of the eye. A watery fluid from your tear glands under the eyelids spreads across this membrane every time you blink. The fluid contains lysozyme – a chemical that destroys bacteria.

?? Did you know …?

The blind spot is a region where the retina is not sensitive to light (no light-sensitive cells are present).

Inside the eye

Figure 1 shows the inside of a human eye. Light enters the eye through a tough transparent layer called the **cornea**. This protects the eye and helps to focus light onto the **retina**. The retina is a layer of light-sensitive cells inside the back of the eye.

The amount of light entering the eye is controlled by the **iris**, which adjusts the size of the **pupil** – the circular opening at the centre of the iris. In bright light, the iris becomes larger and makes the pupil narrow so less light enters the eye. In dim light, the iris becomes smaller and makes the pupil wide so more light enters the eye. The **eye lens** is a converging lens that focuses light to give a sharp image on the retina. Although the image on the retina is inverted, the brain interprets it so that you see it the right way up.

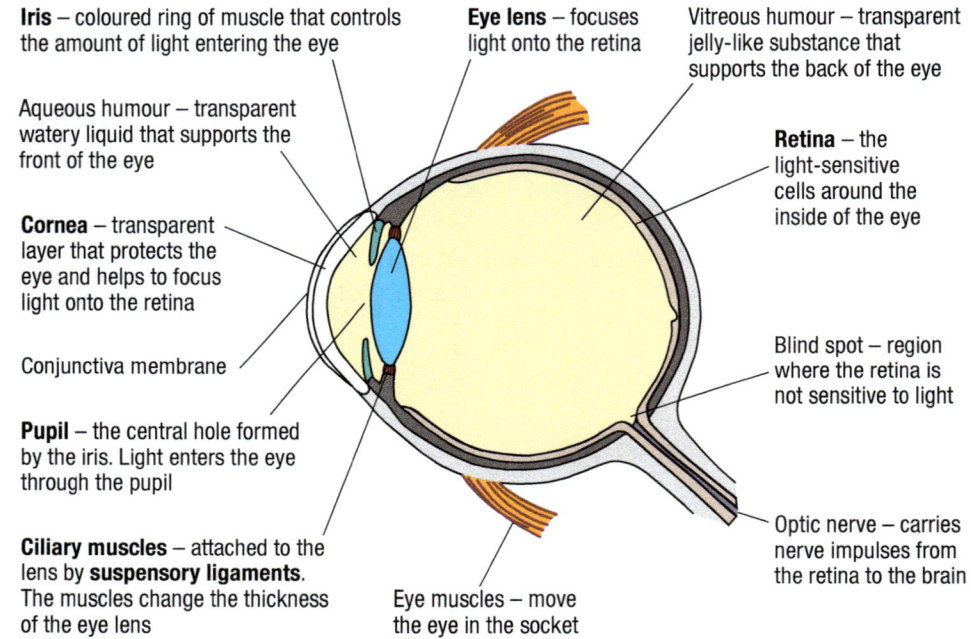

Iris – coloured ring of muscle that controls the amount of light entering the eye

Eye lens – focuses light onto the retina

Vitreous humour – transparent jelly-like substance that supports the back of the eye

Aqueous humour – transparent watery liquid that supports the front of the eye

Retina – the light-sensitive cells around the inside of the eye

Cornea – transparent layer that protects the eye and helps to focus light onto the retina

Conjunctiva membrane

Blind spot – region where the retina is not sensitive to light

Pupil – the central hole formed by the iris. Light enters the eye through the pupil

Optic nerve – carries nerve impulses from the retina to the brain

Ciliary muscles – attached to the lens by **suspensory ligaments**. The muscles change the thickness of the eye lens

Eye muscles – move the eye in the socket

Figure 1 *The human eye*

How does the eye focus on objects at different distances? If you look up from this book and gaze out of a window, your eye lens automatically becomes thinner to keep what you see in focus. The **ciliary muscle** alters the thickness of the eye lens. It is attached to the edge of the lens by the **suspensory ligaments**. The ciliary muscle contains fibres that are parallel to the edge of the lens and fibres joined to the cornea. When the muscle contracts, the fibres shorten and squeeze the eye lens and the other fibres reduce the tension in the ligaments. The two effects make the lens thicker.

The normal human eye has a **range of vision** from 25 cm to infinity. This means it can see clearly any object that is 25 cm or more from the eye. In other words, the normal eye has a **near point** of 25 cm and a **far point** at infinity.

To see a nearby object clearly, the eye lens has to be thicker than if the object is far away. Figure 2 shows this.

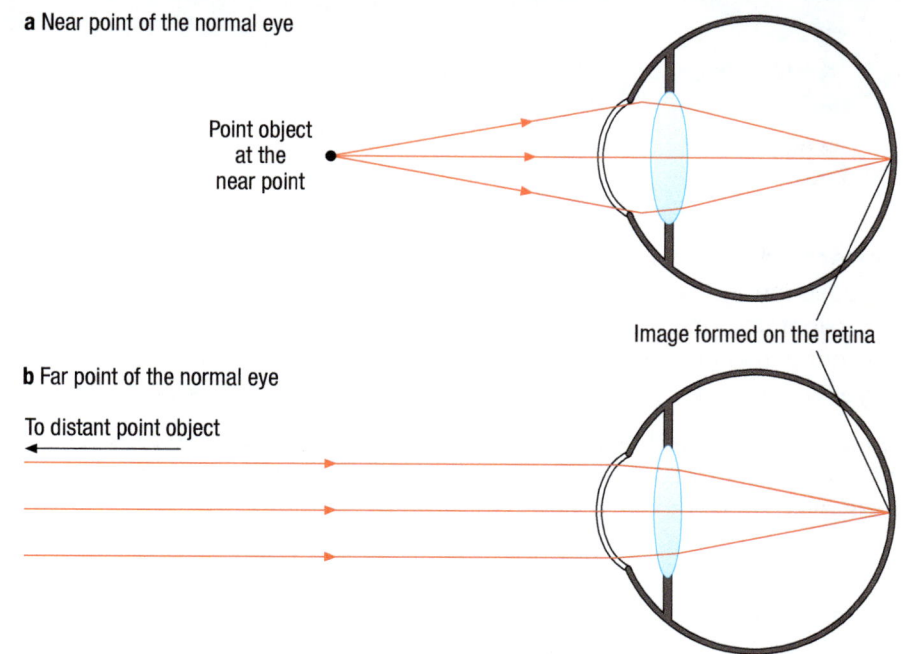

a Near point of the normal eye

Point object at the near point

Image formed on the retina

b Far point of the normal eye

To distant point object

Figure 2 *The normal eye*

Study tip

Make sure that you can correctly label a diagram of the eye, and that you know the function of each part of the eye.

Table 1 *Comparison of the optics of the eye and a camera*

	The eye	The camera
Type of lens	Variable-focus converging lens	Fixed-focus converging lens
Focusing adjustment	Ciliary muscle alters the lens thickness	Adjustment of lens position
Image	Real, inverted, magnification less than 1	
Image detection	Light-sensitive cells on the retina	Photographic film (or CCD sensors in a digital camera)
Brightness control	Iris controls the width of the pupil	Adjustment of aperture stop

Comparison of the eye and the camera

How do the eye and the camera compare as optical instruments? They are similar in that they both contain a converging lens which forms a real image. Look at Table 1 to see how they compare in other ways.

Summary questions

1 a State the function of the following parts of the eye:
  i ciliary muscles   **v** cornea
  ii eye lens   **vi** iris
  iii pupil   **vii** retina.
  iv suspensory ligaments
  b Why does the pupil of the eye appear much wider in darkness than in daylight?

2 A person with normal eyesight who is reading a book looks up to observe a distant object.
  a Describe what happens to the shape of each eye lens in this change.
  b What change takes place in the focal length of each eye lens?

3 **a** Describe the change that takes place in the eye when it adjusts to very bright light.
  b What change is made in a camera when it is adjusted to very bright light?

Key points

- Light is focused onto the retina by the cornea and the eye lens, whose shape is changed by the ciliary muscle.

- The normal human eye has a range of vision from 25 cm to infinity.

- An image is brought to focus:
  – on the retina in an eye by changing the shape of the eye lens
  – on the film in a camera by changing the distance between the film and lens.

11.4 More about the eye

Learning objectives

After this topic, you should know:

- what short sight is and how to correct it

- what long sight is and how to correct it

- why the refractive index of glass is important in making spectacle lenses.

Sight defects

Short sight occurs when an eye cannot focus on distant objects. The **uncorrected** image is formed in front of the retina, as shown in Figure 1. This is because the eyeball is too long or the eye lens is too powerful because it refracts light from the object too much. With the ciliary muscle relaxed, the eye lens is not thin enough to focus the image of a faraway object on the retina of the eye. The eye can focus nearby objects, so the defect is called 'short sight'.

Short sight is corrected by placing a diverging lens of a suitable focal length in front of the eye, as shown in Figure 1. The diverging lens counteracts the excess refraction by the eye lens.

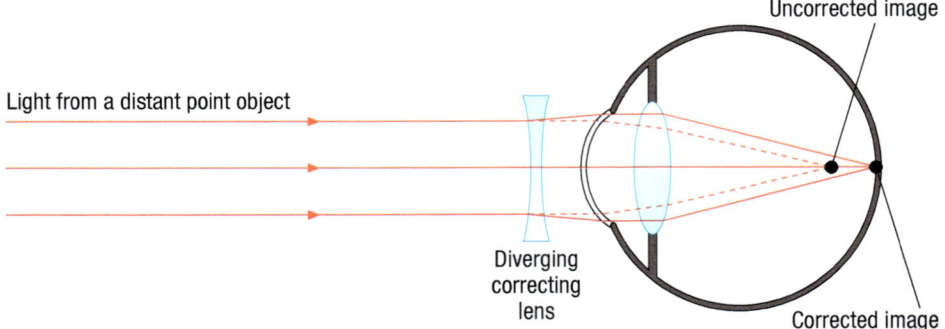

Figure 1 *Short sight and its correction*

Long sight occurs when an eye cannot focus on nearby objects. The uncorrected image is formed behind the retina, as shown in Figure 3. The eyeball is too short or the eye lens is too weak. The eye lens cannot be made thick enough to focus an image on the retina. The eye can focus on distant objects, so the defect is called 'long sight'.

Long sight is corrected by placing a converging lens of a suitable focal length in front of the eye, as shown in Figure 3. The correcting lens makes the rays from the object diverge less. The eye lens can then focus the rays onto the retina. The correcting lens adds to the refraction by the eye lens.

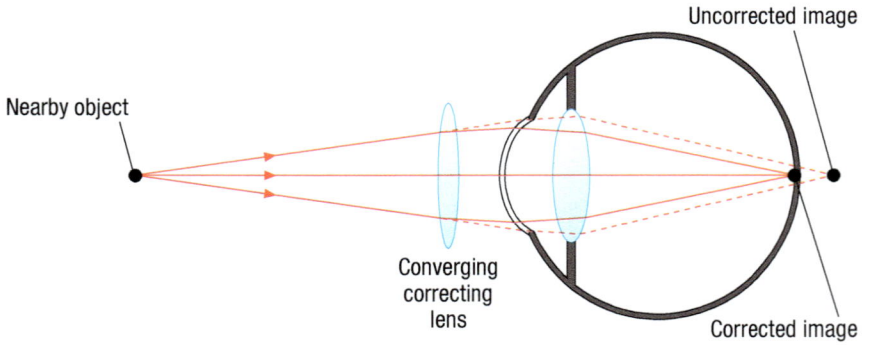

Figure 3 *Long sight and its correction*

Figure 2 *A contact lens corrects the image so it is focused at the retina*

Study tip

Make sure that you know: the difference between short sight and long sight, what causes the defect, and how the defect can be corrected.

Lens makers at work

The eye lens is a remarkable optical device, as it has a variable focal length that depends on its thickness. Lens makers working for opticians need to make contact lenses and spectacle lenses exactly the right shape to achieve the correct focal length for each lens.

The focal length of a lens depends on the refractive index of the lens material and the curvature of the two lens surfaces.

- The larger the refractive index or the greater the curvature of the lens surfaces, the greater the power of the lens (and the shorter its focal length).
- For a lens of a given focal length, the greater the refractive index of the lens material, the flatter and thinner the lens can be manufactured. This is because the lens surfaces need to be less curved if the material has a high refractive index.

Laser treatment

A laser produces a narrow concentrated beam of light. High-power lasers can be used to burn through metal. Low-power lasers are also dangerous because laser light entering the eye will damage the retina. Special protective eye goggles should always be worn when lasers are in use.

Lasers are used in medicine for cutting tissue and for sealing blood vessels (called cauterising). They are also used to correct sight defects. For example, short sight can be corrected by using a special laser called an excimer laser to make part of the cornea slightly thinner. This has the same effect as a diverging correcting lens because it cancels out some of the excess focusing power of the eye lens.

Figure 4 *An eye test*

Summary questions

1 Using his left eye, a student can only see the writing on a board at the front of the class if he sits near the board.
   a What sight defect does he have in this eye?
   b What type of lens should be used to correct this defect?

2 An optician prescribes a convex lens to correct a sight defect.
   a State what the sight defect is and give **one** possible cause of the defect.
   b Explain why the sight defect could **not** be corrected by making the cornea flatter using laser treatment.

3 a Explain what is meant by:
   i   the near point
   ii the far point of an eye.
   b A student's right eye has a near point of 25 cm and a far point of 5.0 m.
   i   State what the sight defect of this eye is.
   ii Describe how the sight defect can be corrected by means of a suitable spectacle lens.

Key points

- A short-sighted eye is an eye that can see only nearby objects clearly. A diverging (concave) lens is used to correct it.

- A long-sighted eye is an eye that can see only distant objects clearly. A converging (convex) lens is used to correct it.

- The higher the refractive index of the lens material, the flatter and thinner the lens can be.

Chapter summary questions

1 The figure shows an incomplete ray diagram of image formation by a lens. The object distance from the lens is 2.5 × the focal length of the lens.

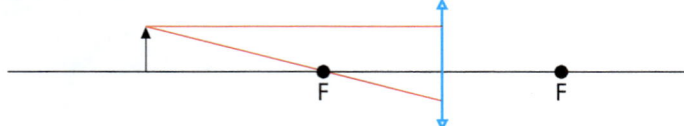

a i What type of lens is shown in this diagram?
ii Copy and complete the ray diagram and label the image.

b Describe the image and state an application of the lens used in this way.

2 An object of height 40 mm is placed perpendicular to the principal axis of a convex lens at a distance of 80 mm from the pole of the lens. The focal length of the lens is 50 mm.

a Draw a scale ray diagram to find the distance from the lens to the image.

b State whether the image is:
i real or virtual
ii upright or inverted.

c Determine the magnification produced by the lens.

3 An object is placed perpendicular to the principal axis of a diverging lens.

a Draw a scale ray diagram to show where the image of the object is formed.

b State whether the image is:
i real or virtual
ii upright or inverted.

4 a Copy and complete the ray diagram in the figure to show how a converging lens is used as a magnifying glass.

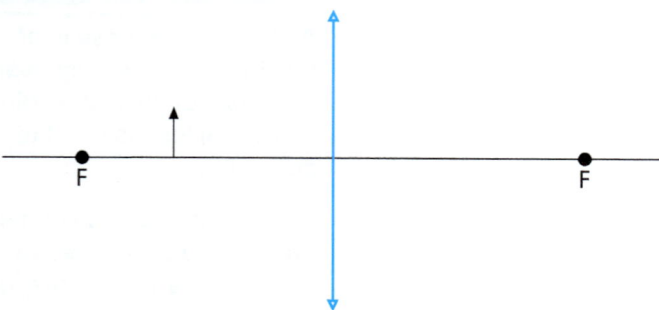

b Determine the magnification of the image.

c State whether the image is:
i real or virtual
ii upright or inverted.

5 a i Explain what is meant by short sight.
ii A short-sighted eye views a distant object. Copy and complete the figure to show the path of two light rays that enter the eye from a point on a distant object.

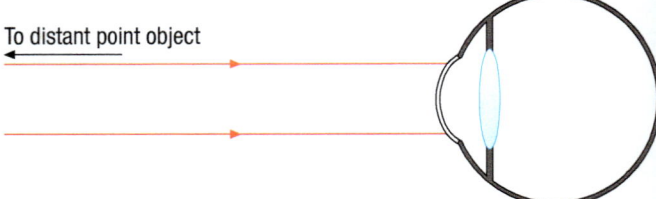

b State the type of lens that is used to correct short sight.

c Short sight may be corrected by eye surgeons using a laser.
i Which part of the eye is treated in this operation, and what is done to it with the laser?
ii Give **one** reason why laser light is dangerous.

6 The figure shows a long-sighted eye when it is viewing a nearby object.

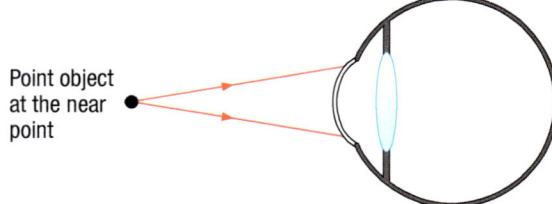

a i Copy and complete the figure to show the path of two light rays that enter the eye from a point on a nearby object.
ii Explain why the eye is unable to see the nearby object clearly.

b i State the type of lens that is used to correct long sight.
ii To correct a long-sighted eye, an optician recommends that a convex lens with a certain focal length be used. Explain why this lens enables distant objects to be seen clearly by this eye.

Practice questions

1 The focal length of a lens determines how powerful it is.

a Draw diagrams to show what is meant by the focal length of:
 i a convex lens (3)
 ii a concave lens. (3)

b An object, AB, is placed 5 cm from a convex lens of focal length 10 cm.

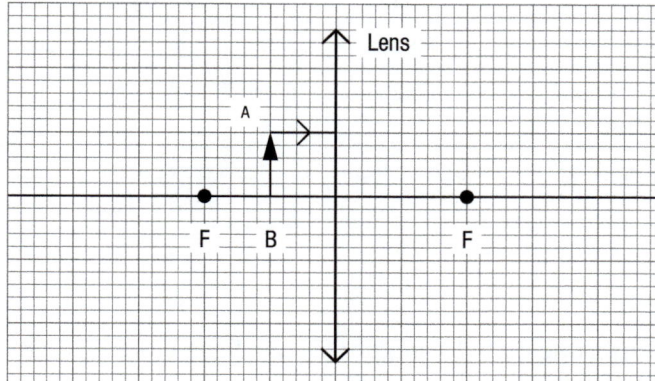

 i Copy and complete the ray diagram, which is drawn to scale, to show how the convex lens forms an image of the object.
 Draw and label the image. (4)
 ii Calculate the magnification. (2)
 iii State **three** differences between the image and the object. (3)

c A convex lens is used in both the camera and the eye.
 i Give **two** other similarities between the camera and the eye. (2)
 ii Describe how the camera and the eye use different methods to produce sharp images of objects at different distances. (4)
 iii Give **two** other differences between the camera and the eye. (2)

2 Devices containing lenses form different types of image.

a Match each device in the left-hand column to the correct image in the right-hand column. (3)

Device		Nature of image
Camera		real, magnified, inverted
Magnifying glass		virtual, magnified, upright
Projector		real, diminished, inverted

b Some students used the following apparatus to determine the focal length of a converging lens.

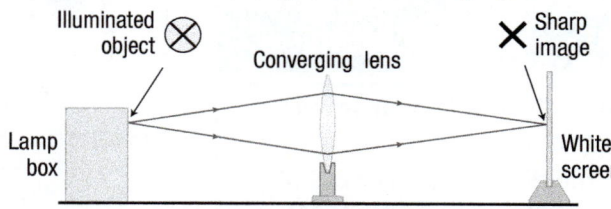

They placed the object 40.0 cm from the lens. Then they moved the screen away from the lens until they saw a sharp image on the screen. They measured the image distance. It was 13.1 cm.

Next they put the screen further away and then moved it towards the lens until they saw a sharp image. This time the image distance was 13.5 cm.
 i What was the resolution of the ruler they were using? (1)
 ii Use their results to determine the focal length of the lens by drawing a scale ray diagram. (4)
 iii They repeated the experiment using six different object distances.
 How should they use their results to obtain a more accurate value for the focal length? (2)
 iv When they made the object distance 5 cm, they could not get a sharp image on the screen. Explain why. (3)

3 The diagram shows a cross-section of an eye.

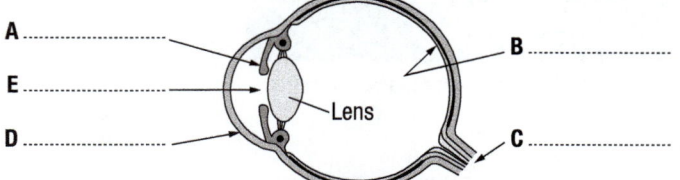

a Label the parts marked **A**, **B**, **C**, **D**, and **E**. (5)

b Describe and explain how the eye adapts to different light conditions and to objects at different distances. (6)

c The diagram shows an eye with a defect.

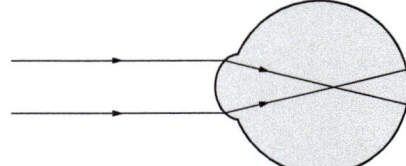

 i What is this eye defect? (1)
 ii What type of lens is used to correct the defect? (1)

12.1 States of matter

Learning objectives

After this topic, you should know:

- **the different properties of solids, liquids, and gases**
- **the arrangement of particles in a solid, a liquid, and a gas**
- **the difference in energy of particles in a solid, a liquid, and a gas.**

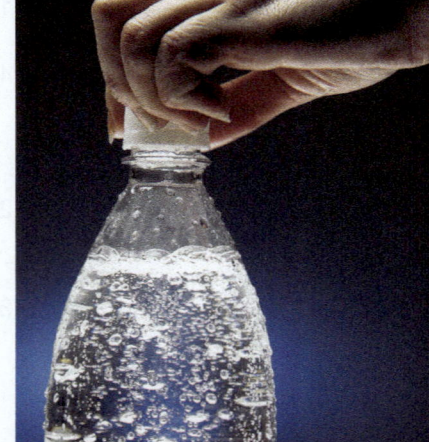

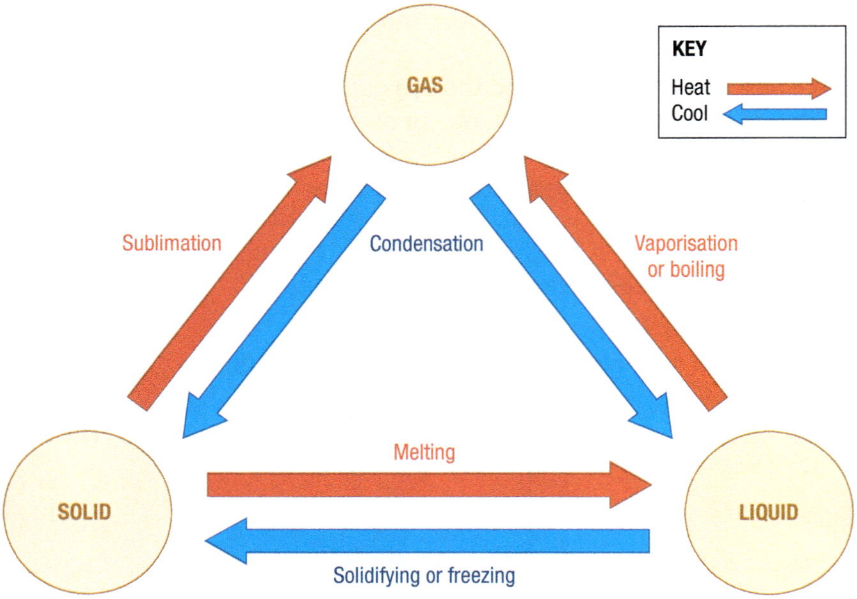

Figure 1 *Spot the three states of matter*

Did you know ... ?

Density is a measure of how much matter is in a certain volume. Lead is 11 times more dense than water, so the mass of a volume of 1 cm^3 of lead is 11 times greater than the mass of 1 cm^3 of water.

Study tip

Make sure you know how particles are arranged in solids, liquids, and gases.

Everything around you is made of matter and exists in one of three states – solid, liquid, or gas. Table 1 summarises the main differences between the three **states of matter**.

Table 1 *States of matter*

	Flow	Shape	Volume	Density
Solid	No	Fixed	Fixed	Much higher than a gas
Liquid	Yes	Fits container shape	Fixed	Much higher than a gas
Gas	Yes	Fills container	Can change	Low compared with a solid or liquid

Change of state

A substance can change from one state to another, as shown in Figure 2. The substance changes state when it is heated or cooled.

For example:

- when water in a kettle boils, the water turns to steam. Steam, also called water vapour, is water in its gaseous state
- when solid carbon dioxide or 'dry ice' warms up, the solid turns into gas directly
- when steam touches a cold surface, the steam condenses and turns to water.

GAS
KEY
Heat
Cool
Sublimation
Condensation
Vaporisation or boiling
Melting
SOLID
LIQUID
Solidifying or freezing

Figure 2 *Changes of state*

Practical

Changing state

Heat some water in a beaker using a Bunsen burner, as shown in Figure 3. Notice that:

- steam or vapour leaves the water surface before the water boils
- when the water boils, bubbles of vapour form inside the water and rise to the surface to release steam.

Switch the Bunsen burner off and hold a cold beaker or cold metal object above the boiling water. Observe the condensation of steam from the boiling water on the cold object.

Safety: Take care with boiling water and wear eye protection.

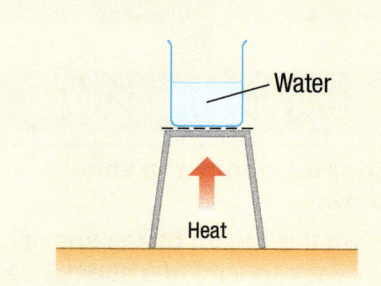

Figure 3 *Changing state*

The kinetic theory of matter

Solids, liquids, and gases are made of particles. Figure 4 shows the arrangement of the particles in a solid, a liquid, and a gas. When the temperature of the substance is increased, the particles move faster.

- The particles in a solid are held next to each other in fixed positions. They vibrate about their fixed positions, so the solid keeps its own shape.
- The particles in a liquid are in contact with each other. They move about at random. So a liquid doesn't have its own shape and it can flow.
- The particles in a gas move about at random much faster. They are, on average, much further apart from each other than the particles in a liquid. So, the density of a gas is much less than that of a solid or liquid.
- The particles in solids, liquids, and gases have different amounts of energy. For a given amount of a substance, its particles have more energy in the gas state than in the liquid state, and more energy in the liquid state than in the solid state.

??? Did you know ...?

Random means unpredictable. Lottery numbers drawn randomly.

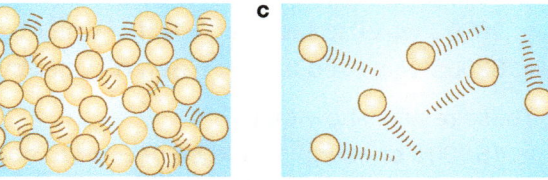

Figure 4 *The arrangement of particles in **a** a solid, **b** a liquid, and **c** a gas*

Summary questions

1. What change of state occurs when:
   a. wet clothing on a washing line dries out?
   b. hailstones form?
   c. snowflakes turn to liquid water?

2. State the scientific word for each of the following changes:
   a. the windows in a bus full of people mist up on a cold day
   b. steam is produced from the surface of water heated in a pan before it boils
   c. ice cubes taken from a freezer thaw out
   d. water put into a freezer gradually turns to ice.

3. Describe the changes that take place in the movement and arrangement of the particles in:
   a. an ice cube when the ice melts
   b. water vapour when it condenses on a cold surface.

Key points

- Flow, shape, volume, and density are the characteristics used to describe each state of matter.

- The particles in a solid are held next to each other in fixed positions. Particles are least energetic in a solid.

- The particles in a liquid move about at random and are in contact with each other. Particles are more energetic in a liquid than in a solid.

- The particles in a gas move about randomly and are much further apart than particles in a solid or liquid. Particles are most energetic in a gas.

12.2 Specific heat capacity

Learning objectives

After this topic, you should know:

- what is meant by the specific heat capacity of a substance

- how the mass of a substance affects how quickly its temperature changes when it is heated

- how to measure the specific heat capacity of a substance.

The metal body of a car can become very hot in strong sunlight. A concrete block of equal **mass** would not become as hot. Metal heats up more easily than concrete. Investigations show that when a substance is heated, its temperature rise depends on:

- the amount of energy supplied to it
- the mass of the substance
- what the substance is.

Practical

Investigating heating

Figure 1 shows how you can use a low-voltage electric heater to heat an aluminium block.

Energy is measured in units called joules (J) or kilowatt-hours (kWh).

Use the energy meter (or joulemeter) to measure the energy supplied to the block. Use the thermometer to measure its temperature rise.

Replace the block with an equal mass of water in a suitable container. Measure the temperature rise of the water when the same amount of energy is supplied to it by the heater.

Your results should show that aluminium heats up more than water.

Safety: Take care with hot objects.

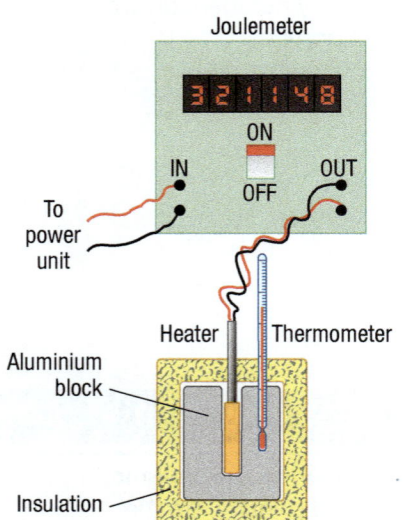

Figure 1 *Heating an aluminium block*

The following results were obtained using two different amounts of water. They show that:

- 1600 J was used to heat 0.1 kg of water by 4 °C
- 3200 J was used to heat 0.2 kg of water by 4 °C.

These results show that:

- 16 000 J of energy would be needed to heat 1.0 kg of water by 4 °C
- 4000 J of energy would be needed to heat 1.0 kg of water by 1 °C.

More accurate measurements would give a value of 4200 J per kg °C for water. This is the **specific heat capacity** of water.

The specific heat capacity of a substance is the energy needed or energy transferred to 1 kg of the substance to raise its temperature by 1 °C.

The unit of specific heat capacity is the joule per kilogram degree Celsius, or J/kg °C.

For a known change of temperature of a known mass of a substance:

$$E = m \times c \times \Delta\theta$$

where:
E is the energy transferred in J; m is the mass in kg;
c is the specific heat capacity in J/kg °C; $\Delta\theta$ is the temperature change in degrees °C.

To find the specific heat capacity you need to rearrange the above equation:

$$c = \frac{E}{m \times \Delta\theta}$$

??? Did you know ...?

Coastal towns are usually cooler in summer and warmer in winter than towns far inland. This is because water has a very high specific heat capacity. Energy from the Sun (or lack of energy) affects the temperature of the sea much less than it affects the temperature of the land.

Practical

Measuring the specific heat capacity of a metal

Use the apparatus shown in Figure 1 to heat a metal block of known mass.

Here are some measurements using an aluminium block of mass 1.0 kg.

Starting temperature = 14 °C
Final temperature = 22 °C
Energy supplied = 7200 J

To find the specific heat capacity of aluminium, the measurements above give:

E = energy transferred = energy supplied = 7200 J
$\Delta\theta$ = temperature change = 22 °C − 14 °C = 8 °C

Inserting these values into the rearranged equation gives:

$$c = \frac{E}{m \times \Delta\theta} = \frac{7200\,J}{1.0\,kg \times 8\,°C} = 900\,J/kg\,°C$$

Note: An ammeter, a voltmeter, and a stopwatch can be used instead of the joulemeter to measure the electrical energy supplied to the block. The circuit for this is shown in Figure 2. The ammeter is used to measure the heater current, and the voltmeter is used to measure the heater voltage.

As you will see in Topic 16.5, the electrical energy supplied E = heater potential difference $V \times$ heater current $I \times$ heating time t.

Safety: Take care with hot objects.

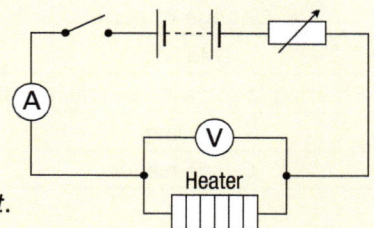

Figure 2 *Circuit diagram*

Table 1 shows the specific heat capacities of some common materials.

Table 1 *Specific heat capacities*

Substance	Water	Oil	Aluminium	Iron	Copper	Lead	Concrete
Specific heat capacity (J/kg °C)	4200	2100	900	390	490	130	850

Summary questions

1 A small bucket of water and a large bucket of water are left in strong sunlight. Which one warms up faster? Give a reason for your answer.

2 Use the information in Table 1 to answer this question.
   a Explain why a mass of lead heats up more quickly than an equal mass of aluminium.
   b Calculate the energy needed:
      i to raise the temperature of 0.20 kg of aluminium from 15 °C to 40 °C.
      ii to raise the temperature of 0.40 kg of water from 15 °C to 40 °C.
      iii to raise the temperature of 0.40 kg of water in an aluminium container of mass 0.20 kg from 15 °C to 40 °C.
   c A copper water tank of mass 20 kg contains 150 kg of water at 15 °C. Calculate the energy needed to heat the water and the tank to 55 °C.

3 State **two** ways in which a storage heater differs from an ordinary electric heater.

12.3 Change of state

Learning objectives

After this topic, you should know:

- what is meant by the melting point and the boiling point of a substance
- what is needed to melt a solid or to boil a liquid
- how to explain the difference between boiling and evaporation.

Melting points and boiling points

When pure ice is heated and melts, its temperature stays at 0 °C until all the ice has melted. When water is heated and boils at atmospheric pressure, its temperature stays at 100 °C.

For any pure substance undergoing a change of state, its temperature stays the same. As shown in Table 1, depending on the change of state, this temperature is called the **melting point** or the **boiling point** of the substance. The temperature at which a liquid changes to a solid is called its freezing point and is the same temperature as the melting point of the solid.

The melting point of a solid and the boiling point of a liquid are affected by impurities in the substance. For example, the melting point of water is lowered if salt is added to the water. This is why salt is added to grit used for gritting roads in freezing weather, so that roads have to be colder before they get icy.

Table 1 *Changes of state*

Change of state	Initial and final state	Temperature
Melting	Solid to liquid	Melting point
Freezing (also called solidification)	Liquid to solid	
Boiling	Liquid to vapour	Boiling point
Condensation	Vapour to liquid	

Required Practical

Measuring the melting point of a substance

The substance in its solid state is placed in a suitable test tube in a beaker of water, as shown in Figure 1a. The water is heated, and the temperature of the substance is measured as it melts. If its temperature is measured every minute, the measurements can be plotted on a graph, as shown in Figure 1b. The melting point is the temperature of the flat section of the graph – this is when the temperature stays the same as the substance melts.

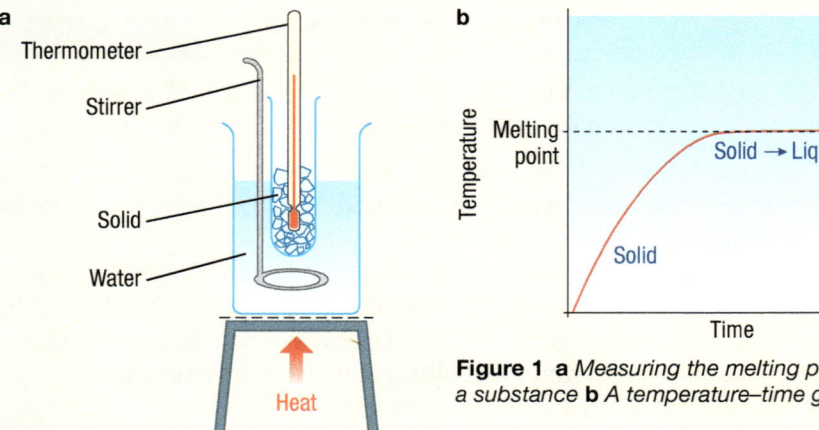

Figure 1 a *Measuring the melting point of a substance* **b** *A temperature–time graph*

The same arrangement without the beaker of water can be used to find the boiling point of a liquid.

Safety: Wear eye protection.

Energy and change of state

Suppose a beaker of ice below 0 °C is heated steadily so that the ice melts and then the water boils. Figure 2 shows how the temperature changes with time. The temperature:

1 increases until it reaches 0 °C, when the ice starts to melt T, then

2 stays constant at 0 °C until all the ice has melted, then

3 increases from 0 °C to 100 °C, when the water in the beaker starts to boil at 100 °C, then

4 stays constant at 100 °C as the water turns to steam.

The energy supplied to a substance when it changes its state is called **latent heat**. 'Latent' means 'hidden'. The energy supplied to melt or boil it is hidden by the substance because its temperature does not change at the melting point or the boiling point.

Most pure substances produce a temperature–time graph with similar features to Figure 2. Note that:

- **fusion** is often used to describe melting because different solids can be joined, or fused, together when they melt
- **evaporation** from a liquid occurs at its surface when the liquid is below its boiling point. At its boiling point, a liquid boils because bubbles of vapour form inside the liquid and rise to the surface to release the gas.

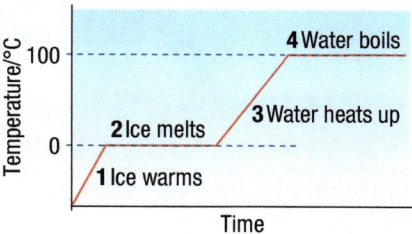

Figure 2 *Melting and boiling of water*

Study tip

Evaporation takes place at any temperature. Boiling occurs only at the boiling point.

 Did you know …?

At high altitude, a liquid boils at a lower temperature than at sea level. This is because the boiling point of a liquid depends on pressure, and atmospheric pressure is lower at high altitude than it is at sea level.

Figure 3 *Boiling water at high altitude*

Summary questions

1 State three differences between evaporation and boiling.

2 A pure solid substance X was heated in a tube and its temperature was measured every 30 seconds. The measurements are given in the table below.

Time (s)	0	30	60	90	120	150	180	210	240	270	300
Temperature (°C)	20	35	49	61	71	79	79	79	79	86	92

a i Use the measurements in the table to plot a graph of temperature on the *y*-axis against time.
  ii Use your graph to find the melting point of X.

b Describe the physical state of the substance as it was heated from 60 °C to 90 °C.

3 Salt water has a lower freezing point than pure water. In icy conditions in winter, gritting lorries are used to scatter a mixture of salt and grit on roads. Explain the purpose of each of the two components of the mixture.

Key points

- For a pure substance:
  – its melting point is the temperature at which it melts (which is the same temperature at which it solidifies)
  – its boiling point is the temperature at which it boils (which is the same temperature at which it condenses).

- Energy is needed to melt a solid or to boil a liquid.

- Boiling occurs throughout a liquid at its boiling point. Evaporation occurs from the surface of a liquid when its temperature is below its boiling point.

12.4 Specific latent heat

Learning objectives

After this topic, you should know:

- what is meant by latent heat of fusion and of vaporisation
- what is meant by *specific* latent heat of fusion and of vaporisation
- how to use the equation $E = m \times L$ in latent heat calculations.

Study tip

Be careful with units when doing calculations, especially joules and kilojoules.

Latent heat of fusion

When a solid substance is heated:

- **below its melting point**, its temperature increases until the melting point is reached. Before it reaches this point, the energy supplied to the solid increases the kinetic energy of its particles. As a result, the particles vibrate more and more and the temperature rises.
- **at its melting point**, it melts and turns to liquid. Its temperature stays constant until all of it has melted. The energy supplied is called **latent heat of fusion**. It is the energy used by the particles to break free from each other.

If the substance in its liquid state is cooled, it will solidify at the same temperature as its melting point. When this happens, the particles bond together into a rigid structure. Latent heat is released as the substance solidifies.

The **specific latent heat of fusion, L_F,** of a substance is the energy needed to melt 1 kg of the substance at its melting point (i.e., without changing its temperature).

The unit of specific latent heat of fusion is the **joule per kilogram (J/kg)**.

If energy E is supplied to a solid at its melting point, and a mass m of the substance melts without change in temperature:

$$\text{specific latent heat of fusion, } L_F = \frac{E}{m}$$

Note that rearranging this equation gives $E = m \times L_F$.

Practical

Measuring the specific latent heat of fusion of ice

In this experiment, a low-voltage heater is used to melt crushed ice in a funnel. The melted ice is collected using a beaker under the funnel, as shown in Figure 1. A joulemeter is used as shown to measure the energy supplied to the heater.

To take account of energy transfer from the surroundings, the mass of ice melted in a certain time must be measured with the heater turned off, then with it turned on. The difference in the two measurements gives the mass of ice melted because of the heater only.

1 With the heater off, water from the funnel is collected in the beaker for a measured time (e.g., 10 minutes). The mass of the beaker and water, m_1, is then measured. The beaker is then emptied for the next stage.

2 With the heater on, the procedure is repeated for exactly the same time. The joulemeter readings before and after the heater is switched on are recorded. After the heater is switched off, the mass of the beaker and the water, m_2, is measured once more.

To calculate the specific latent heat of fusion of ice, note that:

- the mass m of ice melted because of the heater is given by: $m = m_2 - m_1$
- the energy E supplied to the heater = the difference between the joulemeter readings
- the specific latent heat of fusion of ice, $L_F = \dfrac{E}{m} = \dfrac{E}{m_2 - m_1}$.

Instead of using a joulemeter, the energy supplied to the heater can also be measured using the circuit and information in Topic 12.2, Figure 2.

Safety: Take care with the hot immersion heater and wear eye protection.

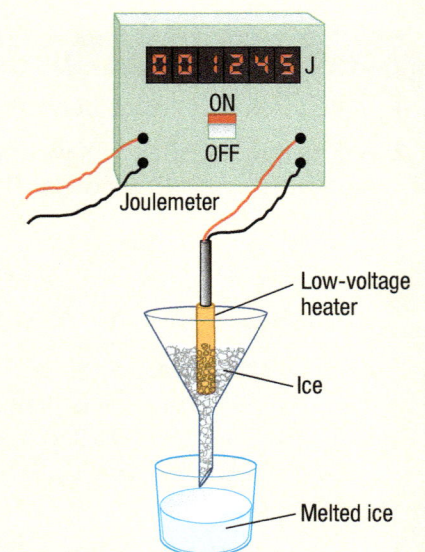

Figure 1 *Measuring the specific latent heat of fusion of ice*

Latent heat of vaporisation

When a liquid is heated:

- **above its freezing point**, its temperature increases until the boiling point is reached. Before it reaches this point, the energy supplied to the liquid increases the kinetic energy of its particles. As a result, the particles move about faster and faster and the temperature rises.
- **at its boiling point**, it turns to vapour. Its temperature stays constant until all of it has boiled. The energy supplied is called **latent heat of vaporisation**. It is the energy used by the particles to break away from the liquid.

If the substance in its gaseous state is cooled, it will condense at the same temperature as its boiling point. When this happens, the particles bond and get much closer together. Latent heat is released as the substance condenses into a liquid.

The **specific latent heat of vaporisation**, L_v, of a substance is the energy needed to change 1 kg of the substance at its boiling point from liquid to vapour.

The unit of specific latent heat of vaporisation is the **joule per kilogram (J/kg)**.

If energy E is supplied to a liquid at its boiling point, and a mass m of the substance boils away without change in temperature:

$$\text{specific latent heat of vaporisation, } L_v = \frac{E}{m}$$

Note that rearranging this equation gives $E = m \times L_v$.

Summary questions

1 In the experiment shown in Figure 1, 0.024 kg of water was collected in the beaker in 300 seconds with the heater turned off. The beaker was then emptied and placed under the funnel again. With the heater on for exactly 300 s, the joulemeter reading increased from zero to 15 000 J and 0.068 kg of water was collected in the beaker.
   a Calculate the mass of ice melted because of the heater being on.
   b Calculate the specific latent heat of fusion of water.

2 a In the experiment shown in Figure 2, the balance reading decreased to 0.144 kg from 0.152 kg in the time taken to supply 18 400 J of energy to the boiling water. Calculate the specific latent heat of vaporisation of water.
   b After the heater in **a** was switched off, the temperature of the water in the beaker gradually decreased. What does this observation tell you about the experiment and the value calculated for the specific latent heat of vaporisation of water?

3 An ice cube of mass 0.008 kg at 0 °C was placed in water at 15 °C in an insulated plastic beaker. The mass of water in the beaker was 0.120 kg. After the ice cube had melted, the water was stirred and its temperature was found to have fallen to 9 °C. The specific heat capacity of water is 4200 J/kg °C.
   a Calculate the energy transferred from the water.
   b Show that when the melted ice warmed from 0 °C to 9 °C, it gained 300 J of energy.
   c Therefore calculate the specific latent heat of fusion of water.

Practical

Measuring the specific latent heat of vaporisation of water

The apparatus shown in Figure 2 can be used. The low-voltage heater is switched on to bring the water to its boiling point. When the water is boiling, the joulemeter reading and the top pan balance reading are measured and then remeasured after a known time (e.g., 5 minutes).

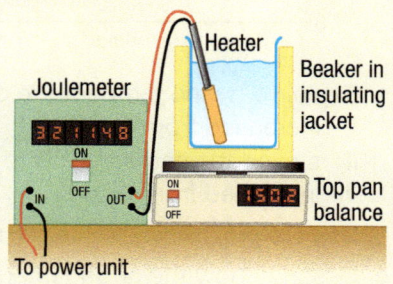

Figure 2 *Measuring the specific latent heat of vaporisation of water*

The energy E supplied during this time = the difference between the joulemeter readings.

The mass m of water boiled away in this time = the difference between the readings of the top pan balance.

The equation $L_v = \frac{E}{m}$ is used to calculate the specific latent heat of vaporisation of water.

Safety: Take care with hot objects and wear eye protection.

Key points

- Latent heat is the energy needed or released when a substance changes its state without changing its temperature.
- Specific latent heat of fusion (or of vaporisation) is the energy needed to melt (or to boil) 1 kg of a substance with no change in temperature.

133

Chapter summary questions

1 A test tube containing a solid substance is heated in a beaker of water. The temperature–time graph shows how the temperature of the substance changed with time as it was heated.

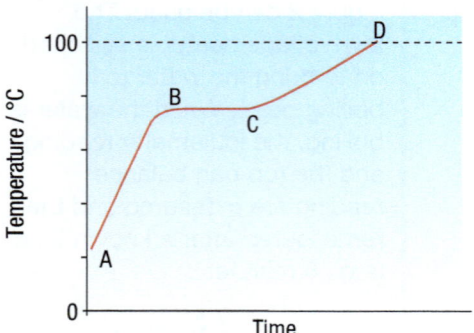

a Explain why the temperature of the substance:
i increased from **A** to **B**
ii stayed the same from **B** to **C**
iii increased from **C** to **D**.

b Use the graph to estimate the melting point of the solid.

c Describe how the arrangement and motion of the particles changes as the temperature increases from **A** to **D**.

2 Geothermal energy is obtained from hot rocks underground by pumping water through the rock.

a i When 400 kg of water is pumped through the rocks at a certain location, the temperature of the water is increased from 20 °C to 80 °C. Calculate the energy transferred to the water. The specific heat capacity of water is 4200 J/kg °C.

ii If the water is pumped through the rocks in a time of 50 seconds, calculate the rate of transfer of energy to the water.

b The temperature of the underground rocks stays constant because the rocks gain energy by heating from the Earth's core. If water had been pumped through the rocks at a much slower rate, discuss the effect this would have had on the temperature of the water flowing out of the rocks.

3 The figure shows a large heater used to heat a big hall. Hot water is pumped through pipes which pass through the heater. Fins attached to the pipes transfer energy by heating to air, which is blown by fans through the heater.

a Water enters the heater at 77 °C and leaves it at a temperature of 15 °C. The water flows through the heater at a rate of 0.050 kg per second. Calculate the energy transferred from the water to the heater each second. The specific heat capacity of water is 4200 J/kg °C.

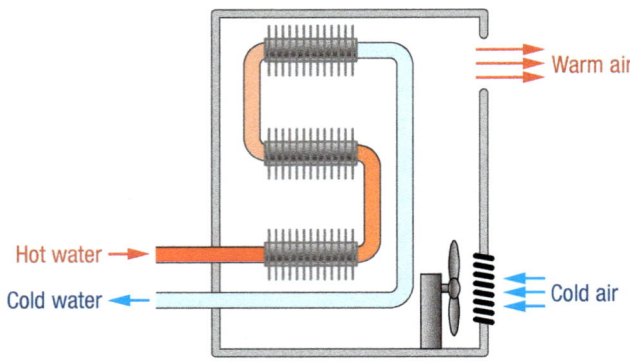

b At a certain fan speed, air enters the heater at 12 °C and leaves it at a temperature of 26 °C. State and explain how the temperature of the air leaving the heater would change if the fan speed is turned up to increase the rate of flow of air through the heater.

4 A plastic beaker containing 0.10 kg of water at 18 °C was placed in a refrigerator for 450 seconds. After this time, the temperature of the water was found to be 3 °C. The specific heat capacity of water is 4200 J/kg °C.

a Calculate the energy transferred from the water.

b Calculate the rate of transfer of energy from the water.

5 A 3.0 kW electric kettle is fitted with a safety cut-out designed to switch it off as soon as the water boils. Unfortunately, the cut-out does not operate correctly and allows the water to boil for 30 seconds longer than it is supposed to.

a How much electrical energy is supplied to the kettle in this time?

b The specific latent heat of vaporisation of water is 2.3 MJ/kg. Estimate the mass of water boiled away in this time.

6 When the brakes of a vehicle are applied, the vehicle decelerates and comes to rest from a velocity of 30 m/s.

a The mass of the vehicle is 1200 kg. Show that the loss of kinetic energy of the vehicle is 0.54 MJ.

b The brakes become hot as a result of the energy transferred to them by the braking force. The brakes contain a total of 16 kg of metal of specific heat capacity 560 J/kg °C. Estimate the increase in temperature of the metal in the brakes.

Practice questions

1 A student put some crushed ice in a beaker and measured the temperature. She then put the beaker on a tripod and gauze and gently heated it with a Bunsen burner. She measured the temperature every minute.

The graph shows how the temperature changed with time.

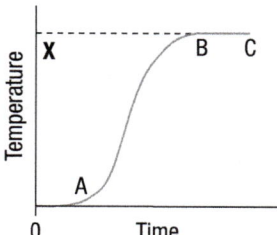

a Name the temperature marked **X** on the graph. (1)

b The crushed ice does not absorb all the energy supplied by the Bunsen burner.
What happens to the energy that is not absorbed by the crushed ice? (3)

c Use the graph to help you explain what happens to the energy that is absorbed. (6)

d Another student carried out a similar experiment. She used a temperature sensor and a data logger instead of a thermometer and a clock.
Give **two** advantages of her method. (2)

2 a A leaflet about saving energy states:

Only boil the water you need.

Explain how doing this saves energy. (2)

b A dishwasher heats 8 kg of water. The normal setting of the dishwasher uses water at 65 °C.
The specific heat capacity of water is 4200 J/kg °C.
Calculate how much energy, in kJ, is saved by using the economy setting of water at 45°C. (3)

3 The diagram shows a storage heater.

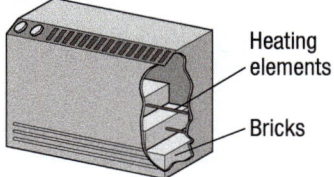

The bricks are heated up during the night when electrical energy is cheaper. During the day, the bricks release energy into the room.

a i The bricks may be made from different materials.

Choose **two** properties that the bricks should have if they are to store as much energy as possible. (1)

| large mass per unit volume |
| small mass per unit volume |
| large specific heat capacity |
| small specific heat capacity |

ii The storage heater contains concrete bricks of mass 12 kg. Concrete has a specific heat capacity of 900 J/kg °C.
During the night 540 kJ of energy are supplied to the bricks.
Calculate the temperature rise of the bricks. (3)

iii Actually the temperature rise will be different from the calculated value.
Explain how the temperature rise will be different. (3)

b The diagram shows an oil-filled heater.

i Give **one** advantage of the oil-filled heater over the storage heater. (1)

ii Give **one** advantage of the storage heater over the oil-filled heater. (1)

4 a The diagram shows one method of determining the specific latent heat of vaporisation of water.

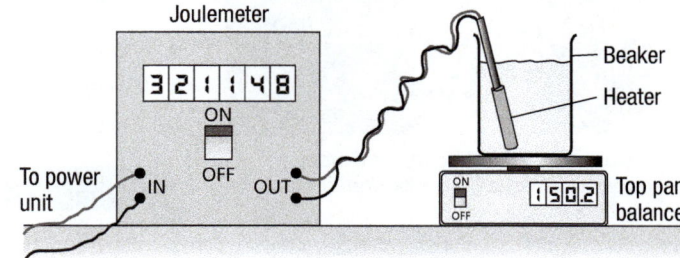

Some students switch on the heater to boil the water. When the water is boiling they take the reading on the top pan balance and set the joulemeter to zero. After heating for a further 10 minutes, they take the joulemeter reading and the new reading on the top pan balance.

Their results are shown below.

1st balance reading	184 g
2nd balance reading	168 g
Joulemeter reading	36 800 J

i Use the students' results to calculate the specific latent heat of vaporisation of water. (4)

ii Suggest how the method could be improved. (1)

13.1 Conduction

Learning objectives

After this topic, you should know:

- which materials make the best conductors
- which materials make the best insulators
- why metals are good conductors
- why non-metals are poor conductors.

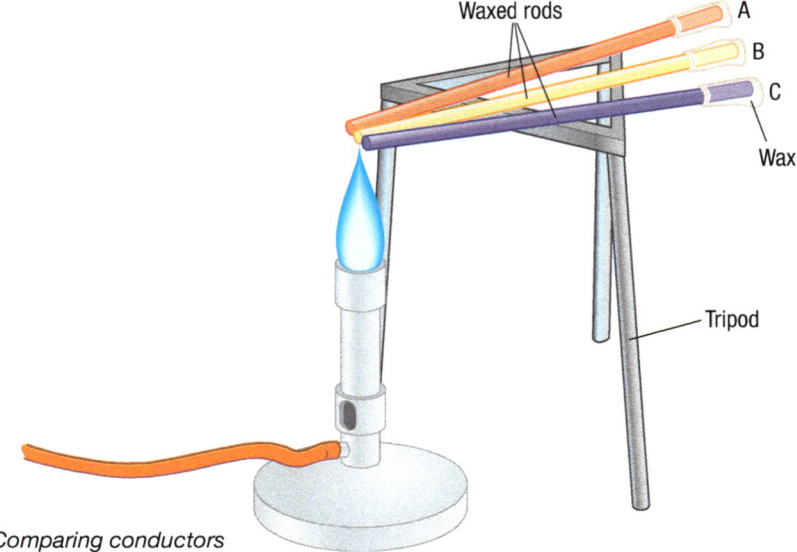

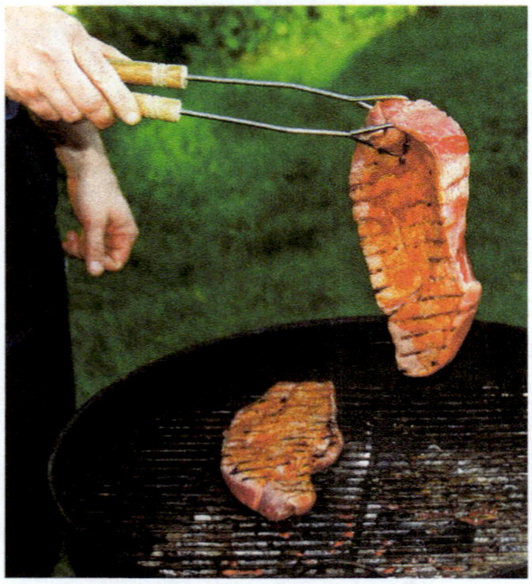

Figure 1 *At a barbecue – the steel cooking utensils have wooden or plastic handles*

If you cook on a barbecue, you need to know which materials are good **conductors** of heat, which ones are good **insulators**. If you use the wrong utensils, you're likely to burn your fingers!

Testing rods of different materials as conductors

The rods need to be the same width and length for a fair test. Each rod is coated with a thin layer of wax near one end. The uncoated ends are then heated together.

Look at Figure 2. The wax melts fastest on the rod that conducts best.

Figure 2 *Comparing conductors*

- Metals conduct energy better than non-metals.
- Copper is a better conductor than steel.
- Wood conducts better than glass.

Practical

Testing sheets of materials as insulators

Use different materials to insulate identical cans or beakers of hot water. The volume of water and its temperature at the start should be the same.

Use a thermometer to measure the water temperature after a fixed time. The results should tell you which insulator was best.

Safety: Take care if using very hot water.

Table 1 shows some results of comparing two different materials using the method in the practical above.

Table 1 *Testing sheets of materials as insulators*

Material	Starting temperature (°C)	Temperature after 300 s (°C)
Paper	40	32
Felt	40	36

Conduction in metals

Metals contain lots of **free electrons**. These electrons move about at random within the metal structure, and hold the positive metal ions (charged atoms) in their fixed positions. The free electrons collide with each other and with the positive ions.

When a metal rod is heated at one end, the free electrons at the hot end gain kinetic energy and move faster.

- These electrons **diffuse** (i.e., spread out) and collide with other free electrons and ions in the cooler parts of the metal.
- As a result, they transfer kinetic energy to these electrons and ions.

So energy is transferred from the hot end of the rod to the cooler end.

In a non-metallic solid, all the electrons are held in the atoms. Energy transfer takes place as the atoms vibrate and shake each other. This is much less effective than energy transfer by free electrons. This is why metals are much better conductors than non-metals.

Figure 3 *Insulating a loft. The air trapped between fibres make fibreglass a good insulator*

?? Did you know …?

Materials such as wool and fibreglass are good insulators. This is because they contain air trapped between the fibres. Trapped air is a good insulator. Materials such as fibreglass are used for insulating lofts and hot water pipes.

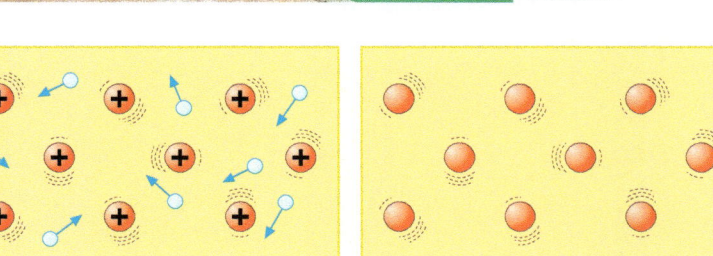

- ⊕ Ion
- ○ Free Electron
- 🔴 Atom

a

b

Figure 4 *Energy transfer in* **a** *a metal* **b** *a non-metal*

Key points

- Metals are the best conductors of energy.
- Materials such as wool and fibreglass are the best insulators.
- Conduction of energy in a metal is caused mainly by free electrons transferring energy inside the metal.
- Non-metals are poor conductors because they do not contain free electrons.

Summary questions

1. a Why do steel pans have handles made of plastic or wood?
   b Which material, felt or paper, is the better insulator? Give a reason for your answer.

2. a Choose a material you would use to line a pair of winter boots. Explain your choice of material.
   b How could you carry out a test on three different lining materials?

3. a Explain why metals are good conductors of energy.
   b Describe how non-metals conduct energy.

13.2

Convection

Learning objectives

After this topic, you should know:

- what convection is
- where convection can occur
- why convection occurs.

??? Did you know ...?

The Gulf Stream is a stream of warm water that flows across the Atlantic Ocean from the Gulf of Mexico to the British Isles. If it ever turned away, UK winters would be much colder.

Figure 1 *A natural glider – some birds use convection to soar high above the ground*

Glider aircraft and birds use convection to stay in the air. **Convection** streams can keep them high above the ground for hours.

Convection happens whenever a **fluid** is heated. A fluid is a gas or a liquid. Look at the diagram in Figure 2. It shows a simple demonstration of convection.

- The hot gases from the burning candle go straight up the chimney above the candle.
- Cold air is drawn down the other chimney to replace the air leaving the box.

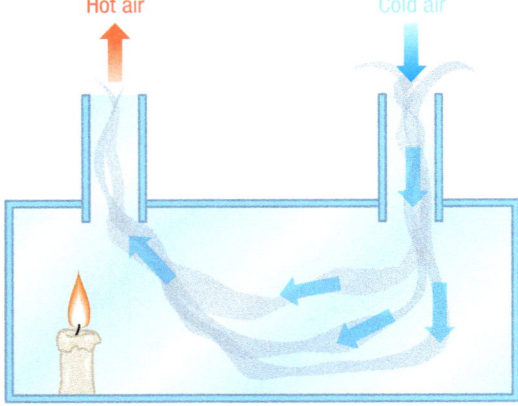

Figure 2 *Convection*

Using convection

Hot water at home

Many homes have a hot water tank. Hot water from the boiler rises and flows into the tank, where it rises to the top. Figure 3 shows the system. When you use a hot water tap at home, you draw off hot water from the top of the tank.

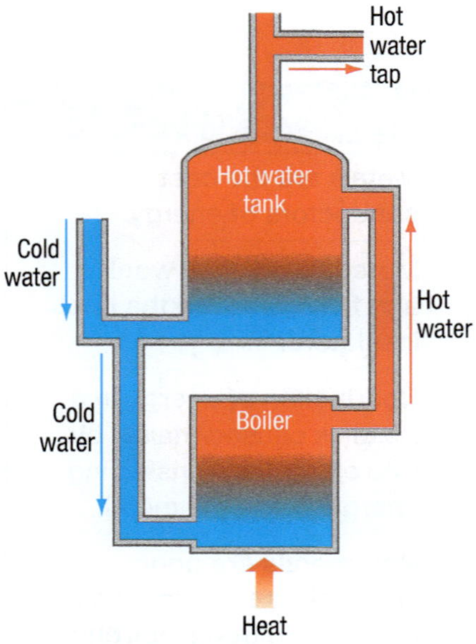

Sea breezes

Sea breezes keep you cool at the seaside. On a sunny day, the ground heats up faster than the sea. So the air above the ground warms up and rises. Cooler air from the sea flows in as a sea breeze to take the place of the rising warm air (see Figure 4).

Figure 3 *Hot water at home*

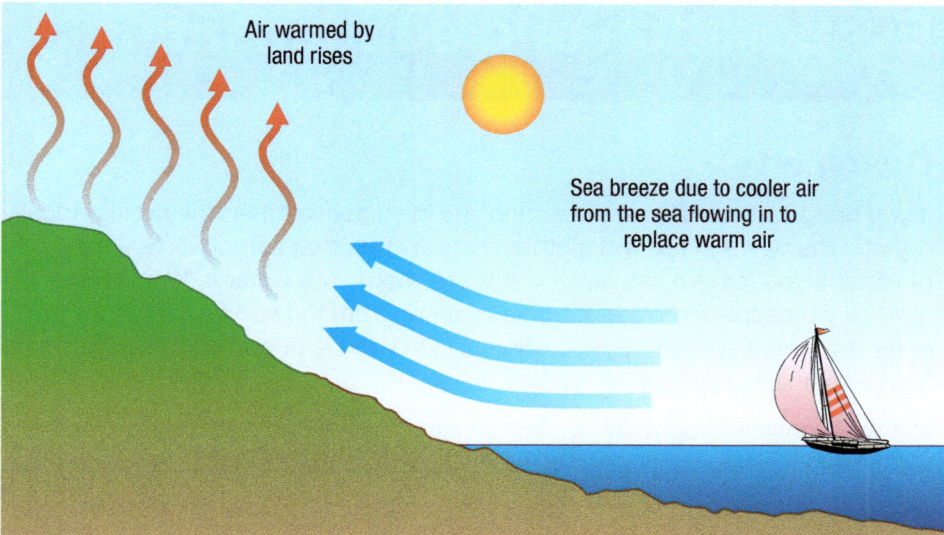

Figure 4 *Sea breezes*

How convection works

Convection takes place:

- only in fluids (liquids and gases)
- because of circulation (convection) within the fluid.

The convection occurs because fluids rise where they are heated. Then they stop rising where they cool down. Convection streams transfer energy from the hotter parts to the cooler parts.

Why do fluids rise when heated?

Most fluids expand when heated. This is because the particles move around more and they move apart, taking up more space. Therefore, the **density** of the fluid decreases, because the same mass of fluid now occupies a bigger volume. So heating part of a fluid makes that part less dense and therefore it rises. As the fluid cools, its particles move together. It becomes more dense and sinks.

Summary questions

1 A pan of water is heated on a gas cooker. Describe how energy is transferred by heating from the gas flame under the pan to all of the water in the pan.

2 The figure shows a convector heater. It has an electric heating element inside and a metal grille on top.

   Hot air

   a What does the heater do to the air inside the heater?

   b Why is there a metal grille on top of the heater?

   c Where does air flow into the heater?

3 **a** Describe how you could demonstrate convection in water using a strongly coloured crystal or a suitable dye. Explain in detail what you would see.

   b A wall-mounted radiator in a room is often fitted under a window. Give one advantage and one disadvantage of fitting such a radiator under a window rather than away from the window.

13.3 Evaporation and condensation

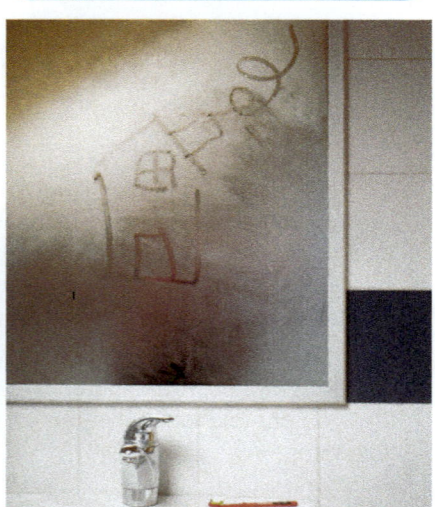

Figure 2 *Condensation*

Study tip

Make sure that you can use the kinetic theory to explain why a liquid cools when it evaporates.

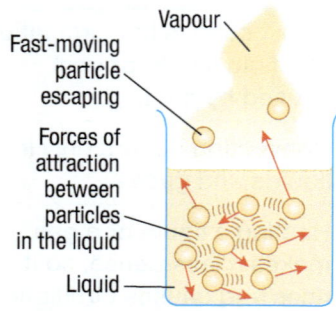

Figure 3 *Explanation of cooling by evaporation*

Drying off

If you hang wet clothes on a washing line in sunny weather, they will gradually dry off. The water in the wet clothes **evaporates**. You can observe evaporation of water if you leave a saucer of water in a room. The water in the saucer gradually disappears. Water particles escape from the surface of the water and enter the air in the room. Evaporation is the process of particles leaving a liquid below its boiling point at the surface.

In a well-ventilated room, the water particles in the air are not likely to re-enter the liquid. Particles continue to leave the liquid until all the water has evaporated.

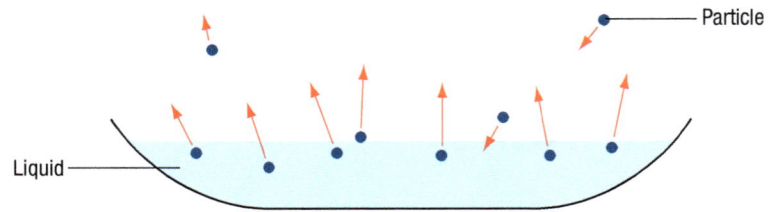

Figure 1 *Water particles escaping from a liquid*

Condensation

In a steamy bathroom, a mirror is often covered by a film of water. There are lots of water particles in the air. Some of them hit the mirror, lose energy, and stay there as the liquid. Water vapour in the air **condenses** on the mirror.

Cooling by evaporation

If you have an injection, the doctor or nurse might numb your skin by dabbing it with a liquid that quickly evaporates. As the liquid evaporates, your skin becomes too cold to feel any pain.

Demonstration

Cooling by evaporation

Watch your teacher carry out this experiment in a fume cupboard.

Why is ether used in this experiment?

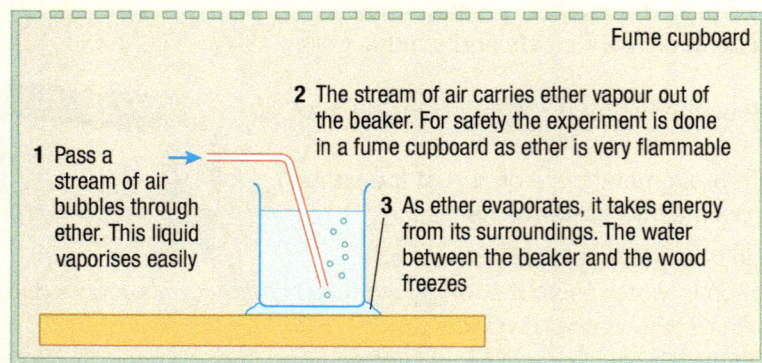

Figure 4 *A demonstration of cooling by evaporation*

Safety: There should be no naked flames in the laboratory.

You can use the kinetic theory of matter, as shown in Figure 3, to explain why evaporation causes this cooling effect.

- Relatively weak attractive forces exist between the particles in the liquid.
- The faster particles, which have more kinetic energy, break away from the attraction of the other particles at the surface and escape from the liquid.
- After they leave, the liquid is cooler because the average kinetic energy of the particles left in the liquid has decreased.

Factors affecting the rate of evaporation

Clothes dry faster on a washing line:

- if the items of wet clothing are spread out on the line. This increases the area of the wet clothing in contact with dry air.
- if the washing line is in sunlight. Wet clothes dry faster the warmer they are.
- if there is a breeze to take away the particles that escape from the water in the wet clothes.

The example of clothes drying shows that the rate of evaporation from a liquid is increased by:

- increasing the surface area of the liquid
- increasing the temperature of the liquid
- creating a draught of air across the liquid's surface.

Factors affecting the rate of condensation

In a steamy kitchen, water can often be seen trickling down a window pane. The glass pane is a cold surface, so water vapour condenses on it. The air in the room is moist, or humid. The bigger the area of the window pane, or the colder it is, the greater the rate of condensation. This example shows that the rate of condensation of a vapour on a surface is increased by:

- increasing the surface area
- reducing the surface temperature.

Did you know ... ?

Air conditioning

An **air-conditioning unit** in a room transfers energy from inside the room to the outside. The unit contains a coolant liquid that easily evaporates. The coolant is pumped around a sealed circuit of pipes that pass through the unit and outside.

- The liquid coolant evaporates in the pipes in the room and cools the room.
- The evaporated coolant condenses in the pipes outside and transfers energy to the surroundings.

Figure 5 *An air-conditioning unit*

Key points

- Evaporation happens when a liquid turns into a gas below its boiling point.
- Condensation happens when a gas turns into a liquid.
- The cooling of a liquid by evaporation is caused by faster-moving particles escaping from the liquid.
- Evaporation can be increased by increasing the surface area of the liquid, by increasing the liquid's temperature, or by creating a draught of air across the liquid's surface.
- Condensation on a surface can be increased by increasing the area of the surface or by reducing the temperature of the surface.

Summary questions

1 Describe in terms of particles what happens in each of the following situations:
   a water droplets form on a cold surface
   b a liquid cools because of evaporation.

2 a In terms of particles, explain why the windows on a bus become misty when there are lots of people on the bus and it's cold outside the bus.
   b A refrigerator usually has a drip tray inside to collect any water that runs to the bottom of the refrigerator. Explain why opening the refrigerator too often might cause the drip tray to fill.

3 Explain the following statements.
   a Wet clothes on a washing line dry faster on a hot day than on a cold day.
   b A person wearing wet clothes on a cold windy day is likely to feel much colder than someone wearing dry clothes.

13.4 Infrared radiation

Learning objectives

After this topic, you should know:

- what infrared radiation is

- what gives out infrared radiation

- how infrared radiation depends on the temperature of an object.

Figure 1 *Keeping watch in darkness*

Study tip

Remember that all bodies emit **and** absorb infrared radiation.

??? Did you know ... ?

A **passive infrared (PIR) detector** in a burglar alarm circuit will trigger the alarm if someone moves in front of the detector. The detector contains sensors that detect infrared radiation from different directions.

∞ links

For more information on electromagnetic waves, look back to Topic 8.1 'The electromagnetic spectrum'.

Seeing in the dark

Special cameras can be used to 'see' animals and people in the dark. These cameras detect **infrared radiation**. Every object gives out (**emits**) infrared radiation because of the motion of their particles.

The higher the temperature of an object, the more infrared radiation it emits in a given time.

Look at the photo in Figure 1. The rhinos are hotter than the ground.

Practical

Detecting infrared radiation

You can use a thermometer with a blackened bulb to detect infrared radiation. Figure 2 shows how to do this. A black surface readily absorbs infrared radiation. See Topic 13.5.

1 The glass prism splits a narrow beam of white light into the colours of the spectrum.

2 The thermometer reading rises when it is placed just beyond the red part of the spectrum. The infrared radiation in the beam is dispersed there. Your eyes cannot detect it, but the thermometer can.

Infrared radiation is beyond the red part of the visible spectrum.

- What would happen to the thermometer reading if the thermometer were moved away from the screen?

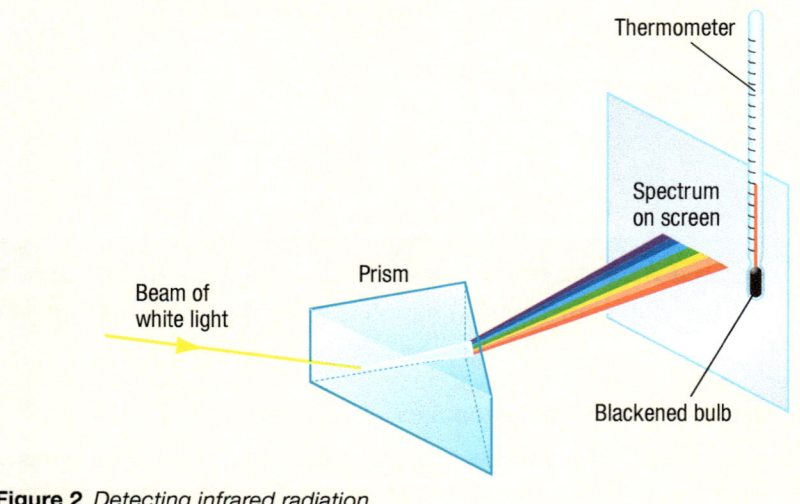

Figure 2 *Detecting infrared radiation*

The electromagnetic spectrum

Radio waves, **microwaves**, **infrared radiation**, and **visible light** are parts of the electromagnetic spectrum. So too are ultraviolet rays and X-rays.

An object that has a constant temperature emits electromagnetic waves across a continuous range of wavelengths. The intensity of the radiation is most intense at a certain wavelength.

The increase in radiation intensity is greater at shorter wavelengths than at lower temperatures. If an object's temperature becomes higher, the radiation it emits is more intense (i.e., it emits more energy per second) at every wavelength, but this increase in intensity is greater at shorter wavelengths. So for a hot object the radiation is more intense at a shorter wavelength. Astronomers can tell that bluish-white stars are hotter than stars like the Sun because they emit more light in the shorter wavelength part of the visible spectrum (i.e., the blue part).

Energy from the Sun

The Sun emits all types of electromagnetic radiation. Fortunately for us, the Earth's atmosphere blocks most of the radiation that would harm us, but it doesn't block infrared radiation from the Sun.

Figure 3 shows a solar furnace. This uses a giant reflector that focuses sunlight.

The temperature at the focus can reach thousands of degrees.

The greenhouse effect

The Earth's atmosphere acts like a greenhouse made of glass. In a greenhouse:

- short-wavelength infrared radiation (and light) from the Sun can pass through the glass and warm the objects inside the greenhouse
- infrared radiation from these warm objects is trapped inside by the glass because the objects emit infrared radiation of longer wavelengths that can't pass out through the glass.

So the greenhouse stays warm.

Some gases in the atmosphere, such as water vapour, methane, and carbon dioxide, act like a greenhouse and are called greenhouse gases. They trap infrared radiation from the Earth. This makes the Earth warmer than it would be if it had no atmosphere.

But the Earth is becoming too warm. If the polar ice caps melt, it will cause sea levels to rise. Reducing our use of fossil fuels will help to reduce the production of greenhouse gases.

Figure 3 *A solar furnace in the Eastern Pyrenees, France*

Summary questions

1 **a** What is infrared radiation?
   b An infrared camera on a satellite shows that more infrared radiation is emitted at night from a city than from the surrounding rural area. What does this tell you about the city compared with the rural area? Give a reason for your answer.

2 **a** Copy and complete the table to show if the object emits infrared radiation, light, or both.

Object	Infrared	Light
A hot iron		
A light bulb		
A TV screen		
The Sun		

   b How can you tell if an electric iron is hot without touching it?

3 **a** Explain why penguins huddle together to keep warm.
   b What can you deduce from Figure 2 about the wavelength of the infrared radiation that passes out of the prism?

Key points

- Infrared radiation is part of the spectrum of electromagnetic waves.
- All objects emit infrared radiation.
- The hotter an object is, the more infrared radiation it emits in a given time.

13.5 Surfaces and radiation

Learning objectives

After this topic, you should know:

- which surfaces are the best emitters of infrared radiation
- which surfaces are the best absorbers of infrared radiation
- which surfaces are the best reflectors of infrared radiation.

Figure 1 An emergency blanket in use

Which surfaces are the best emitters of radiation?

Rescue teams use light, shiny blankets to keep accident survivors warm (see Figure 1). A light, shiny outer surface emits much less radiation than a dark, matt (non-glossy) surface.

Practical

Testing radiation from different surfaces

To compare the radiation from two different surfaces, you can measure how fast two cans of hot water cool. The surface of one can is light and shiny, and the other has a dark, matt surface (see Figure 2).

At the start, the volume and temperature of the water in each can must be the same.

- Why should the volume and temperature of the water be the same at the start?
- Which can will cool faster?

Safety: Take care with hot water.

Thermometer to measure water temperature at intervals as it cools

Figure 2 Testing different surfaces

Your tests should show that:

- Dark, matt surfaces **emit more** infrared radiation than light, shiny surfaces.

Which surfaces are the best absorbers and reflectors of radiation?

When you use a photocopier, infrared radiation from a lamp dries the ink on the paper. Otherwise, the copies would be smudged. Black ink absorbs more infrared radiation than white paper. Just as all objects emit infrared radiation, all objects absorb it.

A light, shiny surface absorbs less radiation than a dark, matt surface. A matt surface has lots of cavities, as shown in Figure 3:

- the radiation reflected from the matt surface hits the surface again
- the radiation reflected from the shiny surface travels away from the surface.

So, the shiny surface absorbs less and reflects more radiation than a matt surface.

In general:

- Light, shiny surfaces **absorb less** infrared radiation than dark, matt surfaces.
- Light, shiny surfaces **reflect more** infrared radiation than dark, matt surfaces.

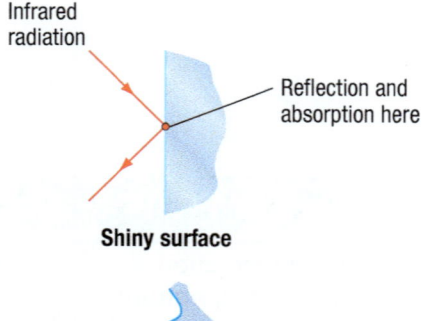

Infrared radiation

Reflection and absorption here

Shiny surface

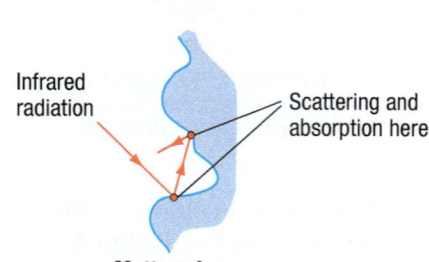

Infrared radiation

Scattering and absorption here

Matt surface

Figure 3 Absorbing infrared radiation

Practical

Absorption tests

Figure 4 shows how you can compare absorption by two different surfaces.

- The front surfaces of the two metal plates are at the same distance from the heater.
- The back of each plate has a coin stuck on with wax. The coin drops off the plate when the wax melts.
- The coin at the back of the matt black surface drops off first. The matt black surface absorbs more radiation than the light, shiny surface.

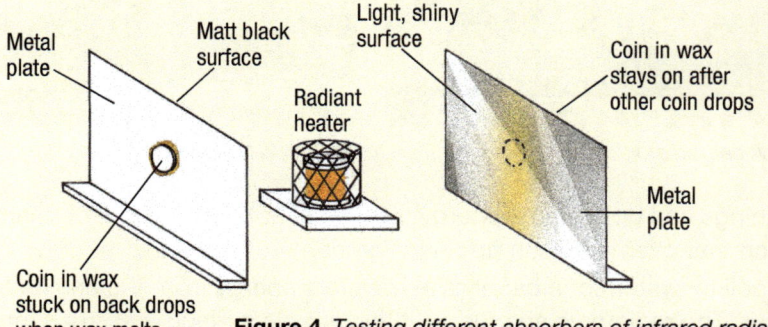

Figure 4 *Testing different absorbers of infrared radiation*

Safety: Take care with hot objects.

Summary questions

1 a Why does ice on a road melt faster in sunshine if sand is sprinkled on it?
  b Why are solar heating panels painted matt black?

2 a A black car and a metallic silver car that are otherwise identical are parked next to each other on a sunny day. Why does the temperature inside the black car rise more quickly than the temperature inside the silver car?
  b Which of the two cars in part **a** would cool down more quickly at night in winter?

3 A metal cube filled with hot water was used to compare the infrared radiation emitted from its four vertical faces, A, B, C, and D. One face was light and shiny, one was light and matt, one was dark and shiny, and one was dark and matt. An infrared sensor was placed opposite each face at the same distance. The sensors were connected to a computer. The results of the test are shown in the graph opposite.

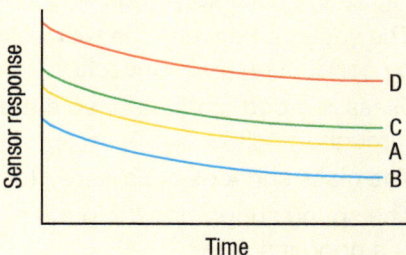

  a Why was it important for the distance from each sensor to the face to be the same?
  b Which face, A, B, C, or D, emits the:
    i least radiation?   ii most radiation?
  c What are the advantages of using data-recording equipment to collect the data in this investigation?

Black body radiation

Scientists are developing blacker and blacker materials. These new materials have very tiny pits in the surface to absorb almost all the light that hits them. They can be used to coat the insides of telescopes so that there are no reflections.

A perfect black body is an object that absorbs all the radiation that hits it. It doesn't reflect any radiation, and it doesn't transmit any radiation (i.e., no radiation passes through it). A good absorber is also a good emitter, so a perfect black body is also the best possible emitter.

The electromagnetic radiation emitted by a perfect black body is called **black body radiation**. Like any other object, when it becomes hotter a black body emits more radiation at all wavelengths, and the increase in intensity is greater at shorter wavelengths. But no other object emits as much radiation as a black body at the same temperature.

Key points

- All objects absorb infrared radiation.
- Dark, matt surfaces emit more infrared radiation than light, shiny surfaces.
- Dark, matt surfaces absorb more infrared radiation than light, shiny surfaces.
- Light, shiny surfaces reflect more infrared radiation than dark, matt surfaces.

13.6 Energy transfer by design

Energy transfer by heating

Learning objectives

After this topic, you should know:

- what design factors affect the rate at which a hot object transfers energy

- how the rate of energy transfer to or from an object can be controlled.

Figure 1 *A car radiator*

Figure 2 *Motorcycle engine fins*

?? Did you know ... ?

Some electronic components get warm when they are working, but if they become too hot they stop working. Such components are often fixed to a metal plate to keep them cool. The metal plate increases the effective surface area of the component. The metal plate is called a **heat sink**.

Figure 3 *A heat sink in a computer*

Lots of things can go wrong if energy transfer is not controlled. For example, a car engine that overheats can go up in flames.

- The cooling system of a car engine transfers energy from the engine to a radiator. The radiator is shaped so it has a large surface area. This increases the rate of energy transfer through convection in the air and through radiation.

- A motorcycle engine is shaped with **fins** on its outside surface. The fins increase the surface area of the engine in contact with air, so the engine transfers energy to its surroundings faster than if it had no fins.

Most cars also have a cooling fan that switches on when the engine is too hot. This increases the flow of air over the surface of the radiator.

The vacuum flask

If you are outdoors in cold weather, a hot drink from a vacuum flask keeps you warm. In the summer the same vacuum flask keeps your drinks cold.

In Figure 4, the liquid you drink is in the double-walled glass container.

- The vacuum between the two walls of the container cuts out energy transfer by conduction and convection between the walls.

- Glass is a poor conductor so there is little energy transfer by conduction through the glass.

- The glass surfaces are silvered to reduce radiation from the outer wall.

- The spring supporting the double-walled container is made of plastic, which is a good insulator.

- The plastic cap stops cooling by evaporation as it stops vapour loss from the flask. In addition, energy transfer by conduction is cut down because the cap is made from plastic.

So why does the liquid in the flask eventually cool down?

The above features decrease but do not totally stop the transfer of energy from the liquid. Energy transfer occurs at a very low rate because some radiation occurs from the silvered glass surface, and some conduction occurs through the cap, spring, and glass walls. The liquid transfers energy slowly to its surroundings so it eventually cools.

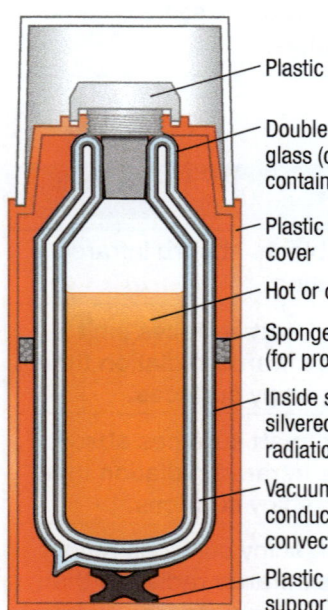

- Plastic cap
- Double-walled glass (or plastic) container
- Plastic protective cover
- Hot or cold liquid
- Sponge pad (for protection)
- Inside surfaces silvered to stop radiation
- Vacuum prevents conduction and convection
- Plastic spring for support

Figure 4 *A vacuum flask*

Factors affecting the rate of energy transfer

The bigger the **temperature difference** between an object and its surroundings, the faster the rate at which energy is transferred. In addition, the examples on the opposite page show that the rate at which an object transfers energy depends on its design. The design factors that are important are:

- the nature of the surface that is in contact with the object
- the object's volume and surface area.

Also important are:

- the object's mass
- the material the object is made from.

These factors affect how quickly the object's temperature changes (and therefore the rate of transfer of energy to or from it) when it loses or gains energy.

⚭ links

For more information on factors affecting energy transfer, look back to Topic 12.2 'Specific heat capacity'.

Practical

Investigating the rate of energy transfer

You can plan an investigation using different beakers and hot water to find out what affects the rate of cooling.

- Write a question that you could investigate.
- Identify the independent, dependent, and control variables in your investigation.

Safety: Take care with hot water.

Summary questions

1 Hot water is pumped through a radiator like the one in the photo below. Describe how energy is transferred from the hot water to the surrounding air.

2 An electronic component in a computer is attached to a heat sink.
   a i Explain why the heat sink is necessary.
   ii Why is a metal plate used as the heat sink?
   b Plan a test to show that double glazing is more effective at preventing energy transfer than single glazing.

3 **a** Explain, in detail, how the design of a vacuum flask reduces the rate of energy transfer.
   b A freezer is defrosted. Explain why a bowl of warm water placed in the freezer speeds up the defrosting process.

?? **Did you know … ?**

Foxy survivors

A desert fox has much larger ears than an arctic fox. Blood flowing through the ears transfers energy from inside the body to the air surrounding the surface of the ears. Big ears have a much larger surface area than little ears so they transfer energy to the surroundings more quickly.

A desert fox has big ears so it keeps cool by transferring energy quickly to its surroundings.

An arctic fox has little ears so it transfers energy more slowly to its surroundings. This helps keep it warm.

a

b

Figure 5 *Fox ears* **a** *A desert fox* **b** *An arctic fox*

Key points

- The rate of energy transferred to or from an object depends on:
  - the shape, size, and type of material of the object
  - the nature of the surface that the object is in contact with
  - the temperature difference between the object and its surroundings.

13.7 Expansion by heating

After this topic, you should know:

- how solids, liquids, and gases expand when heated
- some applications and hazards of expansion by heating.

Comparing the expansion by heating of liquids and gases

Most substances expand when heated. Hot air expands and makes a hot-air balloon fill up and take-off. A burner under the balloon causes the balloon to fill with hot air, which then lifts the balloon. The expansion of a substance due to increasing its temperature is called **expansion by heating**.

Figure 1 *Filling a hot-air balloon*

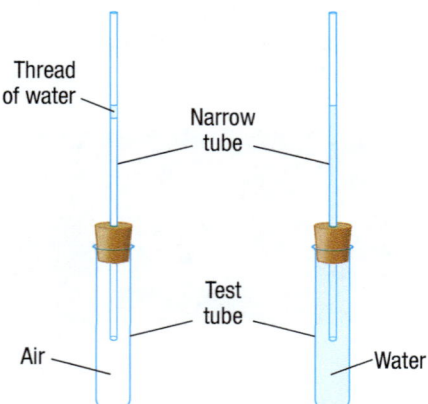

Figure 2 *Comparing the expansion by heating of air and water*

Gases expand much more than solids or liquids. Usually, the volume of a solid increases by no more than about 0.01% for a temperature increase of 1 °C. Most liquids expand slightly more. Gases at constant pressure expand on heating about 30 times more than liquids and solids do.

Figure 2 shows how you can compare the expansion by heating of air with that of water. The air in the air-filled test tube is trapped by a thread of water in the narrow tube which is open at both ends. In each case, expansion of the air or water in the test tube causes the level of the water in the narrow tube to move towards the top end of the narrow tube. To ensure both test tubes are heated exactly the same amount, they could be placed next to each other in strong sunlight, or in the same beaker of warm water. The results show that air expands much more than water for the same temperature rise. The same result applies to a comparison between any gas and any liquid.

A **radiator thermostat** in a car makes use of the expansion of oil when it becomes warm. Figure 3 shows how it works. The valve stays closed until it has been warmed up by energy from the engine. The metal tube is forced out of the oil chamber as the oil becomes hotter and expands. The movement of the metal tube opens the valve, allowing the hot water to flow to the radiator.

A **liquid in glass thermometer** makes use of the expansion of a liquid such as alcohol or mercury. When the thermometer bulb becomes warm, the liquid in the bulb expands into the long narrow tube joined to the bulb. The temperature reading is given by the position on the thermometer scale of the end of the thread of liquid in the tube.

Study tip

Make sure you know some examples of expansion by heating being a nuisance, and some examples of expansion by heating being useful.

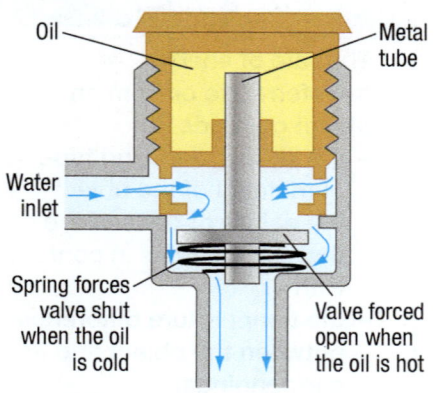

Figure 3 *A radiator thermostat*

Expansion of solids by heating

A jar with a tight-fitting lid can be difficult to open. Warming the lid in warm water may help to open the jar. The increase of temperature of the lid causes it to expand just enough to unscrew the lid. As with liquids, the expansion of a solid by heating is much less than that of a gas. Even so, the expansion of a solid by heating has some interesting applications and consequences.

Expansion gaps are needed in buildings, bridges, and railway tracks to allow for expansion by heating. Outdoor temperatures can change by as much as 50 °C between summer and winter. A 100 m concrete bridge span would expand by about 5 cm if its temperature increased by 50 °C. Without expansion gaps, the bridge would buckle. The gaps are usually filled with soft material such as rubber to prevent any debris falling in.

Steel tyres are fitted on train wheels by heating the tyre so it expands when fitting it on the wheel. As the tyre cools, it contracts so it fits very tightly on the wheel.

Bimetallic strips are used as thermostats to operate safety cut-out switches or valves. A bimetallic switch consists of a strip of two different metals such as brass and steel bonded together. When the temperature of the strip rises, one metal expands more than the other (e.g., brass expands more than steel) so the strip bends. This can be used to switch on or off an electrical device. Figure 5 shows a bimetallic strip in a fire alarm. When heated, the strip bends towards the contact screw, and when it touches it, the circuit is completed and the bell rings.

In a heater thermostat, when the strip becomes hot, it bends away from the contact screw and loses contact with it. As a result, the heater is switched off.

Figure 4 *Expansion gaps*

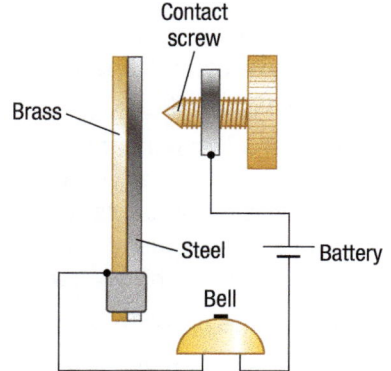

Figure 5 *A bimetallic strip in an alarm circuit*

Summary questions

1 The figure below shows a thermostat in a gas oven. When the oven overheats, the brass tube expands more than the Invar rod. Explain why this change reduces the flow of gas through the valve.

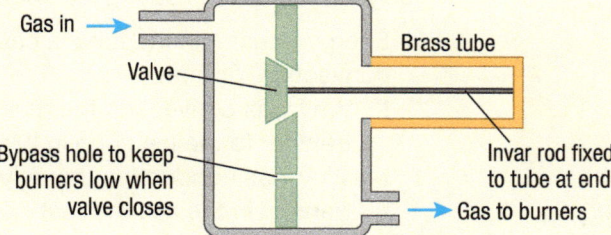

2 Explain why expansion gaps are:
   a needed between concrete sections in buildings
   b usually filled with soft material such as rubber.

3 a Explain why a bimetallic strip bends when it is heated.
   b A bimetallic thermostat is used to switch off the electricity supply to a heater if the heater overheats. With the aid of a diagram, describe such a thermostat and explain how it works.

Key points

- Gases expand on heating much more than solids and liquids.

- Applications of expansion by heating include glass thermometers and thermostats containing liquid.

- Expansion by heating must be allowed for in buildings and bridges by using expansion gaps.

Chapter summary questions

1 a i Why could seasonal variations in temperature cause a flat roof surface to develop cracks?

ii What type of surface is better for a flat roof – matt or smooth, dark or shiny? Explain your answer.

b A solar heating panel is used to heat water. Some panels have a transparent cover and a matt black base. Others have a matt black cover and a shiny base.

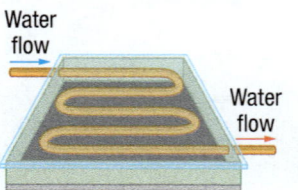

Panel X with a transparent cover and a matt black base

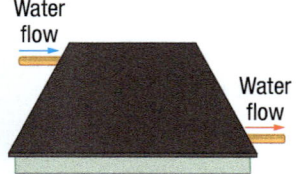

Panel Y with a matt black cover and a shiny base

Which of these two designs, Panel X or Panel Y, do you think is better? Give reasons for your answer.

2 A heat sink is a metal plate or clip fixed to an electronic component to stop it overheating.

a When the component becomes hot, how does energy transfer from:

i where it is in contact with the plate to the rest of the plate?

ii the plate to the surroundings?

b What is the purpose of the metal fins on the plate?

c Heat sinks are made from metals such as copper or aluminium. Copper is approximately three times as dense as aluminium and its specific heat capacity is about twice as large. Discuss how these physical properties are relevant to the choice of whether to use copper or aluminium for a heat sink in a computer.

3 a The figure shows a central heating radiator fixed to a wall and consisting of two parallel panels separated by an air gap.

i Describe how energy transfer by heating to the surroundings from hot water takes place when the hot water passes through the radiator pipes.

ii Explain the advantage of having two panels with an air gap between the panels.

b An electric storage heater contains bricks that become hot when the heater is switched on and cool down slowly when the heater is switched off. Explain in terms of conduction, convection, and radiation how the bricks store energy and release it to the surroundings.

4 a Explain why woolly clothing is very effective at keeping people warm in winter.

b Wearing a hat in winter is a very effective way of keeping your head warm. Describe how a hat helps to reduce energy loss from your head.

c Keeping your ears warm is important too. Explain why energy transfer from your ears to the surroundings can be significant in winter.

5 Marathon runners at the end of a race are often supplied with a shiny emergency blanket. These silvered blankets are very light in weight because they are made from plastic film with a reflective metallic coating inside.

a At the end of a race, the runners continue sweating. Explain why this continued sweating might cause them to become cold without an emergency blanket.

b i What form of energy transfer is reduced by the reflective coating inside the emergency blanket?

ii Explain why an emergency blanket helps to stop the runners becoming cold.

6 A glass tube containing water with a small ice cube floating at the top was heated at its lower end. The time taken for the ice cube to melt was measured. The test was repeated with a similar ice cube weighted down at the bottom of the tube of water. The water in this tube was heated near the top of the tube. The time taken for the weighted ice cube to melt was much longer than for the floating ice cube.

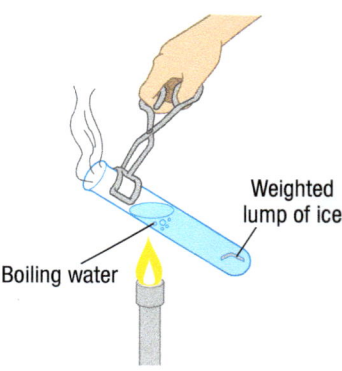

Weighted lump of ice

Boiling water

a Energy transfer in the tube is caused by conduction, convection, or both.

i Why was convection the main cause of energy transfer to the ice cube in the first test?

ii Why was conduction the only cause of energy transfer in the second test?

b Which of the following conclusions about these tests is true?

1 Energy transfer caused by conduction does not take place in water.

2 Energy transfer in water is mainly caused by convection.

3 Energy transfer in water is mainly caused by conduction.

Practice questions

1 a The diagram shows apparatus used to find out which of four materials is the best conductor of heat.

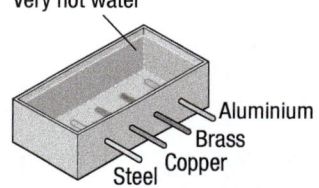

Very hot water

Aluminium
Brass
Copper
Steel

i Which two variables listed should be controlled? (2)

colour of rods
diameter of rods
length of rods
material of rods

ii Describe how the experiment can tell you which of the four materials is the best conductor of heat. (2)

b Water in a house is often heated by an electric immersion heater.

The diagram shows a metal tank containing water and a heating element.

When the heating element is switched on, the water near the element becomes hot.

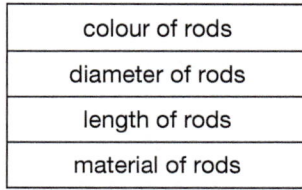

Water in tank

Metal wall of tank

Heating element

i Explain the process that transfers energy through the whole tank of water. (6)

ii Energy is lost to the surroundings by transfer through the metal wall of the tank.
Explain the process that transfers energy through the metal wall of the tank. (5)

c Metal pipes expand when they are heated. This can cause problems in a central heating system.

i Give another example of where expansion by heating can be a problem, and state how the problem can be overcome. (2)

ii Expansion by heating can also be useful.
The diagram shows a bimetallic strip in a fire alarm.

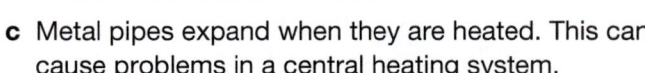
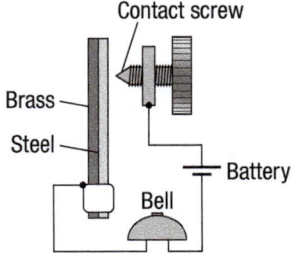

Contact screw

Brass
Steel
Battery
Bell

Explain how the fire alarm works. (6)

2 The diagram shows an outdoor swimming pool. Water is heated in pipes in a solar panel. The heated water is pumped into the pool.

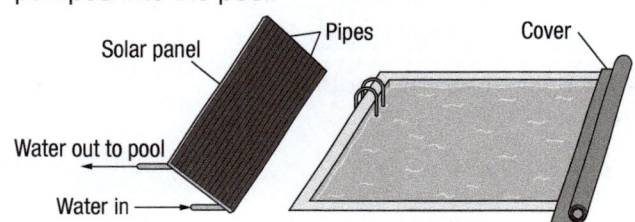

Solar panel
Pipes
Cover
Water out to pool
Water in

The pipes in the solar panel are painted black. The panel underneath the pipes is shiny.

Copy and complete **a** and **b** using words from the list:
absorbers   emitters   insulators   reflectors

a The pipes are painted black because black surfaces are good .... of radiation. (1)

b The panel underneath the pipes is shiny because shiny surfaces are good .... of radiation. (1)

c What other feature of the solar panels can ensure that they absorb as much energy from the Sun as possible? (1)

3 Hot water in a mug cools by transferring energy to its surroundings.

a i Which of the following factors affect the rate at which hot water transfers energy to its surroundings? (4)

Choose **four** of the factors.

nature of the surface of the mug
specific latent heat of ice
specific latent heat of steam
surface area of the mug
temperature of the water
temperature of surroundings

ii State **two** other factors that control the rate at which the hot water's temperature falls. (2)

b A man spilled some petrol on his hand whilst refuelling his car. He noticed that the petrol soon disappeared from his hand, and he also noticed that his hand felt cold.
Explain why. (4)

c The diagram shows a washing basket of wet clothes. The wet clothes dry quicker when they are hung on a washing line outside on a sunny day rather than a line inside.

Wet clothes
Basket

i Explain why. (3)

ii Describe how the particles of water remaining on the wet clothes differ in behaviour from the particles leaving the wet clothes on the washing line. (3)

14.1 Electrical charges

Learning objectives

After this topic, you should know:

- what happens when insulating materials are rubbed together

- what is transferred when objects become charged

- what happens when charged objects are brought together.

Have you ever stuck a balloon on a ceiling? All you need to do is to rub the balloon on your clothing and touch it to the ceiling. The rubbing action charges the balloon with **static electricity**. In other words, the balloon becomes electrically charged. The charge on the balloon attracts it to the ceiling.

Practical

The Van de Graaff generator

A Van de Graaff generator can make your hair stand on end. The dome charges up when the generator is switched on. Massive sparks are produced if the charge on the dome builds up too much.

 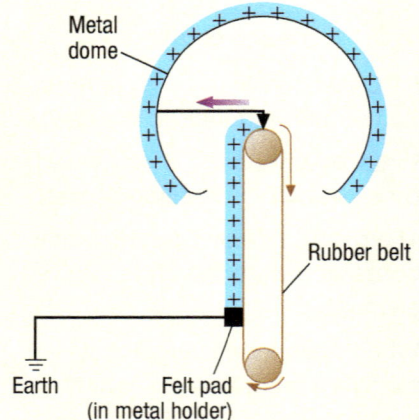

Figure 1 *The Van de Graaff generator*

The Van de Graaff generator charges up because:
- the belt rubs against a felt pad and becomes charged
- the belt carries the charge onto an insulated metal dome
- sparks are produced when the dome can no longer hold any more charge.
- Why should you keep away from a Van de Graaff generator?

Safety: Do not take part in this experiment if you have a heart condition. All electronic equipment should be switched off and kept well away from the generator.

Inside the atom

Protons and **neutrons** make up the nucleus of the atom. Electrons move around in the space surrounding the nucleus.

- A proton has a positive charge.
- An electron has an equal negative charge.
- A neutron is uncharged.

An uncharged atom has equal numbers of electrons and protons. Electrons can be transferred to or from an atom. A charged atom is called an **ion**.

1 Adding electrons to an uncharged atom makes it negative (because the atom then has more electrons than protons).

2 Removing electrons from an uncharged atom makes it positive (because the atom has fewer electrons than protons).

Did you know ... ?

Take off a woolly jumper and listen out! You can hear it crackle as tiny sparks from static electricity are created. If the room is dark, you can even see the sparks.

You can get charged up just by sitting in a plastic chair. If this happens, you may feel a slight shock from static electricity when you stand up.

Study tip

Objects become electrically charged by gaining electrons (so becoming more negative) or losing electrons (so becoming more positive).

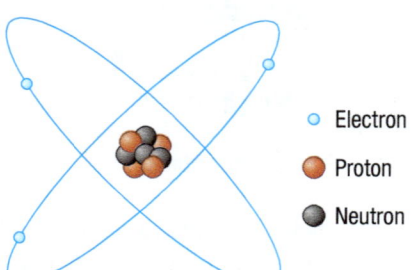

- Electron
- Proton
- Neutron

Figure 2 *Inside an atom*

Charging by friction

Substances that can't conduct electricity are called **insulators**. Some insulators become charged by rubbing them with a dry cloth.

- Rubbing a polythene rod with a dry cloth transfers electrons to the surface atoms of the rod from the cloth. So the polythene rod becomes negatively charged.
- Rubbing a perspex rod with a dry cloth transfers electrons from the surface atoms of the rod to the cloth. So the perspex rod becomes positively charged.

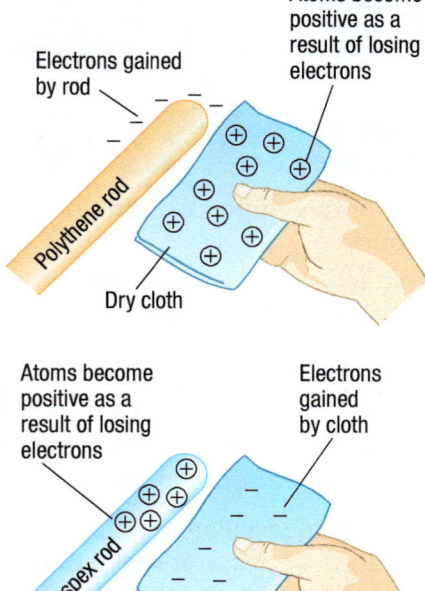

Figure 3 *Charging by friction*

Practical

The force between two charged objects

Two charged objects exert a force on each other. Figure 4 shows how you can investigate this force.

- What happens?

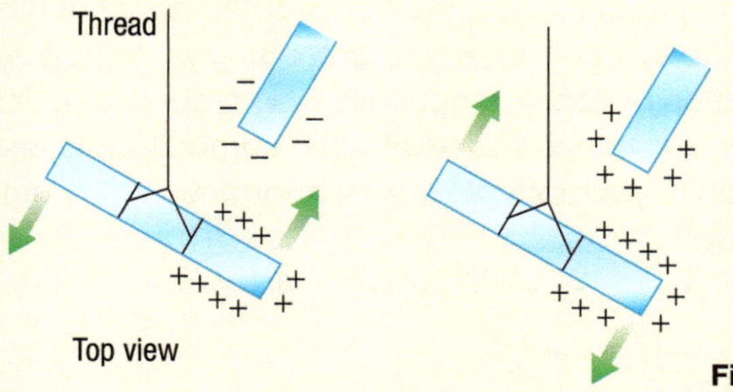

Figure 4 *The law of force for charges*

Your results in the experiment above should show that:

- two objects with the same type of charge (i.e., like charges) repel each other
- two objects with different types of charge (i.e., unlike charges) attract each other.

Like charges repel. Unlike charges attract.

Summary questions

1 a In terms of electrons, explain why:
   i   a polythene rod becomes negatively charged when rubbed with a dry cloth
   ii a perspex rod becomes positively charged when rubbed with a dry cloth.
   b Glass is charged positively when it is rubbed with a cloth. Does glass gain or lose electrons when it is charged?

2 When rubbed with a dry cloth, perspex becomes positively charged. Polythene and ebonite become negatively charged. State whether or not attraction or repulsion takes place when:
   a a charged perspex rod is held near a charged polythene rod
   b a charged perspex rod is held near a charged ebonite rod
   c a charged polythene rod is held near a charged ebonite rod.

3 a After two rods X and Y made of different insulating materials are charged, they are found to repel each other. What does this tell you about the charge on the two rods?
   b Given a rod R that is known to charge positively, how would you determine the type of charge on X and on Y?

Key points

- **Certain insulating materials become charged when rubbed together.**

- **Electrons are transferred when objects become charged:**
  - **Insulating materials that become positively charged when rubbed lose electrons.**
  - **Insulating materials that become negatively charged when rubbed gain electrons.**

- **Like charges repel, unlike charges attract.**

14.2 Electric circuits

Learning objectives

After this topic, you should know:

- how electric circuits are shown as diagrams

- the difference between a battery and a cell

- what determines the size of an electric current

- how to calculate the size of an electric current from the charge flow and the time taken.

An electric torch can be very useful in a power cut at night. But it needs to be checked to make sure it works. Figure 1 shows what is inside a torch. The figure shows how the torch lamp is connected to the switch and the two cells.

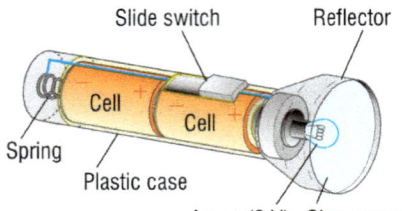

Figure 1 *An electric torch*

A circuit diagram shows how the components in a circuit are connected together without having to draw the actual components. Each component has its own symbol. Figure 2 shows the symbols for some of the components you will meet in this course. The function of each component is also described. You need to recognise these symbols and remember what each component is used for – otherwise you'll get mixed up in your exams. More importantly, you could get a big shock if you mix them up!

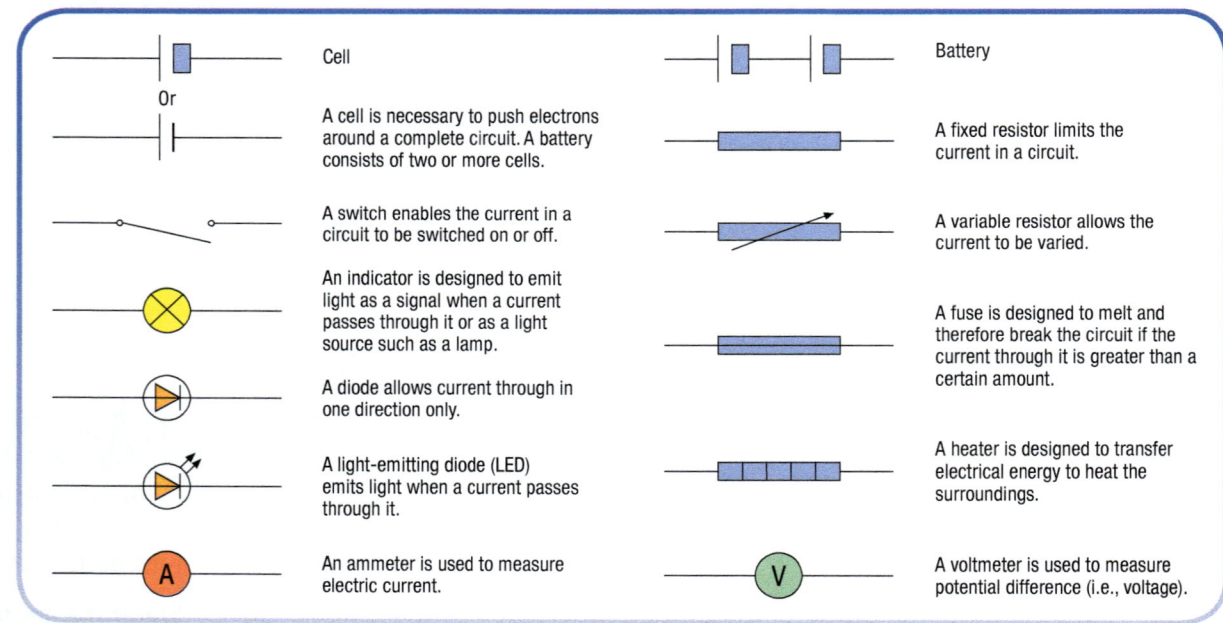

Cell	A cell is necessary to push electrons around a complete circuit. A battery consists of two or more cells.
Battery	A fixed resistor limits the current in a circuit.
A switch enables the current in a circuit to be switched on or off.	A variable resistor allows the current to be varied.
An indicator is designed to emit light as a signal when a current passes through it or as a light source such as a lamp.	A fuse is designed to melt and therefore break the circuit if the current through it is greater than a certain amount.
A diode allows current through in one direction only.	
A light-emitting diode (LED) emits light when a current passes through it.	A heater is designed to transfer electrical energy to heat the surroundings.
An ammeter is used to measure electric current.	A voltmeter is used to measure potential difference (i.e., voltage).

Figure 2 *Components and symbols*

Electric current

An electric current is a flow of charge. When an electric torch is on, millions of **electrons** pass through the torch lamp and through the cell every second. Each electron carries a negative charge. Metals contain lots of electrons that move about freely between the positively charged metal ions. These electrons stop the ions moving away from each other. The electrons pass through the lamp because its filament is made of a metal. The electrons transfer energy from the cell to the torch lamp.

The size of an electric current is the rate of flow of electric charge. This is the flow of charge per second. The greater the number of electrons that pass through a component, the bigger the current passing through it.

Study tip

Direction matters

The direction of the current in a circuit is marked from + to – round the circuit. This convention was agreed long before electrons were discovered.

Electric charge is measured in **coulombs (C)**. Electric current is measured in **amperes (A)**, sometimes abbreviated as amps.

An electric current of 1 ampere is a rate of flow of charge of 1 coulomb per second. If a certain amount of charge flows steadily through a wire or a component in a certain time:

$$\textbf{current (amperes, A)} = \frac{\textbf{charge flow (coulombs, C)}}{\textbf{time taken (seconds, s)}}$$

You can write the equation above using symbols as:

$$I = \frac{Q}{t}$$

where:
I = current in A
Q = charge in C
t = time taken in s.

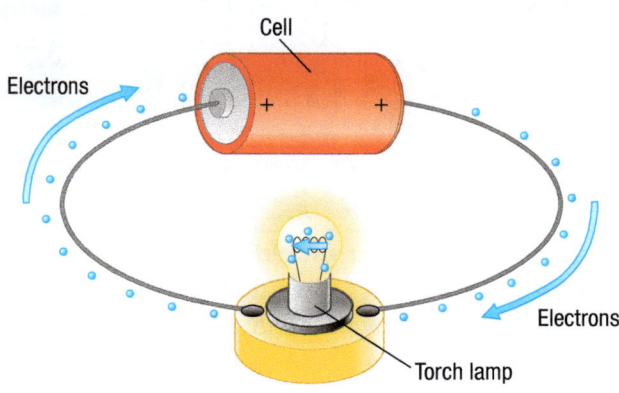

Figure 3 *Electrons on the move*

Worked example

A charge of 8.0 C passes through a lamp in 4.0 seconds. Calculate the current through the lamp.

Solution

$$I = \frac{Q}{t} = \frac{8.0\,C}{4.0\,s} = \textbf{2.0 A}$$

Practical

Circuit tests

Connect a variable resistor in series with a torch lamp and a cell, as shown in Figure 5.

Adjust the slider of the variable resistor. This alters the amount of current flowing through the lamp and therefore affects its brightness.

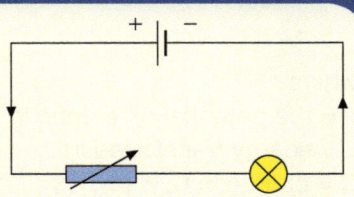

Figure 5 *Using a variable resistor*

- In Figure 5, the torch lamp goes dim when the slider is moved one way. What happens if the slider is moved back again?
- What happens if you include a diode in the circuit?

??? Did you know ... ?

You would damage a portable radio if you put the batteries in the wrong way round. A diode in the battery circuit prevents this damage. The diode allows current through only when it is connected as shown in Figure 4.

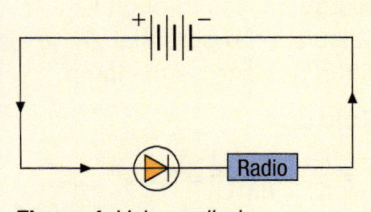

Figure 4 *Using a diode*

Summary questions

1 Name the numbered components in the circuit diagram in the figure opposite.

2 **a** Redraw the circuit diagram in Question 1 with a diode in place of the switch so it allows current through.
   b What further component would you need in this circuit to change the current in it?
   c When the switch is closed in the figure for Question 1, a current of 0.25 A passes through the lamp. Calculate the charge that passes through the lamp in 60 seconds.

3 **a** What is a light-emitting diode?
   b What is a variable resistor used for?

Key points

- Every component has its own agreed symbol. A circuit diagram shows how components are connected together.

- A battery consists of two or more cells connected together.

- The size of an electric current is the rate of flow of charge.

- Electric current = $\dfrac{\text{charge flow}}{\text{time taken}}$

14.3 Potential difference and resistance

Potential difference

Look at the circuit in Figure 1. The battery forces electrons to pass through the ammeter and the torch lamp.

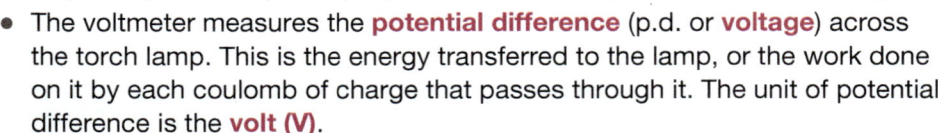

Figure 1 *Using an ammeter and a voltmeter*

- The ammeter measures the current through the torch lamp. It is connected in **series** with the lamp so the current through them is the same. The ammeter reading gives the current in amperes (or milliamperes (mA) for small currents, where 1 mA = 0.001 A).

- The voltmeter measures the **potential difference** (p.d. or **voltage**) across the torch lamp. This is the energy transferred to the lamp, or the work done on it by each coulomb of charge that passes through it. The unit of potential difference is the **volt (V)**.

- The voltmeter is connected in **parallel** with the torch lamp so it measures the potential difference across it. The voltmeter reading gives the potential difference in volts (V).

When charge flows steadily through a component:

$$\textbf{potential difference across the component (volts, V)} = \frac{\textbf{energy transferred (joules, J)}}{\textbf{charge (coulombs, C)}}$$

You can write the equation above using symbols as:

$$V = \frac{E}{Q}$$

where:
V = the potential difference in V
E = energy transferred in J
Q = charge in C.

Worked example

The energy transferred to a lamp is 24.0 J when 8.0 C of charge passes through it. Calculate the potential difference across the lamp.

Solution

$V = \dfrac{E}{Q} = \dfrac{24.0\,\text{J}}{8.0\,\text{C}} = \textbf{3.0 V}$

Resistance

Electrons passing through a torch lamp have to push their way through lots of vibrating ions in the metal filament. The ions resist the passage of electrons through the torch lamp.

The **resistance** of an electrical component is defined as:

$$\textbf{resistance (ohms, } \Omega\textbf{)} = \frac{\textbf{potential difference (volts, V)}}{\textbf{current (amperes, A)}}$$

The unit of resistance is the **ohm**. The symbol for the ohm is the Greek capital letter Ω (omega). Note that a resistor in a circuit limits the current. For a given potential difference, the larger the resistance of a resistor, the smaller the current is.

You can write the definition above as:

$$R = \frac{V}{I}$$

where:
R = resistance in Ω
V = potential difference in V
I = current in A.

Maths skills

Rearranging the equation

$R = \dfrac{V}{I}$ gives $V = I \times R$ or $I = \dfrac{V}{R}$

The last equation shows that for a fixed potential difference across a component, the greater the resistance, the smaller the current.

The equation tells us that the current I is **inversely proportional** to the resistance R. For example, if R was ten times greater, I would be ten times smaller. See the 'Using data' section for more about inverse proportionality.

Study tip

Make sure you can rearrange the equation $V = I \times R$.

Practical

Investigating the resistance of a wire

Does the resistance of a wire change when the current through it is changed? Figure 2 shows how you can use a variable resistor to change the current through a wire. Make your own measurements and use them to plot a current–potential difference graph like the one in Figure 2.

- Discuss how your measurements compare with the ones in Table 1.
- Calculate the resistance of the wire you tested.

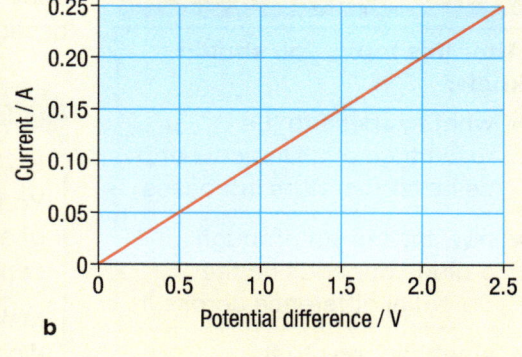

Table 1 *These results are plotted in Figure 2*

Current (A)	0	0.05	0.10	0.15	0.20	0.25
Potential difference (V)	0	0.50	1.00	1.50	2.00	2.50

Figure 2 *Investigating the resistance of a wire.* **a** *Circuit diagram* **b** *A current–potential difference graph for a wire*

Current–potential difference graphs

The graph in Figure 2 is a straight line through the origin. This means that the current is directly proportional to the potential difference. In other words, the resistance (= potential difference ÷ current) is constant. This was first discovered for a wire at constant temperature by Georg Ohm and is known as **Ohm's law**:

The current through a resistor at constant temperature is directly proportional to the potential difference across the resistor.

A wire is known as an **ohmic conductor** because its resistance is constant. As shown in Figure 3, reversing the potential difference makes no difference to the shape of the line. The resistance is the same whichever direction the current is in.

Note: The gradient of the line depends on the resistance of the resistor. The greater the resistance of the resistor, the less steep the line.

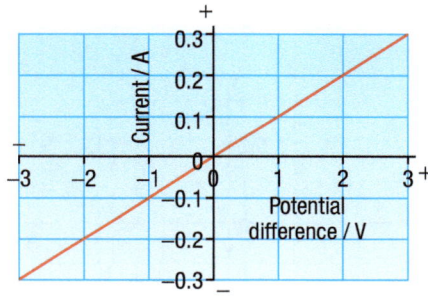

Figure 3 *A current–potential difference graph for a resistor*

Summary questions

1 a The current through a wire is 0.5 A when the potential difference across it is 4.0 V. Calculate the resistance of the wire.
b Calculate the resistance of the wire that gave the results in the graph in Figure 2.

2 Copy and complete the table. Calculate each missing value using the equation $V = I \times R$ or a rearrangement of it.

Resistor	Current (A)	Potential difference (V)	Resistance (Ω)
W	2.0	12.0	
X	4.0		20
Y		6.0	3.0

3 A torch lamp lights normally when the current through it is 0.060 A and the potential difference across it is 3.0 V.
a Calculate the resistance of the torch lamp when it lights normally.
b When the torch lamp lights normally, calculate:
  i the charge passing through the torch lamp in 300 s
  ii the energy delivered to the torch lamp in this time.

Key points

- The potential difference across a component (V)
  $$= \frac{\text{energy transferred (J)}}{\text{charge (C)}}$$

- Resistance (Ω)
  $$= \frac{\text{potential difference (V)}}{\text{current (A)}}$$

- Ohm's law states that the current through a resistor at constant temperature is directly proportional to the potential difference across the resistor.

- Reversing the current through a component reverses the potential difference across it.

14.4 Component characteristics

Learning objectives

After this topic, you should know:

- what happens to the resistance of a filament lamp as its temperature increases

- how the current through a diode depends on the potential difference across it

- what happens to the resistance of a thermistor as its temperature increases and of an LDR as the light level increases.

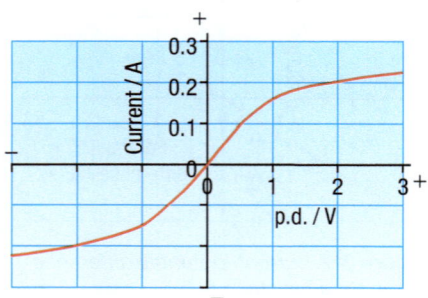

Figure 1 *A current–potential difference graph for a filament lamp*

Study tip

Try to learn the current–potential difference graphs for a resistor, a filament lamp, and a diode.

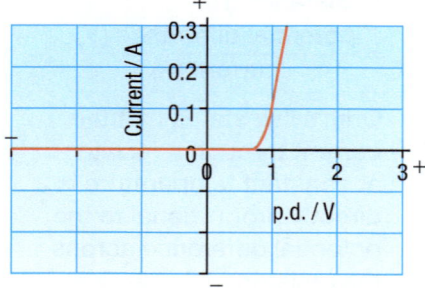

Figure 2 *A current–potential difference graph for a diode*

Have you ever switched a light lamp on only to hear it 'pop' and fail? Electrical appliances can fail at very inconvenient times. Most electrical failures happen because too much current passes through a component in the appliance.

Practical

Investigating different components

You can use the circuit in Figure 2 in Topic 14.3 to find out if the resistance of a component depends on the current. You can also see if reversing the component in the circuit has any effect.

Make your own measurements using a resistor, a filament lamp, and a diode.

Plot your measurements on a current–potential difference graph. Plot the 'reverse' measurements on the negative section of each axis.

Using current–potential difference graphs

A filament lamp

Figure 1 shows the graph for a torch lamp (i.e., a low-voltage filament lamp).
- The line **curves** away from the current axis. So, the current is *not* directly proportional to the potential difference. The filament lamp is a non-ohmic conductor.
- **The resistance (= potential difference ÷ current) increases as the current increases.** So, the resistance of a filament lamp increases as the filament temperature increases. This is because the ions in the metal filament vibrate more as the temperature increases. So they resist the passage of the electrons through the filament more. The resistance of any metal increases as its temperature increases.
- Reversing the potential difference makes no difference to the shape of the curve. The resistance is the same for the same current, regardless of its direction.

The diode

Figure 2 shows a graph for a diode.
- In the 'forward' direction, the line curves towards the current axis. So the current is not directly proportional to the potential difference. A **diode** is a non-ohmic conductor.
- In the reverse direction, the current is virtually zero. So the diode's resistance in the reverse direction is much higher than in the forward direction.

Note that a **light-emitting diode (LED)** emits light when a current passes through it in the forward direction.

Practical

Thermistors and light-dependent resistors

Thermistors and light-dependent resistors are used in sensor circuits. A **thermistor** is a temperature-dependent resistor. The resistance of a **light-dependent resistor (LDR)** depends on how much light shines on it. For example:

- a thermistor may be used in a thermostat used to control temperature
- an LDR is used in a sensor circuit that switches an electric light on and off.

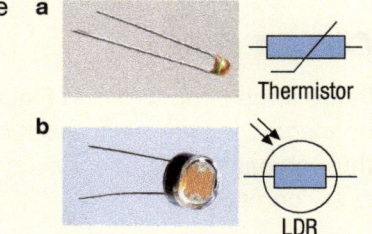

Figure 3 a *A thermistor and its circuit symbol* **b** *An LDR and its circuit symbol*

Test a thermistor and then an LDR in series with a battery and an ammeter.

- What did you find out about each component tested?

Current–potential difference graphs for a thermistor and an LDR

For a thermistor, Figure 4 shows the current–potential difference graph at two different temperatures.

- At constant temperature, the line is straight so its resistance is constant.
- If the temperature is increased, its resistance decreases.

For a light-dependent resistor, Figure 5 shows the current–potential difference graph in bright light and in dim light. If the brightness of the light is increased, the resistance of the LDR decreases.

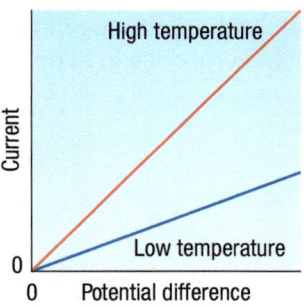

Figure 4 *Thermistor graph*

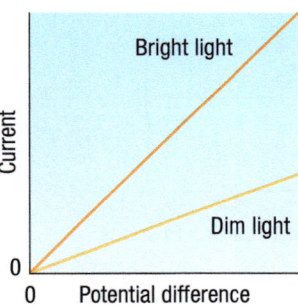

Figure 5 *LDR graph*

Summary questions

1 a Identify the type of component that has a resistance that:
  i decreases as its temperature increases
  ii depends on which way round it is connected in a circuit
  iii increases as the current through it increases.
  b Calculate the resistance of the filament lamp that gave the graph in Figure 1 at a current of:
  i 0.1 A   **ii** 0.2 A.

2 A thermistor is connected in series with an ammeter and a 3.0 V battery, as shown in the figure opposite.
  a At 15 °C, the current through the thermistor is 0.2 A and the potential difference across it is 3.0 V. Calculate its resistance at this temperature.
  b State and explain what happens to the ammeter reading if the thermistor's temperature is increased.

3 a The thermistor in the figure for Question **2** is replaced by a light-dependent resistor (LDR). State and explain what happens to the ammeter reading when the LDR is covered.
  b For the diode that gave the graph in Figure 2, describe how:
  i the current
  ii the resistance
  changes when the potential difference is increased steadily from zero.

Did you know ... ?

When a filament light lamp fails, it usually happens when you switch it on. Because the resistance is low when the lamp is off, a large current passes through it when you switch it on. If the current is too large, it burns the filament out.

Key points

- The resistance of a filament lamp increases as the filament temperature increases.
- For a diode, the 'forward' resistance is low, and the 'reverse' resistance is high.
- The resistance of a thermistor decreases as its temperature increases.
- The resistance of an LDR decreases as the light intensity on it increases.

159

14.5 Series circuits

Learning objectives

After this topic, you should know:

- about the current, potential difference, and resistance for each component in a series circuit
- about the potential difference of several cells in series.

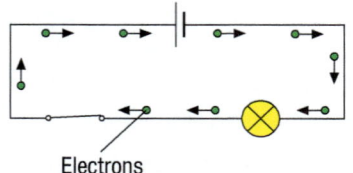

Electrons

Figure 1 *A torch lamp circuit*

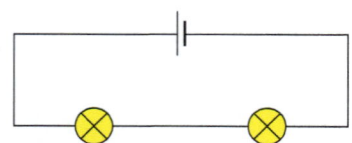

Figure 2 *Lamps in series*

Table 1 *Results of voltage tests*

Filament lamp	Voltmeter V_1 (volts)	Voltmeter V_2 (volts)
Normal	1.5	0.0
Dim	0.9	0.6
Very dim	0.5	1.0

Circuit rules

In the torch lamp circuit in Figure 1, the lamp, the cell, and the switch are connected in series with each other. The same number of electrons passes through each component every second. So the same current passes through every component.

> **In a series circuit, the same current passes through each component.**

In Figure 2, each electron from the cell passes through two lamps. The electrons are pushed through each lamp by the cell. The **potential difference** (or **voltage**) of a cell or any other source of potential difference is a measure of the energy transferred from the source by each electron that passes through it. Since each electron in the circuit in Figure 2 goes through both lamps, the potential difference of the cell is shared between the lamps. This rule applies to any series circuit.

> **In a series circuit, the total potential difference of the voltage supply is shared between the components.**

Cells in series

What happens if you use two or more cells in series in a circuit? As long as you connect the cells so they act in the same direction, each electron gets a push from each cell. So an electron would get the same push from a battery of three 1.5 V cells in series as it would from a single 4.5 V cell.

In other words, **as long as the cells act in the same direction**:

> **The total potential difference of cells in series is the sum of the potential difference of each cell.**

Required Practical

Investigating potential differences in a series circuit

Figure 3 shows how to test the potential difference rule for a series circuit. The circuit consists of a filament lamp in series with a variable resistor and a cell. You can use the variable resistor to see how the voltmeter readings change when you change the current. Make your own measurements.

- How do they compare with the data in Table 1?

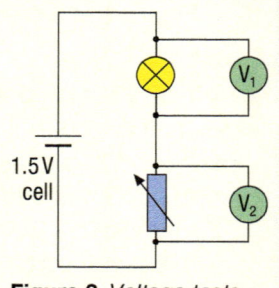

Figure 3 *Voltage tests*

The results in Table 1 show that the voltmeter readings for each setting of the variable resistor add up to 1.5 V. This is the potential difference of the cell. The share of the cell's potential difference across each component depends on the setting of the variable resistor.

The resistance rule for components in series

In Figure 3, suppose the current through the lamp is 0.1 A when the lamp is dim.

Using data from Table 1:

- the resistance of the lamp would then be $9\,\Omega$ (= 0.9 V ÷ 0.1 A)
- the resistance of the variable resistor at this setting would be $6\,\Omega$ (= 0.6 V ÷ 0.1 A).

If you replaced the lamp and the variable resistor by a single resistor, what should its resistance be for the same current of 0.1 A? You can calculate this because you know the potential difference across it would be 1.5 V (from the cell). So the resistance would need to be $15\,\Omega$ (= 1.5 V ÷ 0.1 A). This is the sum of the resistance of the two components. The rule applies to any series circuit.

The total resistance of components in series is equal to the sum of the resistance of each component.

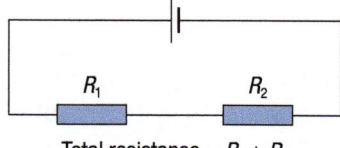

Total resistance = $R_1 + R_2$

Figure 4 *Resistors in series*

Summary questions

1. **a** In Figure 2, if the potential difference of the cell is 1.2 V and the potential difference across one lamp is 0.8 V, what is the potential difference across the other lamp?

   b In Figure 3, the lamp lights normally when the resistance of the variable resistor is $5.0\,\Omega$ and the potential difference across the variable resistor is 1.0 V. Calculate the current through the lamp and the potential difference across it.

2. A 1.5 V cell is connected to a $3.0\,\Omega$ resistor and $2.0\,\Omega$ resistor in series with each other.

   a Draw the circuit diagram for this arrangement.

   b Calculate:

   i the total resistance of the two resistors

   ii the current through the resistors.

3. For the circuit in the figure below, each cell has a potential difference of 1.5 V.

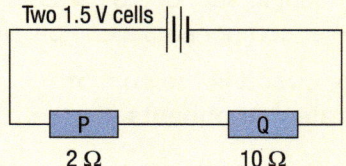

Two 1.5 V cells

P   Q

$2\,\Omega$ $10\,\Omega$

   a Calculate:

   i the total resistance of the two resistors

   ii the total potential difference of the two cells.

   b Show that the current through the battery is 0.25 A.

   c Calculate the potential difference across each resistor.

   d If a $3\,\Omega$ resistor R is connected in series between the two resistors, calculate:

   i their total resistance

   ii the current through the resistors

   iii the potential difference across each resistor.

14.6 Parallel circuits

??? Did you know ...?

A bypass is a parallel route. A heart bypass is another route for the flow of blood. A road bypass is a road that passes a town centre instead of going through it. For components in parallel, charge flows separately through each component. The total flow of charge is the sum of the flow through each component.

Study tip

Remember that when components are connected in parallel, there is the same potential difference across each component.

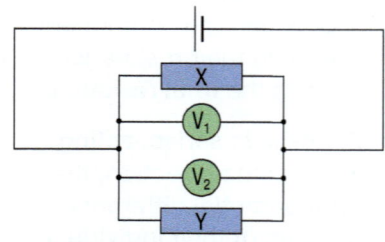

Figure 2 *Components in parallel*

Practical

Investigating parallel circuits

Figure 1 shows how you can investigate the current through two lamps in parallel with each other. You can use ammeters in series with the lamps and the cell to measure the current through each component.

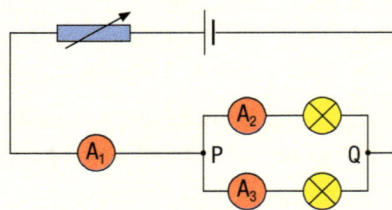

Figure 1 *At a junction*

Set up your own circuit. Adjust the variable resistor and collect your data.

- How do your measurements compare with the ones for different settings of the variable resistor shown in Table 1?
- Discuss whether your own measurements show the same pattern.

Look at Table 1.

Table 1 *Some results of investigating parallel circuits*

Ammeter A_1 (A)	Ammeter A_2 (A)	Ammeter A_3 (A)
0.50	0.30	0.20
0.30	0.20	0.10
0.18	0.12	0.06

In each case, the reading of ammeter A_1 is equal to the sum of the readings of ammeters A_2 and A_3.

This shows that the current through the cell is equal to sum of the currents through the two lamps. This rule applies wherever components are in parallel.

The total current through the whole circuit is the sum of the currents through the separate components.

Potential difference in a parallel circuit

Figure 2 shows two resistors X and Y in parallel with each other. A voltmeter is connected across each resistor. The voltmeter across resistor X shows the same reading as the voltmeter across resistor Y. This is because each electron from the cell passes either through X or through Y. So it delivers the same amount of energy from the cell, whichever resistor it goes through. In other words:

For components in parallel, the potential difference across each component is the same.

Calculations on parallel circuits

Components in parallel have the same potential difference across them. The current through each component depends on the resistance of the component.

- The bigger the resistance of the component, the smaller the current through it. The resistor that has the largest resistance passes the smallest current.
- You can calculate the current using the equation:

$$\text{current (A)} = \frac{\text{potential difference (V)}}{\text{resistance } (\Omega)}$$

Worked example

The circuit diagram in Figure 3 shows three resistors, $R_1 = 1\,\Omega$, $R_2 = 2\,\Omega$, and $R_3 = 6\,\Omega$, connected in parallel to a 6 V battery.

Calculate:
a the current through each resistor
b the current through the battery.

Solution

a $I_1 = \dfrac{V_1}{R_1} = \dfrac{6\,V}{1\,\Omega} = \mathbf{6\,A}$

$I_2 = \dfrac{V_2}{R_2} = \dfrac{6\,V}{2\,\Omega} = \mathbf{3\,A}$

$I_3 = \dfrac{V_3}{R_3} = \dfrac{6\,V}{6\,\Omega} = \mathbf{1\,A}$

b The total current through the battery $= I_1 + I_2 + I_3 = 6\,A + 3\,A + 1\,A = \mathbf{10\,A}$.

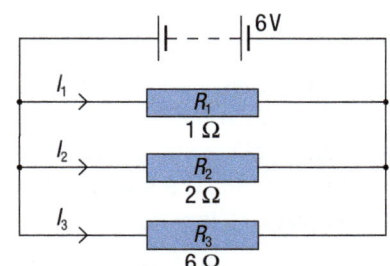

Figure 3 *A parallel circuit*

Summary questions

1 a In Table 1, if ammeter A_1 reads 0.40 A and A_2 reads 0.1 A, what would A_3 read?

b A 3 Ω resistor and a 6 Ω resistor are connected in parallel in a circuit. Which resistor passes more current?

c In the circuit shown in Figure 3, what would be the resistance of a single resistor that could replace the three parallel resistors across the 6 V battery and allow the same current to pass through the battery?

2 A 1.5 V cell is connected across a 3 Ω resistor in parallel with a 6 Ω resistor.

a Draw the circuit diagram for this circuit.

b Show that the current through:

   i the 3 Ω resistor is 0.50 A   **ii** the 6 Ω resistor is 0.25 A.

c Calculate the current passing through the cell.

3 a The circuit diagram shows three resistors, $R_1 = 2\,\Omega$, $R_2 = 3\,\Omega$, and $R_3 = 6\,\Omega$, connected to each other in parallel and to a 6 V battery. Calculate:

   i the current through each resistor

   ii the current through the battery.

b If the 6 Ω resistor in the figure was replaced by a 4 Ω resistor, calculate the battery current now.

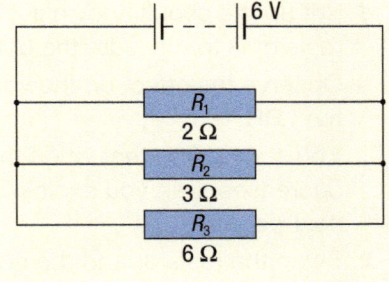

Key points

- For components in parallel:
  - the total current is the sum of the currents through the separate components
  - the potential difference across each component is the same.

- In parallel circuits, the bigger the resistance of a component, the smaller the current through it.

- To calculate the current through a resistor in a parallel circuit:

$$\text{current} = \frac{\text{potential difference}}{\text{resistance}}$$

14.7   Sensor circuits

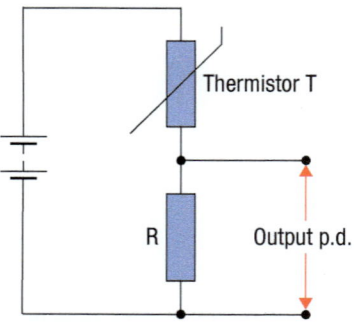

Figure 1 *Using a thermistor*

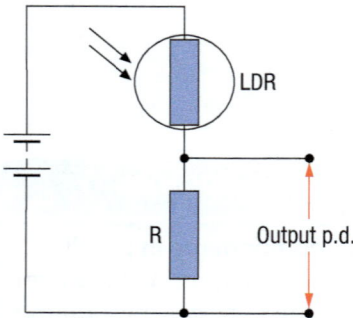

Figure 2 *Using an LDR*

Sensor circuits are designed to respond to a change in the surroundings, such as a change in temperature or light intensity. Many sensor circuits include a resistor such as a thermistor or an LDR that is sensitive to an external change.

A **sensor circuit** is made up of two or more resistors in series connected to a source of fixed p.d. as shown in Figure 1. The fixed p.d. is shared (divided) between the resistors. In the sensor circuit, one of the resistors is sensitive to a change in the surroundings – it could be a thermistor or a light-dependent resistor (LDR).

A temperature sensor and a light sensor

A temperature sensor includes a thermistor, which you learnt about in Topic 14.4. You can see the symbol for a thermistor in Figure 1.

Figure 1 shows a sensor circuit in which a thermistor T and a resistor R are connected in series to a battery. When the temperature of the thermistor increases, its resistance decreases, so its share of the battery p.d. decreases. Because of this, the share of the battery p.d. across resistor R increases, so the output p.d. increases.

A light sensor includes an LDR, which you also learnt about in Topic 11.4. You can see the symbol for an LDR in Figure 2.

Figure 2 shows a sensor circuit in which an LDR and a resistor R are connected to a battery. When the brightness of the incident light (i.e., the light intensity) on the LDR increases, its resistance decreases, so its share of the battery p.d. decreases. Because of this, the share of the battery p.d. across resistor R increases, so the output p.d. increases.

In both of these sensor circuits, the resistor R can be replaced with a variable resistor that is adjusted so that it produces a specific output p.d. when a specific temperature or light intensity is reached. The output p.d. would then:

- increase if the temperature or the light intensity increased
- decrease if the temperature or the light intensity decreased.

The circuits above could be used to switch a device (called the output device) such as buzzer or an LED (in series with a resistor).

Practical

Investigate a light sensor

1 Set up the circuit shown in Figure 2 and connect a voltmeter across the resistor R to measure the output p.d.

   Observe the effect on the voltmeter reading of covering and uncovering the LDR.

   You should find that when the LDR is covered, the voltmeter reading decreases. Can you explain why this happens? See Question 2 on the next page.

2 Swap the resistor and the LDR over in the circuit and repeat the test. Observe and explain what happens to the voltmeter reading this time when the LDR is covered.

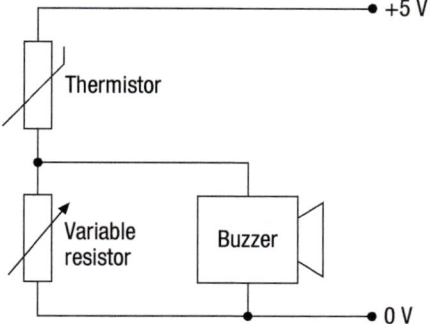

Figure 3 *A temperature-operated buzzer*

A temperature-operated alarm

Figure 3 shows a circuit that is used to switch a buzzer on if the temperature of a thermistor increases above a particular level. When the thermistor is cold, its resistance is high, so the output p.d. across the buzzer is too small to switch it on.

Suppose the variable resistor is adjusted so that the buzzer is only just switched off. If the thermistor temperature then increases, its resistance decreases, and the p.d. across the buzzer increases and switches the buzzer on.

By swapping over the thermistor and the resistor in series with it, the circuit can be made to switch a buzzer on when the thermistor temperature decreases.

The relay

A **relay** is an electrically operated switch used in electric circuits to switch machines on or off. It contains an electromagnet and a switch that is open when there is no current in the electromagnet. The switch closes when a small current, for example from the output of a sensor circuit, passes through the electromagnet. The electrical symbol for a relay is shown in Figure 4.

Study tip

By replacing the thermistor in Figure 3 with an LDR, the circuit could be used to switch a buzzer or an LED on when the LDR is exposed to bright light.

links

You will learn more about the relay in Topic 15.2 'Electromagnets'.

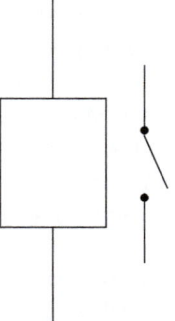

Figure 4 *The symbol for a relay*

Summary questions

1 A potential divider consisting of a 20 Ω resistor in series with a 30 Ω resistor is connected to a 5.0 V battery.
 a i Sketch the circuit diagram
  ii Explain why the p.d. across the 30 Ω resistor is greater than the p.d. across the other resistor.
 b Calculate the current through the battery and the p.d. across the 30 Ω resistor.

2 Describe and explain why, in Figure 2, the output p.d. decreases when the LDR is covered completely so that it is in darkness.

3 The thermistor and the resistor in Figure 1 are swapped over in the circuit so that the output p.d. is across the thermistor.
 a Draw the new circuit diagram.
 b Explain how the output p.d. in the new circuit changes when the temperature of the thermistor is reduced.

4 a Explain the function in Figure 3 of the variable resistor.
 b A lamp is used as the output device in Figure 3 instead of the buzzer.
  i Suggest what further changes would need to be made to switch the lamp on automatically at night.
  ii Draw a circuit diagram to show the changes.

Key points

- A sensor circuit responds to a change in the surroundings.

- The output p.d. of a sensor circuit changes when there is a change in the surroundings.

- A sensor circuit containing a thermistor can be used to switch a device on when the temperature increases too much.

- A sensor circuit containing an LDR can be used to switch a device on when the light intensity decreases.

Chapter summary questions

1 State and explain how the resistance of a filament lamp changes when the current through the filament is increased.

2 Match each component in the list to each statement **a** to **d** that describes it.

diode   filament lamp   resistor   thermistor

a Its resistance increases if the current through it increases.

b The current through it is proportional to the potential difference across it.

c Its resistance decreases if its temperature is increased.

d Its resistance depends on which way round it is connected in a circuit.

3 a Sketch a circuit diagram to show two resistors **P** and **Q** connected in series to a battery of two cells in series with each other.

b In the circuit in part **a**, resistor **P** has a resistance of 4 Ω, resistor **Q** has a resistance of 6 Ω, and each cell has a potential difference of 1.5 V. Calculate:
  i the total potential difference of the two cells
  ii the total resistance of the two resistors
  iii the current in the circuit
  iv the potential difference across each resistor.

4 a Sketch a circuit diagram to show two resistors **R** and **S** in parallel with each other connected to a single cell.

b In the circuit in part **a**, resistor **R** has a resistance of 2 Ω, resistor **S** has a resistance of 4 Ω, and the cell has a potential difference of 2 V. Calculate:
  i the current through resistor **R**
  ii the current through resistor **S**
  iii the current through the cell in the circuit.

5 The figure shows a light-dependent resistor (LDR) in series with a 200 Ω resistor, a 3.0 V battery, and an ammeter.

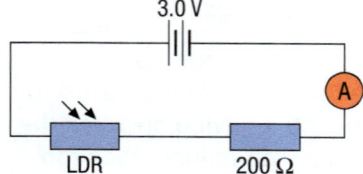

a With the LDR in daylight, the ammeter reads 0.010 A.
  i Calculate the potential difference across the 200 Ω resistor when the current through it is 0.010 A.

ii Show that the potential difference across the LDR is 1.0 V when the ammeter reads 0.010 A.

iii Calculate the resistance of the LDR in daylight.

b i If the LDR is then covered, explain whether the ammeter reading increases, decreases, or stays the same.

ii Explain how the resistance of the LDR can be calculated from the current *I*, the battery potential difference *V*, and the resistance *R* of the LDR.

6 In the figure to Question **5**, the LDR is replaced by a 100 Ω resistor and a voltmeter connected in parallel with this resistor.

a Draw the circuit diagram for this circuit.

b Calculate:
  i the total resistance of the two resistors in the circuit
  ii the current through the ammeter
  iii the voltmeter reading
  iv the potential difference across each resistor.

7 The figure shows a light-emitting diode (LED) in series with a resistor and a 3.0 V battery.

a The LED in the circuit emits light. The potential difference across it when it emits light is 0.6 V.
  i Explain why the potential difference across the 1000 Ω resistor is 2.4 V.
  ii Calculate the current in the circuit.

b The current through the LED must not exceed 15 mA or else it will be damaged. If the resistor in the figure is replaced by a different resistor **R**, what should be the minimum resistance of **R**?

c If the LED in the circuit is reversed, what would be the current in the circuit? Give a reason for your answer.

8 a Design a temperature sensor that will switch an electric heater on if the temperature is too low.

b Explain how your circuit works.

Practice questions

1 a Some of the particles in an atom are charged.

Copy the table and use words from the list to complete it. (2)

negative   positive   uncharged

Particle	Charge
Electron	i)
Neutron	ii)
Proton	iii)

b A plastic ruler is rubbed with a duster. The ruler becomes negatively charged.

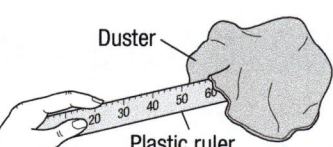
Duster
Plastic ruler

Explain what happens to the ruler and to the duster. (4)

c The diagram shows how electrical charges are used when spraying paint onto car doors.

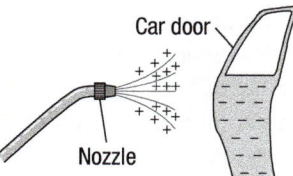
Car door
Nozzle

As the paint drops leave the nozzle, they become positively charged. The car door is given a negative charge.

Explain how the paint drops become electrically charged and how electrical charges improve the paint spraying process. (6)

2 Circuit diagrams help people when they are setting up apparatus for an experiment.

a i Draw a circuit diagram to help someone who wants to determine the resistance of a short length of resistance wire. The circuit should have a switch and a variable resistor. (4)

ii Explain why the circuit should have a switch. (3)

iii Explain why the circuit should have a variable resistor. (2)

b The circuit diagram shows one way of connecting two **identical** heating coils, **A** and **B**, in an electric cooker hob.

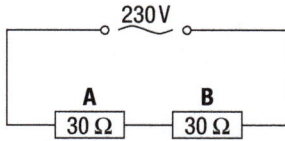
230V
A   30 Ω   B   30 Ω

i Calculate the total resistance of the heating coils. (1)

ii Calculate the potential difference across heating coil **A**. (1)

c The circuit diagram shows another way of connecting two **identical** heating coils, **C** and **D**, in an electric cooker hob.

The current flowing through heating coil **C** is 7.7 A.

i What is the potential difference across heating coil **C**? (1)

ii Calculate the resistance of heating coil **C**. (1)

iii Calculate the total current flowing in this circuit. (1)

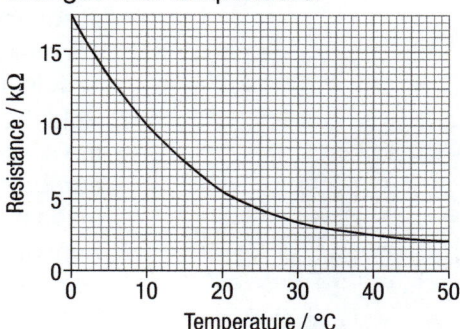
230V
C
D   30 Ω

3 Thermistors can be used as temperature sensors.

a The graph shows how the resistance of a thermistor changes with temperature.

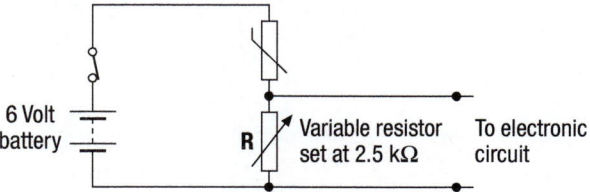
Resistance / kΩ
Temperature / °C

Use the graph to find the resistance of the thermistor at 33 °C. (1)

b The thermistor is connected in the circuit below.

6 Volt battery
R   Variable resistor set at 2.5 kΩ   To electronic circuit

Calculate the current through the circuit, in mA, when the temperature of the thermistor is 33 °C. (3)

c An electronic circuit is connected across resistor **R** as shown in the circuit diagram.

The potential difference across resistor **R** is used to power the electronic circuit.

Calculate the potential difference across resistor **R** when the temperature is:

i 33 °C (2)

ii 0 °C. (3)

iii Explain how this circuit could be used. (4)

15.1 | Magnetic fields

Figure 1 *Using a compass*

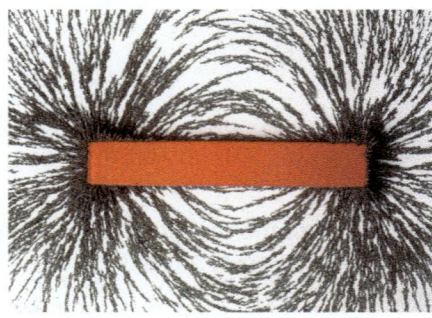

a

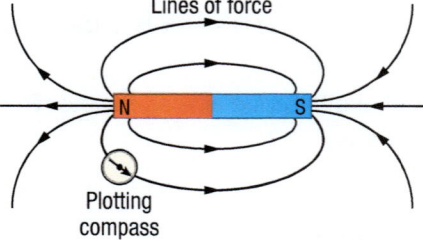

b

Figure 3 *The magnetic field of a bar magnet using **a** Iron filings **b** A plotting compass (a small compass used to show magnetic field lines)*

About magnets

A magnetic compass is a tiny magnetic needle pivoted at its centre. Because the Earth's magnetic field is concentrated at the magnetic north and south poles, one end of the compass always points north and the other south. The end of the compass that points north is the 'north-seeking' pole (usually called the magnet's **north pole** (N)), and the other end is the 'south-seeking' pole (the magnet's **south pole** (S)).

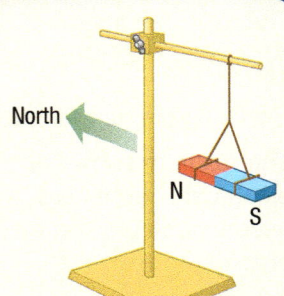
The tests above show the general rule that:

like poles repel, unlike poles attract.

Magnetic materials

Any iron or steel object can be magnetised (or demagnetised if it's already magnetised). Steel is called a ferrous material because it contains iron. Any ferrous material can be magnetised or demagnetised. Only a few non-ferrous materials can be magnetised and demagnetised. Cobalt and nickel are two examples. Oxides of these metals are used to make ceramic magnets and to coat magnetic tapes and discs.

Magnetic fields

If a sheet of paper is placed over a bar magnet and iron filings are sprinkled onto the paper, the filings form a pattern of lines. The space round the magnet is called a **magnetic field**. Any other magnetic material placed in this space experiences a force caused by the first magnet.

In Figure 3:

- the iron filings form lines as shown in Figure 3a that end at or near the poles of the magnet. These lines are lines of force, also called **magnetic field lines**.

- a plotting compass placed in the magnetic field would align itself along a line of force, pointing in a direction away from the N-pole of the magnet and towards the magnet's S-pole, as shown in Figure 3b. For this reason, the direction of a line of force is always from the north pole of the magnet to the south pole.

The further the plotting compass is from the magnet, the less effect the magnet has on the plotting compass. This is because the greater the distance from the magnet, the weaker the magnetic field.

Practical

Plotting a magnetic field

Mark a dot near the north pole of the bar magnet. Place the tail of the compass needle above the dot and mark a second dot at the tip of the needle. Move the compass so the tail of the needle is above the second dot and mark a new dot at the tip. Figure 4 shows the idea. Repeat the procedure until the compass reaches the S-pole of the magnet. Draw a line through the dots and mark direction from the N-pole to the S-pole. Repeat the procedure for further lines.

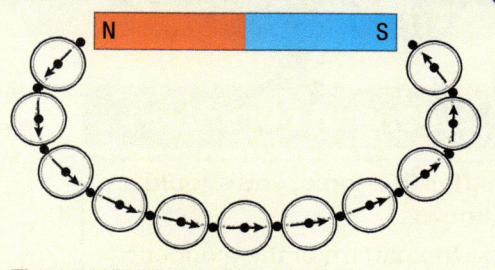

Figure 4 *Plotting a magnetic field*

Induced magnetism

An unmagnetised magnetic material can be magnetised by placing it in a magnetic field. The magnetic field is said to **induce** magnetism in the material. For example, Figure 5 shows that an unmagnetised iron rod placed in line with a bar magnet becomes a magnet with poles at each end.

- The nearest poles of the rod and the bar magnet have opposite polarity.
- Because of this opposite polarity, a force of attraction acts between the iron rod and the bar magnet.
- If the iron rod is moved out of the magnetic field, it will lose most of its magnetism.

Notice the field lines are parallel between opposite poles. The field here is said to be **uniform** because the lines are in the same direction.

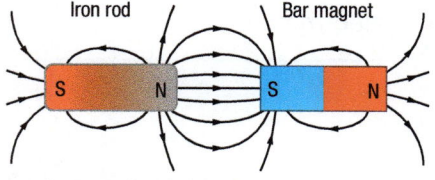

Figure 5 *Magnetic induction. Induced magnetism will cause a force of attraction between any unmagnetised magnetic material placed near one end of a bar magnet. The force is always an attractive force whichever end of the bar magnet is nearest to the material. An unmagnetised steel rod is harder to magnetise than iron. So the steel rod would need to be placed in a stronger magnetic field. Also, after steel is magnetised and then removed from the magnetic field, it keeps more of its magnetism than iron does*

Summary questions

1 A bar magnet XY is freely suspended in a horizontal position so that end X points north and end Y points south.
   a State the magnetic polarity of: **i** end X   **ii** end Y.
   b End P of a second bar magnet PQ placed near end X of bar magnet XY repels end X and attracts end Y. State the magnetic polarity of end P and explain this observation.

2 The tip of an iron nail is held in turn near each end of a plotting compass needle. State whether the tip of the nail is an N-pole, an S-pole, or is unmagnetised in each of the following possible observations:
   a the N-pole of the compass needle is repelled by the tip of the nail and the S-pole is attracted by it
   b the N-pole of the compass needle is attracted by the tip of the nail and the S-pole is repelled by it
   c the N-pole of the plotting compass is attracted by the tip of the nail and the S-pole is also attracted by it.

3 a Sketch the pattern of the magnetic field lines around a bar magnet. On your diagram, label the north and south pole of the bar magnet and indicate the direction of the field lines.
   b The figure shows a bar magnet XY and a plotting compass near end Y of the bar magnet. The needle of the plotting compass points towards end Y of the bar magnet.

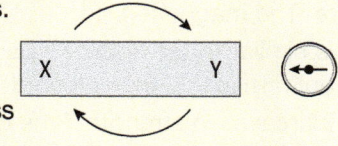

   i State the magnetic polarity of each end of the bar magnet.
   ii If the bar magnet was rotated gradually about its centre through 180°, describe and explain the effect on the direction of the plotting compass needle.

Key points

- Like poles repel, and unlike poles attract.
- The magnetic field lines of a bar magnet curve round from the north pole of the bar magnet to the south pole.
- In a uniform field the lines of a magnetic field are parallel to each other.

15.2 Electromagnets

Learning objectives

After this topic, you should know:

- the pattern of the magnetic field around a wire carrying a current

- how the strength and direction of this field varies with position

- how an electromagnet works

- some applications of electromagnets.

When an electric current passes along a wire, a magnetic field is set up around the wire. Figure 1 shows how the pattern of the magnetic field around a long straight wire can be seen using iron filings or a plotting compass. The lines of force are a series of circles, which are centred on the wire. The field is strongest near the wire. The direction of the field is reversed if the direction of the current is reversed. You can use the corkscrew rule shown in Figure 1 to remember the direction of the magnetic field for each direction of the current.

Plotting the magnetic field near a current-carrying wire

Set up the arrangement shown in Figure 2. To eliminate magnetism caused by nearby iron objects, use a wooden stand (or any non-ferrous object) to support the cardboard sheet so it is horizontal. Use the plotting compass to plot magnetic field lines near the wire as explained in Topic 15.1. You should find that the field lines are concentric circles (in the plane of the card) centred on and perpendicular to the wire.

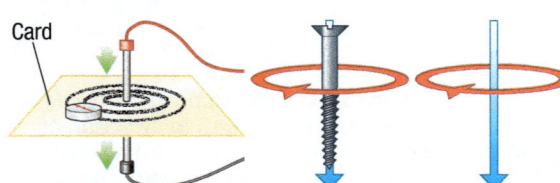

Use the arrangement shown in Figure 2 to observe the effect of:

1 **reversing the current**: You should find that the direction of the plotting compass reverses. This shows that the magnetic field lines reverse direction when the direction of the current is reversed.

2 **moving the plotting compass away from the wire**: You should find it tends to point towards magnetic north. This is because the magnetic field caused by the wire decreases in strength further from the wire, so the Earth's magnetic field has more effect.

Figure 1 *The magnetic field near a long straight wire and the corkscrew rule. The field direction is the same direction as the rotation direction of the screw driven in the same direction as the current*

Study tip

Remember the current direction is from + to – around the circuit. Refer to Topic 14.2 'Electric circuits'.

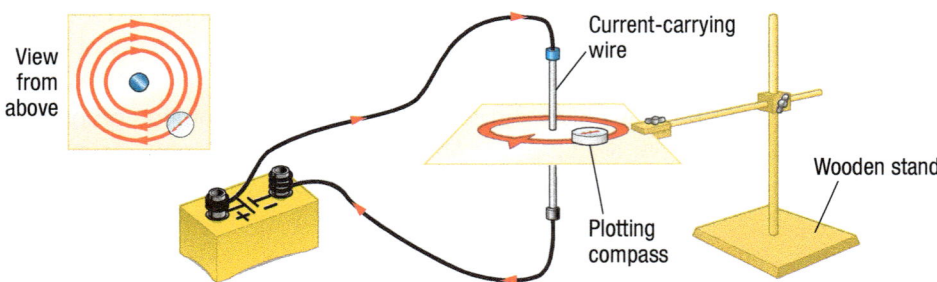

Figure 2 *Plotting magnetic field lines*

Solenoids

A solenoid is a long coil of insulated wire as shown in Figure 3. A magnetic field is produced in and around the solenoid when a current is passed through the wire. The magnetic field:

- is similar to the field of a bar magnet outside the solenoid, and is strong and uniform inside the solenoid

- increases in strength if the current is increased, and reverses its direction if the current is reversed.

Figure 3 shows how you can find the polarity of each end of the solenoid from the direction of the current.

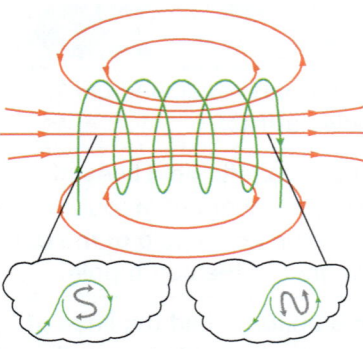

Figure 3 *The magnetic field of a solenoid*

Electromagnets

An **electromagnet** consists of insulated wire wrapped around an iron bar (the core). When a current is passed through the wire, its magnetic field magnetises the iron bar. When the current is switched off, the iron bar loses most of its magnetism.

Electromagnets are used in many devices.

1 **The scrapyard crane**
   Scrap vehicles are lifted in a scrap yard using powerful electromagnets attached to cranes. The steel frame of a vehicle sticks to the electromagnet when current passes through the coil of the electromagnet. When the current is switched off, the frame drops off.

2 **The electric bell** (Figure 4)
   When the bell is connected to the battery, the iron armature is pulled on to the electromagnet. This opens the make-and-break switch, and the electromagnet is switched off. The armature springs back and the make-and-break switch closes again so the cycle repeats.

3 **The relay** (Figure 5)
   A relay is used to switch an electrical machine such as a motor on or off. When current passes through the electromagnet, the armature is pulled onto the electromagnet and closes the switch gap. So a small current (through the electromagnet) switches on a much larger current.

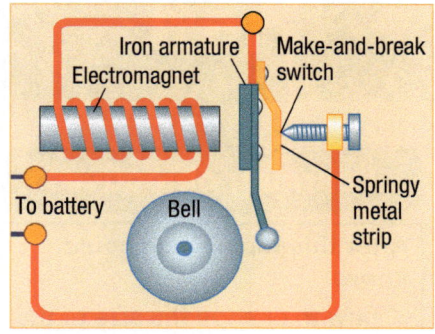
Figure 4 *An electric bell*

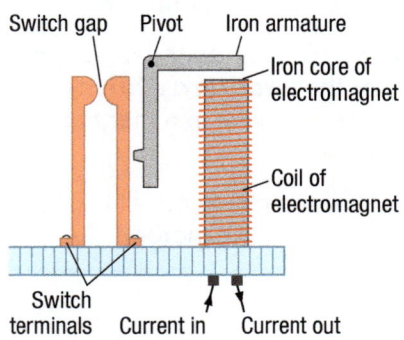
Figure 5 *The construction of a relay*

Required Practical

Investigating the strength of an electromagnet

Use an electromagnet to hold a flat iron bar supporting a known weight. Measure the smallest current needed to hold the load (the bar and the weight).
1 Increase the weight and repeat the measurement for different known weights.
2 Repeat the measurements with more turns on the electromagnet.
3 Plot a graph of your measurements and use it to draw conclusions about your investigation.

Summary questions

1 a Sketch the pattern of the magnetic field lines near a vertical wire carrying current upwards.
  b Explain why iron, and not steel, is used for the core of an electromagnet.

2 List the statements **A–E** below in the correct order to explain how the relay in Figure 6 works. Statement **E** is third in the correct order.
  A The device connected to the switch terminals is switched on.
  B The iron core of the electromagnet is magnetised.
  C A current passes through the coil.
  D The switch gap is closed.
  E The armature is attracted to the core of the electromagnet.

3 The construction of a buzzer is like that of the electric bell, except that the buzzer does not have a striker or a bell.
  a Explain why the armature of the buzzer vibrates when the buzzer is connected to a battery.
  b Why does the buzzer vibrate at a higher frequency than the electric bell?

Key points

- The magnetic field lines around a wire are circles centred on the wire in a plane perpendicular to the wire.

- Increasing the current makes the magnetic field stronger.

- Reversing the current reverses the magnetic field.

- An electromagnet is made up of a coil of insulated wire wrapped round an iron core.

- Electromagnets are used in scrapyard cranes, circuit breakers, electric bells, and relays.

15.3 The motor effect

Learning objectives

After this topic, you should know:

- how to change the size of the force on a current-carrying wire in a magnetic field
- how to reverse the direction of the force on a current-carrying wire in a magnetic field
- how the motor effect is used in an electric motor.

You probably use electric motors several times every day. A hairdryer, an electric shaver, a refrigerator pump, and a computer hard drive are just a few examples of electrical appliances that contain an electric motor. The electric motor works because a force can act on a wire in a magnetic field when a current passes through the wire. This is called the **motor effect**.

Practical

Investigating the motor effect

Figure 1 shows how you can investigate the motor effect. You should find that a force acts on the wire unless the wire is parallel to the magnetic field lines.

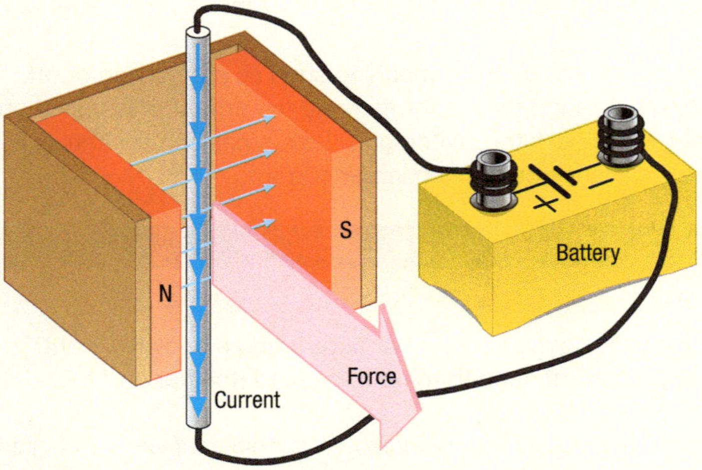

Figure 1 *Investigating the motor effect*

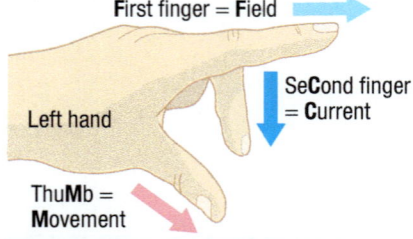

Figure 2 *Fleming's left-hand rule. Hold the fingers at right angles to each other. You can use this rule to work out the direction of the force (i.e., movement) on the wire*

Force factors

Your investigations should show that:

- The size of the force can be increased by:
  - increasing the current
  - using a stronger magnet.
- The size of the force depends on the angle between the wire and the magnetic field lines. The force is:
  - greatest when the wire is perpendicular to the magnetic field
  - zero when the wire is parallel to the magnetic field.
- The direction of the force is always at right angles to the wire and the field lines. Also, the direction of the force is reversed if the direction of the current or the magnetic field is reversed. Figure 2 shows **Fleming's left-hand rule**, which shows how these directions are related to each other.

Study tip

Make sure you can use Fleming's left-hand rule to decide the direction of the force on a current-carrying conductor.

The electric motor

An electric motor is designed to use the motor effect. You can control the speed of the motor by changing the current. Also, you can reverse the direction of turn of the motor by reversing the current.

The simple motor shown in Figure 3 consists of a rectangular coil of insulated wire (the armature coil) that is forced to rotate. The coil is connected via two metal or graphite brushes to the battery. The brushes press against a metal **split-ring commutator** fixed to the coil.

When a current is passed through the coil, the coil spins because:

- a force acts on each side of the coil due to the motor effect
- the force on one side is in the opposite direction to the force on the other side.

The split-ring commutator reverses the current around the coil every half-turn of the coil. Because of this, as the coil spins the current in the half beside each pole of the magnet remains in the same direction, so the force on the coil remains in the same direction.

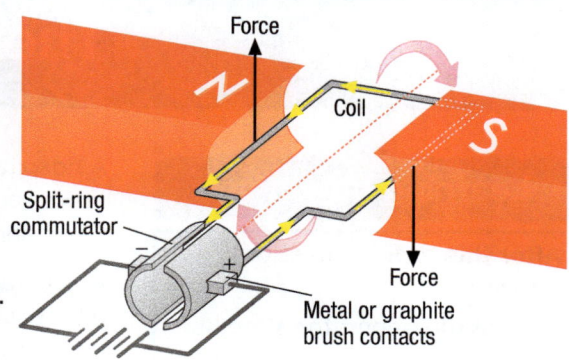

Figure 3 *The electric motor*

Practical

Make and test a simple electric motor like the one in Figure 3.

 Did you know ...?

Graphite is a form of carbon that conducts electricity and is very slippery. It causes very little friction when in contact with the rotating commutator.

The loudspeaker

A loudspeaker is designed to make a diaphragm attached to a coil vibrate when alternating current passes through the coil.

- When a current passes through the coil, a force caused by the motor effect makes the coil move.
- Each time the current changes its direction, the force reverses its direction. So the coil is repeatedly forced backwards and forwards. This motion makes the diaphragm vibrate so that sound waves are created.

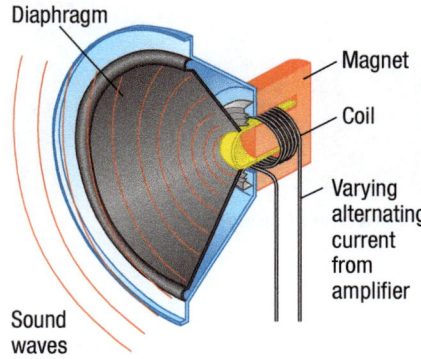

Figure 4 *A loudspeaker*

Summary questions

1 **a** Explain why the coil of a simple electric motor rotates continuously when the motor is connected to a battery.

   b Why does a loudspeaker not produce sound when direct current is passed through it?

2 **a** Explain why a simple electric motor connected to a battery reverses its spin if the battery connections are reversed.

   b Discuss whether or not an electric motor would run faster if the coil was wound on:
   i a plastic block
   ii an iron block, instead of a wooden block.

3 **a** A force is exerted on a straight wire when a current is passed through it and it is at right angles to the lines of a magnetic field. Describe how the force changes if the wire is turned through 90° until it is parallel to the field lines.

   b A loudspeaker contains a small coil in a magnetic field. The coil is attached to a diaphragm. Explain why the loudspeaker produces sound waves when an alternating current passes through the coil.

Key points

- In the motor effect, the force is:
  - increased if the current or the strength of the magnetic field is increased
  - at right angles to the direction of the magnetic field and to the wire
  - reversed if the direction of the current or the magnetic field is reversed
  - zero if the wire is parallel to the magnetic field.

- The motor effect in an electric motor forces the coil to turn when a current is in the coil.

15.4 The generator effect

Figure 1 *A standby generator*

A hospital has its own electricity generator always on standby in case of a power-cut. Patients' lives would be put at risk if the mains electricity supply failed and there was no standby generator.

A generator contains coils of wire that spin in a magnetic field. A potential difference, or voltage, is created, or **induced**, across the ends of the wire when it cuts across the magnetic field lines. This process is called **electromagnetic induction**. If the wire is part of a complete circuit, the induced potential difference makes an electric current pass around the circuit.

Practical

Investigating a simple generator

Connect some insulated wire to an ammeter as shown in Figure 2. Move the wire between the poles of a U-shaped magnet and observe the ammeter. You should discover that the ammeter pointer deflects as a current is generated when the wire cuts across the magnetic field. This is because a potential difference is induced in the wire when it cuts across the lines of the magnetic field. This effect is called the **generator effect**.

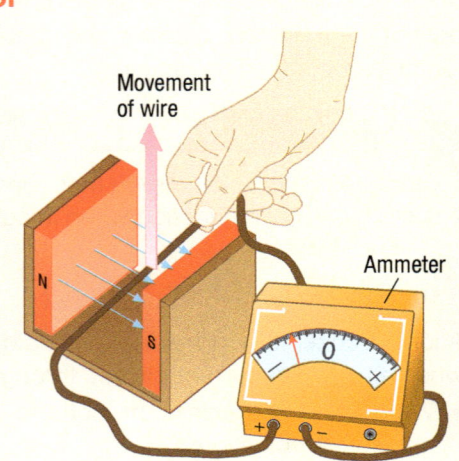

Figure 2 *The generator effect*

Carry out tests to see what difference is made by:

1. holding the wire stationary in the magnetic field
2. moving the magnet instead of the wire
3. moving the wire faster across the magnetic field
4. reversing the direction of motion of the wire.

In the tests above, you should find that:

- no current is generated when the wire is stationary
- a current is generated when the magnet instead of the wire is moved
- a larger current is generated when the wire moves faster
- the current is reversed when the direction of motion is reversed.

A generator test

Look at Figure 3. It shows a coil of insulated wire connected to a centre-reading ammeter. When one end of a bar magnet is pushed into the coil, the ammeter pointer deflects.

This is because:

- the movement of the bar magnet causes an induced potential difference in the coil
- the induced potential difference causes a current, because the coil is part of a complete circuit.

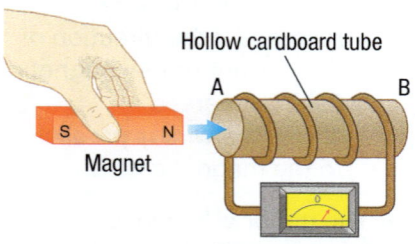

Meter pointer deflects when the magnet is pushed into the coil

Figure 3 *Testing the generator effect*

In Figure 3, if the bar magnet is withdrawn from the coil, the ammeter pointer deflects in the opposite direction. This is because the induced potential difference acts in the opposite direction, so the induced current is in the opposite direction. The direction of the induced current also depends on which way round the polarity of the magnet is. For example, in Figure 3, the north pole of the bar magnet is shown entering end A of the coil.

Table 1 shows the results of different tests on the generator effect. It gives the direction of current as viewed from end A of the coil.

Table 1 *Testing the generator effect*

Magnetic pole entering or leaving the coil	Pushed in or pulled out	Direction of current	Induced polarity of Y	Magnet and coil
North pole	In	Anticlockwise	North pole	Repel
North pole	Out	Clockwise	South pole	Attract
South pole	In	Clockwise	South pole	Repel
South pole	Out	Anticlockwise	North pole	Attract

The induced current creates a magnetic field in and around the coil, but only when the magnet is moving. This induced magnetic field always opposes the original change, which in Table 1 is the movement of the magnet. So, work has to be done by the person moving the magnet. The electrical energy generated is the result of the work done by the person moving the magnet.

In Figure 3, moving the coil instead of the magnet has the same effect as described in Table 1. This is because the motion of the coil and the magnet relative to each other is unchanged. So the magnetic field in the coil due to the magnet changes in the same way.

Note that you can use the solenoid rule in Figure 4 to work out from the current direction the magnetic polarity at end A of the solenoid.

Figure 4 *The solenoid rule*

Summary questions

1 **a** When a wire is moved between the poles of a U-shaped magnet as shown in Figure 2, explain why a current passes through the ammeter.
   b State and explain what would be observed when the wire is moved between the poles more slowly in the opposite direction.

2 A coil of wire is connected to a centre-reading ammeter. A bar magnet is inserted into the coil, making the ammeter pointer flick briefly.
   What would you observe if:
   a the magnet was then held at rest in the coil?
   b the coil had more turns of wire wrapped around the tube?
   c the magnet was withdrawn rapidly from the coil?

3 Look at Figure 5.
   a Explain why the ammeter pointer deflects when the switch is closed.
   b Explain why the pointer does not deflect when there is a constant current in coil X.

Practical

A magnetic puzzle

Wind two separate coils X and Y on a cardboard tube as shown in Figure 5. Connect coil X in series with a battery and a switch. Connect coil Y in a separate circuit to an ammeter.

1 Close the switch and observe that the ammeter pointer deflects briefly. This happens because a magnetic field is created in and around coil X when the switch is closed. This magnetic field has the same effect on Y as pushing a magnet into it, so a potential difference is induced in coil Y.

2 Repeat test **1** with an iron bar in the tube. You should find that the deflection of the ammeter pointer is much bigger.

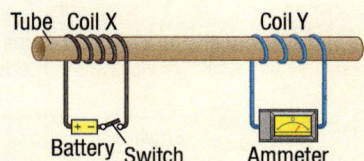

Figure 5 *A magnetic puzzle*

Key points

- The generator effect is the effect of inducing a potential difference using a magnetic field.

- When a wire cuts the lines of a magnetic field, a potential difference is induced across the ends of the wire.

- The faster a wire cuts across the lines of a magnetic field, the greater the induced potential difference. When a direct current electromagnet is used, it needs to be switched on or off to induce a potential difference.

15.5 Alternators and dynamos

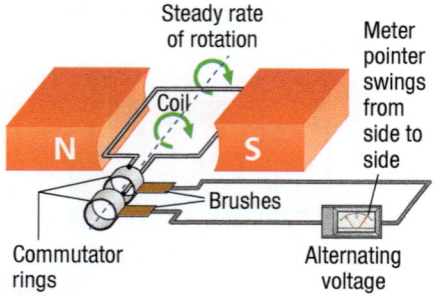

Figure 1 *The a.c. generator*

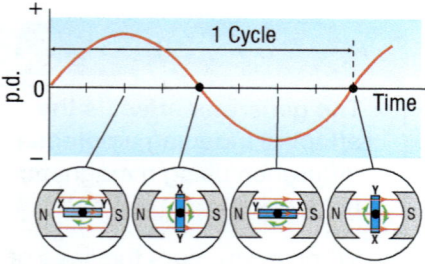

Figure 2 *Alternating voltage*

∞ links

For more information on displaying alternating voltages on an oscilloscope, see Topic 16.1 'Alternating current'.

The alternator

A simple **alternator** is an alternating-current (a.c.) generator. It is made up of a rectangular coil that is forced to spin in a uniform magnetic field, as shown in Figure 1. The coil is connected to a centre-reading meter via metal brushes that press against two metal slip rings. The slip rings and brushes provide a continuous connection between the coil and the meter.

When the coil turns steadily in one direction, the meter pointer deflects first one way, then the opposite way, then back again. This carries on as long as the coil keeps turning in the same direction. The current in the circuit repeatedly changes its direction through the meter because the induced potential difference in the coil repeatedly changes its direction. The induced potential difference and the current **alternate** – they repeatedly change direction.

The induced potential difference varies as the coil rotates, as shown in Figure 2. In one complete rotation of the coil (or one full cycle), the induced potential difference increases from zero to a maximum value, then decreases to zero, reverses and increases to a negative maximum, and then becomes zero again. Both the positive and the negative maximum values are called the **peak value**.

- **The size of the induced potential difference is greatest** when the plane of the coil is parallel to the direction of the magnetic field. At this position, the sides of the coil parallel to the axis of rotation (labelled X and Y in Figure 2) cut directly across the magnetic field lines. So the induced potential difference is at its peak value.

- **The size of the induced potential difference is zero** when the plane of the coil is perpendicular to the magnetic field lines. At this position, the sides of the coil move parallel to the field lines and do not cut through them. So, the induced potential difference is zero.

The faster the coil rotates:

- **the greater the frequency (i.e., the number of cycles per second) of the alternating current.** This is because each full cycle of the alternating potential difference takes the same time as one full rotation of the coil.

- **the larger the peak value of the alternating current.** This is because the sides of the coil move faster and therefore cut the field lines at a faster rate, so the peak value of the induced potential difference is greater.

The peak value can also be increased by using a magnet with a stronger magnetic field, and by using a coil with a larger area and with more turns of wire on it.

As you will see in Topic 16.1 'Alternating current', an alternating voltage can be displayed on an oscilloscope screen. If an a.c. generator (alternator) is rotated faster, the screen display will show more waves on the screen (because the frequency of the induced potential difference will be greater), and the waves will be taller (because the peak value of the induced potential difference will be greater).

The direct current dynamo

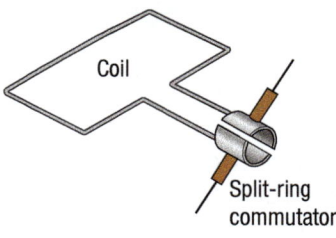

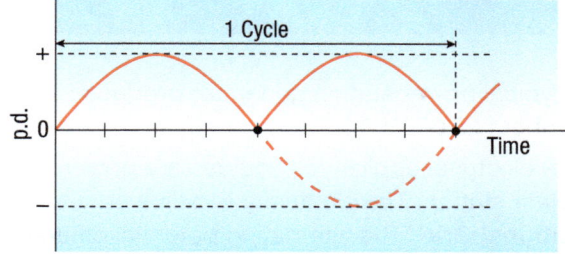

1 Cycle

Figure 3 *The d.c. generator*

A **dynamo** is a direct-current (d.c.) generator. A **simple dynamo** is the same as an alternating current (a.c) generator except that the d.c. generator has a split-ring commutator, as shown in Figure 3, instead of two separate slip rings. As the coil spins, the split-ring commutator reconnects the coil the opposite way round in the circuit every half-turn. This happens each time the coil is perpendicular to the magnetic field lines. As a result, the induced potential difference does not reverse its direction as it does in the a.c. generator. The induced potential difference varies from zero to a maximum value twice each cycle, and never changes polarity.

The moving-coil microphone

A microphone generates a varying potential difference because of the sound waves that reach it. The moving-coil microphone contains a small coil of insulated wire attached to a diaphragm, as shown in Figure 5. The coil is positioned between the poles of a permanent magnet. The pressure variations of the sound waves on the diaphragm make it vibrate, causing the coil to cross through the magnetic field lines. Because of this, an alternating potential difference is induced in the coil. The alternating potential difference has the same frequency as the sound waves, and its peak value has an order of magnitude of millivolts.

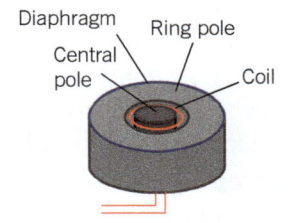

Figure 5 *Inside a microphone*

Key points

- A simple a.c. generator is made up of a coil that spins in a uniform magnetic field.

- The waveform, seen using an oscilloscope, of the a.c. generator's induced potential difference is zero when the sides of the coil move parallel to the field lines, and at its peak value a quarter of a cycle later.

- A simple d.c. generator has a split-ring commutator instead of two slip rings.

- In a moving-coil microphone, sound waves make the coil vibrate in the magnetic field, so an alternating voltage is generated.

Summary questions

1 An alternating current generator has a coil that spins between the poles of a U-shaped magnet.
   a Explain why an alternating voltage is induced in the coil.
   b Describe how the alternating voltage would differ if the coil is made to spin faster.

2 Figure 2 shows how the alternating voltage produced by an a.c. generator changes with time.
   a How would the graph differ if the coil was rotated more slowly?
   b Give reasons for your answer to **a**.

3 a State the function of the split-ring commutator in a simple d.c. generator.
   b Draw a graph to show how potential difference varies with time for a simple d.c. generator.

15.6 Transformers

Learning objectives

After this topic, you should know:

- why transformers work only with a.c.

- what the core of a transformer is made from

- how switch mode transformers differ from ordinary transformers.

A typical power station generator produces an alternating potential difference of about 25 000 V. Mains electricity in homes is 230 V.

The electrical appliances you use are powered by electricity generated at a power station. The electricity is delivered via a network of cables called the **National Grid**. The alternating potential difference of the cables (the grid potential difference) is typically 132 000 V. A **transformer** is used to change the alternating potential difference between the power station and the cables, and between the cables and your home.

How a transformer works

A transformer has two coils of insulated wire, both wound around the same soft iron core, as shown in Figure 1. This core is easily magnetised and demagnetised. The primary coil is connected to an alternating current supply. When alternating current passes through the primary coil, an alternating potential difference is induced in the secondary coil.

This happens because:

- alternating current passing through the primary coil produces an alternating magnetic field
- the lines of the alternating magnetic field pass through the secondary coil
- the magnetic field is changing.

This creates an alternating potential difference between the terminals of the secondary coil. An alternating potential difference is induced in the secondary coil.

If a bulb is connected across the secondary coil, the induced potential difference causes an alternating current in the secondary circuit, so the bulb lights up. Electrical energy is therefore transferred from the primary coil to the secondary coil. This happens even though they are **not** electrically connected in the same circuit.

- A **step-up transformer** makes the potential difference across the secondary coil higher than the potential difference across the primary coil. Its secondary coil has more turns than its primary coil.
- A **step-down transformer** makes the potential difference across the secondary coil lower than the potential difference across the primary coil. Its secondary coil has fewer turns than its primary coil.

For example, if a low-voltage supply is needed for a device, a step-down transformer is used to step the mains potential difference down from 230 V.

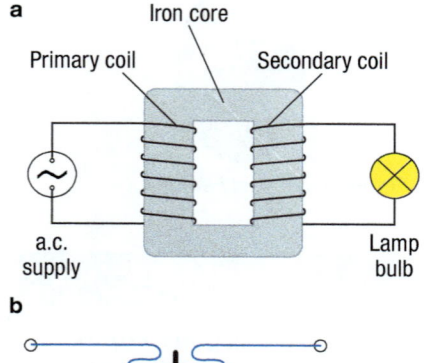

Figure 1 *A transformer* **a** *In a circuit* **b** *Circuit symbol*

Practical

Make a model transformer

Wrap a coil of insulated wire around the iron core of a model transformer to form the primary coil. Connect the coil to a 1 V a.c. supply. Then connect a second length of insulated wire to a 1.5 V torch bulb. If you wrap enough turns of the second wire around the iron core, the bulb should light up.

- Test whether cores made from different materials affect the transformer.

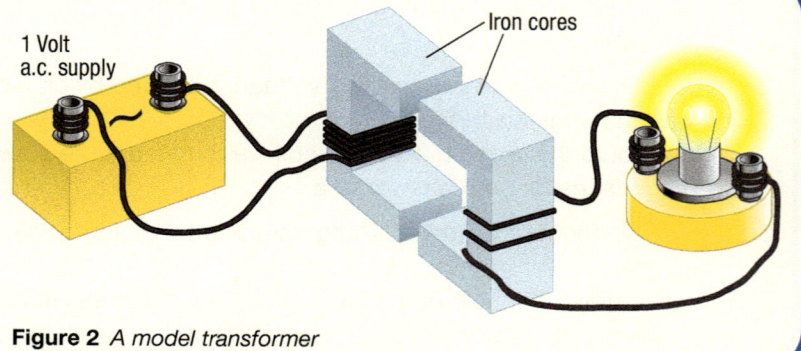

Figure 2 *A model transformer*

Practical transformers

Transformers only work with alternating current. With a direct current, there is no changing magnetic field, so the secondary potential difference is zero.

In the type of transformer described on the previous page, the core of the transformer guides the field lines in a loop through the coils. But the field must be changing to induce a potential difference in the secondary coil.

Figure 3 shows a practical transformer. The primary and secondary coils are both wound together around the same part of the core.

Switch mode transformers

A **switch mode transformer** works in a different way to the traditional transformer described above. It operates at frequencies between 50 000 Hz (50 kHz) and 200 000 Hz (200 kHz). Its main features listed below make it very suitable for use in a mobile-phone charger.

- It is lighter and smaller than a traditional transformer that works at 50 Hz.
- It uses very little power when no electrical device is connected across its output terminals (i.e., no load is applied to it).

A mobile-phone charger has three main circuits. Figure 4 shows what each circuit does. A switch mode transformer has a ferrite core. This is much lighter than an iron core and, unlike an iron core, can work at high frequency. The three circuits work together to convert the mains potential difference (at 230 V and 50 Hz in Europe) to a much lower direct potential difference.

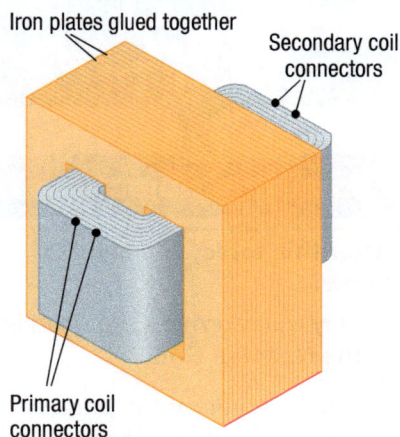

Iron plates glued together
Secondary coil connectors
Primary coil connectors

Figure 3 *A practical transformer*

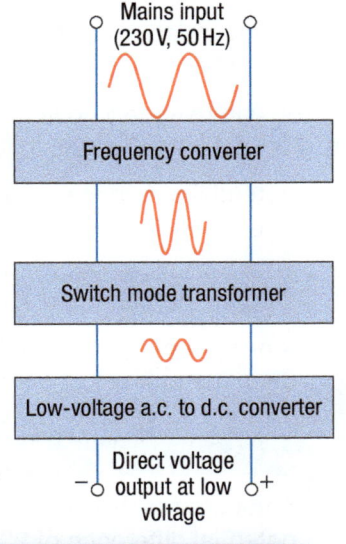

Mains input (230 V, 50 Hz)

Frequency converter

Switch mode transformer

Low-voltage a.c. to d.c. converter

Direct voltage output at low voltage

Figure 4 *Block diagram of a mobile-phone charger*

Study tip

Make sure you can explain how a transformer works.

Summary questions

1 a Explain how a transformer works.
  b A step-down transformer contains a 200-turn coil and a 4000-turn coil wound on the same iron core.
    i Which coil is the primary coil?
    ii Permanent magnets are made from steel, not iron. Explain why the transformer would not work if the core was made of steel instead of iron.

2 a Why does a transformer not work with direct current?
  b Why is it important that the coil wires of a transformer are insulated?
  c Why is the core of a transformer made of iron?

3 a A laptop computer can operate with a 14 V battery or with a mains transformer.
    i What is the benefit of having a dual power supply?
    ii Does the transformer step up or step down the potential difference applied to it?
  b Why is a switch mode transformer lighter than an ordinary transformer?

Key points

- A traditional transformer works only on a.c. because a changing magnetic field is necessary to induce a.c. in the secondary coil.

- A transformer has an iron core. A switch mode transformer has a ferrite core.

- A switch mode transformer is lighter and smaller than an ordinary transformer. It operates at high frequency.

Motors, generators, and transformers

15.7 Transformers in action

Learning objectives

After this topic, you should know:

- why transformers are used in the National Grid

- how the ratio of the primary potential difference to the secondary potential difference depends on the number of turns on each coil

- the difference between a step-up and a step-down transformer

- about the efficiency of a transformer.

Electricity is supplied to homes from power stations via the National Grid. Figure 2 shows how the grid system is used to supply industry as well as homes.

The higher the grid potential difference, the greater is the efficiency of transferring electrical power through the grid.

This is why transformers are used to step up the potential difference from a power station to the grid potential difference, and to step the grid potential difference down to the mains voltage. The grid potential difference is at least 132 000 V. What difference would it make if the grid potential difference were much lower? Much more current would be needed to deliver the same amount of power. The grid cables would therefore heat up more and waste more energy.

The transformer equation

The secondary potential difference of a transformer depends on the primary potential difference and the number of turns on each coil.

You can use the following equation to calculate any one of these factors if you know the other factors:

$$\frac{\text{potential difference across primary, } V_P}{\text{potential difference across secondary, } V_S} = \frac{\text{number of turns on primary, } n_p}{\text{number of turns on secondary, } n_s}$$

- **For a step-up transformer**, the number of secondary turns, n_s, is greater than the number of primary turns, n_p. Therefore V_S is greater than V_P.
- **For a step-down transformer**, the number of secondary turns, n_s, is less than the number of primary turns, n_p. Therefore V_S is less than V_P.

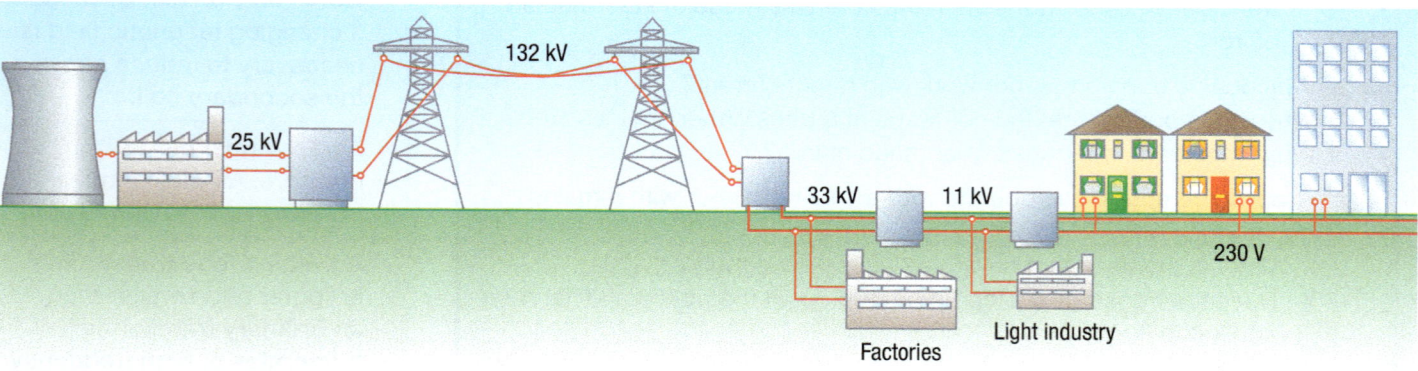

Figure 1 A power transformer under inspection

Worked example

A transformer is used to step a potential difference of 230 V down to 10 V. The secondary coil has 60 turns. Calculate the number of turns of the primary coil.

Solution

$V_P = 230\,V$, $V_S = 10\,V$, $n_s = 60$ turns

Using $\dfrac{V_P}{V_S} = \dfrac{n_P}{n_s}$ gives $\dfrac{230\,V}{10\,V} = \dfrac{n_P}{60}$

Therefore: $n_P = \dfrac{230\,V \times 60}{10\,V}$

$= \mathbf{1380\ turns}$

Figure 2 The grid system

Transformer efficiency

Transformers are almost 100% efficient. When a device is connected to the secondary coil (see Figure 3), almost all the electrical power supplied to the transformer is delivered to the device. If you know how much electrical power a device needs to work normally (i.e., the transformer output power), this tells you how much electrical power needs to be supplied to the transformer (i.e., the transformer input power).

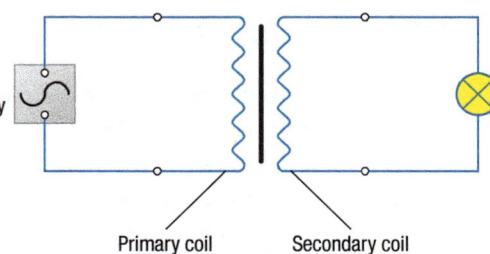

Figure 3 *Transformer efficiency*

- power supplied to transformer (i.e., input power) = primary current, I_p × primary potential difference, V_p

- power delivered by transformer (i.e., output power) = secondary current, I_s × secondary potential difference, V_s

- If you can assume 100% efficiency (power supplied = power delivered), then:

primary potential difference × primary current = secondary potential difference × secondary current

$$V_P \times I_P = V_S \times I_S$$

Worked example

A step-down transformer is used to step down an alternating potential difference of 230 V to 12 V to supply electricity to a 12 V, 2.5 A lamp. Calculate the current in the primary coil of the transformer when the device is switched on. Assume that the transformer is 100% efficient.

Solution

Make I_p the subject of the equation $V_p \times I_p = V_s \times I_s$, and substitute in the known values.

$$I_p = \frac{V_s \times I_s}{V_p} = \frac{12 \times 2.5}{230}$$
$$= 0.13 \text{ A}$$

Summary questions

1 **a** A transformer with 60 turns in the secondary coil is used to step a potential difference of 120 V down to 6 V. Calculate the number of turns on the primary coil.

   b A step-up transformer is required to step an alternating potential difference of 20 V to 230 V. If a coil with 100 turns is used as the primary coil, how many turns should the secondary coil have?

   c A 230 V, 60 W bulb lights normally when it is connected to the secondary coil of a transformer and a 10 V a.c. supply is connected to the primary coil. Assume the transformer is 100% efficient. Calculate:

   i the primary current   **ii** the bulb current.

2 A transformer with a secondary coil of 100 turns is used to step a potential difference down from 240 V to 12 V.

   a Calculate the number of turns on the primary coil of this transformer.

   b A 12 V, 36 W bulb is connected to the secondary coil. Assume the transformer is 100% efficient. Calculate the current in:

   i the bulb   **ii** the primary coil.

3 Two separate cables A and B deliver the same amount of electrical power to two factories. A is at a higher potential difference than B.

   a What can you say about the current in cable A compared with the current in cable B?

   b Cable A has the same resistance as cable B. Why is less power wasted in A than in B?

Key points

- Transformers are used to step potential differences up or down.

- The transformer equation is:

$$\frac{\text{primary potential difference, } V_P}{\text{secondary potential difference, } V_S} = \frac{n_P}{n_S}$$

where: n_P = number of primary turns and n_S = number of secondary turns.

- For a step-down transformer, n_S is less than n_P.

- For a step-up transformer, n_S is greater than n_P.

- For a 100% efficient transformer:

$$V_P \times I_P = V_S \times I_S$$

where: I_P = primary current, I_S = secondary current.

Chapter summary questions

1 a Two identical bar magnets are placed end-to-end on a sheet of paper on a table. There is a gap between them and unlike poles are facing each other.

   i Draw the arrangement and the pattern of the magnetic field lines in the gap.

   ii A plotting compass is placed in the gap at equal distances from the two magnets, a short distance from the midpoint of the gap. On your drawing, show the plotting compass in this position and show the direction in which it points.

b Copy and complete **i** and **ii** using the words below. Each word may be used more than once.

   field   force   lines   current

   i A vertical wire is placed in a horizontal magnetic field. When a .... is passed through the wire, a .... acts on the wire.

   ii A force acts on a wire in a magnetic field when a .... passes along the wire and the wire is not parallel to the .... of the ....

2 a The figure opposite shows the construction of a relay. Explain why the switch closes when a current passes through the coil of the electromagnet.

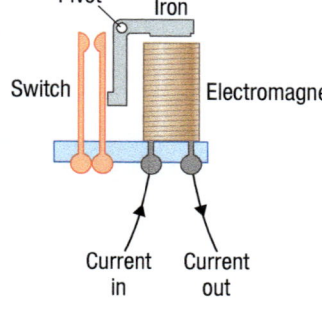

b The diagram below shows the relay coil in a circuit that is used to switch on the starter motor of a car. Explain why the motor starts when the ignition switch is closed.

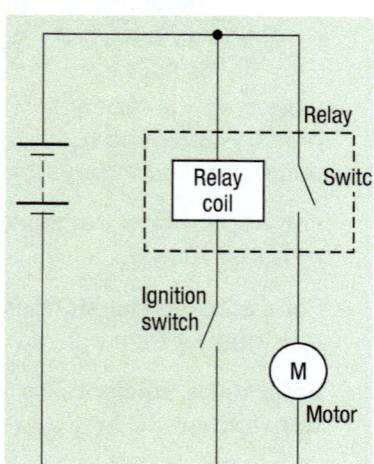

3 This figure shows a rectangular coil of wire in a magnetic field viewed from above. When a direct current passes clockwise round the coil, a downward force acts on side **X** of the coil.

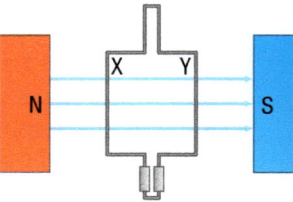

a What is the direction of the force on side **Y** of the coil?

b What can you say about the force on each side of the coil that is parallel to the magnetic field lines?

c What is the effect of the forces on the coil?

4 a i The arrangement shown in the figure for Question **3** could be used to generate a direct current if the coil is made to spin and the split-ring commutator is connected to a lamp. Draw a graph to show how the current would vary with time if the coil is turned steadily.

   ii How you would modify the arrangement shown in the figure for Question **3** to generate an alternating current?

b Explain why a transformer does not work on direct current.

5 a Cables at a potential difference of 100 000 V are used to transfer 1 000 000 W of electrical power in a part of a grid system.

   i Calculate the current in the cable.

   ii If the potential difference had been 10 000 V, how much current would be needed to transfer the same amount of power?

b Explain why power is transmitted through the National Grid at a high potential difference rather than a low potential difference.

6 A transformer has 50 turns in its primary coil and 500 turns in its secondary coil. It is to be used to light a 120 V, 60 W bulb connected to the secondary coil. Assume the transformer is 100% efficient.

a Calculate the primary potential difference.

b Calculate the current in the bulb.

c Calculate the current in the primary coil.

7 A transformer has 3000 turns on its primary coil. An alternating potential difference of 240 V is to be connected to the primary coil, and a 12 V bulb is to be connected to the secondary coil.

a Calculate the number of turns the secondary coil should have.

b What would be the current through the primary coil if the current through the lamp is to be 3.0 A? Assume the transformer is 100% efficient.

Practice questions

1 There are many appliances that transfer energy.

a Some students set up the apparatus shown here to investigate the motor effect.

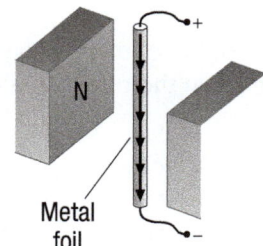

Metal foil

They wanted to see how the force depended on the current flowing through the strip of metal foil.

i To make the test fair, which variable in the list should they keep constant? (1)

current
size of the force
strength of the magnetic field

ii The students changed the apparatus so that they could measure the force acting on the strip of metal foil. Their results are shown in the table.

Current (A)	0.0	0.1	0.2	0.3	0.4
Force (N)	0.0	0.03	0.06	0.09	0.12

Explain what you can conclude from the data in the table. (2)

b The diagrams show two positions of a rectangular coil of wire in a magnetic field.

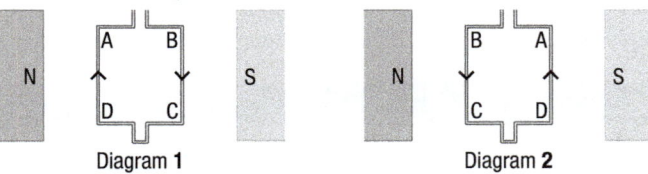

Diagram **1**   Diagram **2**

i Describe and explain what happens to the coil in diagram **1**. (3)

ii Describe and explain what happens to the coil in diagram **2**. (3)

iii To keep the coil spinning round, what has to be done to the current every half-cycle? (1)

2 a The diagram shows a simple generator.

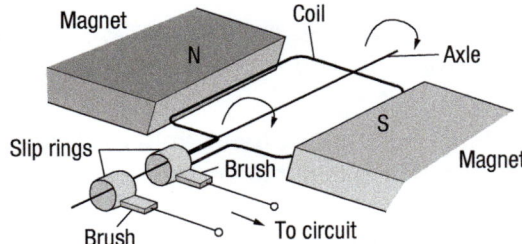

Explain how this generator produces a current in the circuit. (6)

b The graph shows how the current from the generator varies with time.

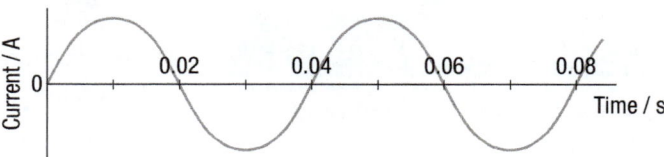

i Electric currents are either alternating or direct. Explain how the graph shows that the current from the generator is an alternating current. (2)

ii Calculate the frequency of the current. (3)

3 The diagram shows a shaver socket in a bathroom and the transformer inside it.

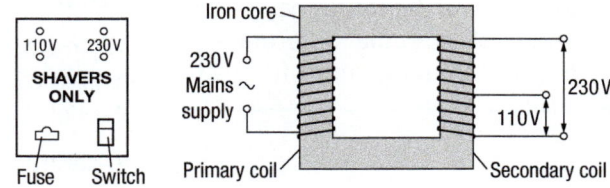

a What material is used for the core of a transformer?

Choose **one** of the materials in the list. (1)

brass   copper   iron   steel   tungsten

b The primary coil of the transformer has 5000 turns. The coil is connected to the 230 V mains supply.
   i How many turns are needed on the secondary coil to give an output of 230 V? (1)
   ii Calculate the number of turns needed on the secondary coil to give an output of 110 V. (3)
   iii Explain how the transformer supplies an alternating current to the shaver. (5)

c A student wanted to run a 2.4 V motor from a 12 V car battery. He connected the battery to the input leads of a transformer. He found that the output from the transformer was zero.

Explain why. (2)

d Power lines supply electricity to a school via a transformer.

The input to the transformer is 415 V. The transformer changes this to 230 V for use in the school.

The power input to the transformer is 48 kW.

Calculate the current that the transformer delivers to the school. (You may assume that the transformer is 100% efficient.) (3)

16.1   Alternating current

Learning objectives

After this topic, you should know:

- what direct current is and what alternating current is

- what is meant by the live wire and the neutral wire of a mains circuit

- how to use an oscilloscope to measure the frequency and peak potential difference of an alternating current.

In a battery-powered torch, the current always goes around the circuit in the same direction. The current in the circuit is a **direct current** (d.c.) – it is in one direction only.

When you switch a light on at home, you use **alternating current** (a.c.) because mains electricity is an a.c. supply. An alternating current repeatedly reverses its direction. It flows one way then the opposite way in successive cycles. Its **frequency** is the number of cycles each second. In the UK, the mains frequency is 50 cycles per second (or 50 Hz).

A light bulb works with either alternating current or direct current.

Mains circuits

Every mains circuit has a **live wire** and a **neutral wire**. The current through a mains appliance alternates as the mains supply provides an alternating potential difference between these two wires.

Study tip

Make sure that you can interpret oscilloscope traces from an a.c. supply.

The neutral wire is **earthed** at the local substation. The potential difference between the live wire and earth is usually referred to as the potential or voltage of the live wire. The live wire is dangerous because its voltage repeatedly changes from + to – and back every cycle. It reaches over 300 V in each direction, as shown in Figure 1.

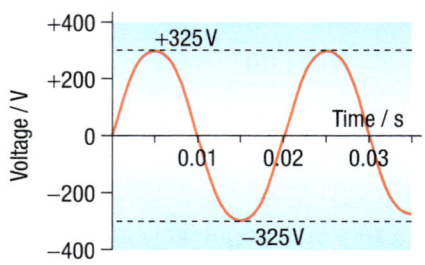

Figure 1 *Mains voltage against time*

Practical

The oscilloscope

You use an **oscilloscope** to show how an alternating potential difference changes with time.

1 Connect a signal generator to an oscilloscope, as shown in Figure 2.
   - The trace on the oscilloscope screen shows electrical waves. They are caused by the potential difference increasing and decreasing continuously. Adjusting the Y-gain control changes how tall the waves are. Adjusting the time base control changes how many waves fit across the screen.

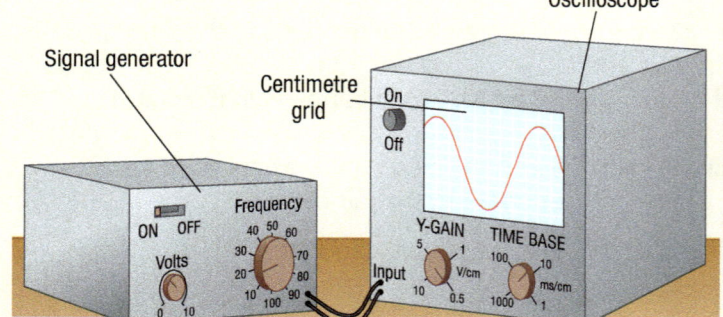

Figure 2 *Using an oscilloscope*

   - The highest potential difference is reached at each peak. The **peak potential difference** (**peak voltage**) is the difference in volts between the peak and the middle level of the waves. Increasing the potential difference of the a.c. supply makes the waves on the screen taller.
   - Increasing the frequency of the a.c. supply increases the number of cycles you see on the screen. So the waves on the screen get squashed together.
     - How would the trace change if the potential difference of the a.c. supply were reduced?
2 Connect a battery to the oscilloscope. You should see a flat line at a constant potential difference.
     - What difference is made by reversing the battery?

Measuring an alternating potential difference

You can use an oscilloscope to measure the peak potential difference and the frequency of a low-voltage a.c. supply. For example, in Figure 2:

- Suppose each peak is 4.2 cm above the centreline, which is at zero potential difference. The Y-gain control at 0.5 V/cm shows that each centimetre of height represents a potential difference of 0.5 V. So the peak potential difference is 2.1 V (= 0.5 V/cm × 4.2 cm).
- Suppose the time base control at 10 milliseconds per centimetre (ms/cm) shows that each centimetre across the screen is a time interval of 10 ms. So the time taken for one cycle is 80 ms (= 10 ms/cm × 8 cm). The frequency is therefore 12.5 Hz (= 1/80 ms or 1/0.08 s).

$$\text{The frequency of a.c. supply} = \frac{1}{\text{the time taken for one cycle}}$$

More about mains circuits

Look at Figure 1 again. It shows how the potential of the live wire varies with time.

- The live wire alternates between +325 V and −325 V. For the same current, this provides the same power as a direct voltage of 230 V. So the voltage of the mains is described as 230 V.
- Each cycle in Figure 1 takes 0.02 seconds. The frequency of the mains supply (the number of cycles per second) is therefore $\frac{1}{0.02 \text{ seconds}}$ = 50 Hz.

Summary questions

1 Choose the correct potential difference from the list for each appliance **a** to **d**.

   1.5 V   12 V   230 V   325 V

   a a car battery   **c** a torch cell
   b the mains voltage   **d** the maximum potential of the live wire.

2 In Figure 2, how would the trace on the screen change if the frequency of the a.c. supply was:
   a increased?   **b** reduced?

3 In Figure 2, what is the frequency if one cycle measures 4 cm across the screen for the same time base setting?

4 **a** How does an alternating current differ from a direct current?
   b The figure shows a diode and a resistor in series with each other connected to an a.c. supply. Explain why the current in the circuit is a direct current, not an alternating current.

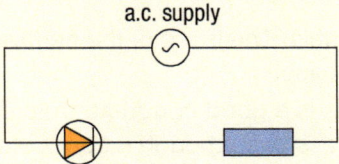
a.c. supply

   c **i** Sketch a graph to show how the current varies with time.
   ii How would your graph differ if a resistor of greater resistance had been used?

Key points

- Direct current (d.c.) is current in one direction only. Alternating current (a.c.) repeatedly reverses its direction.

- A mains circuit has a live wire that is alternately positive and negative every cycle and a neutral wire at zero volts.

- The peak potential difference of an a.c. supply is the maximum voltage measured from zero volts.

- To measure the frequency of an a.c. supply, you measure the time period of the waves then use the formula:
  frequency = $\frac{1}{\text{time taken for 1 cycle}}$

<table>
<tr><td>16.2</td><td># Cables and plugs</td></tr>
</table>

16.2

Cables and plugs

Household electricity

Learning objectives

After this topic, you should know:

- what the casing of a mains plug or socket is made from and why

- what is inside a mains cable

- why a three-pin plug includes an earth pin

- why appliances with plastic cases are not earthed.

When you plug a heater with a metal case into a wall socket, the metal case is automatically connected to earth through a wire referred to as the earth wire. This stops the metal case becoming live if the live wire breaks and touches the case. If the case did become live and you touched it, you would be electrocuted.

Plastic materials are good insulators. An appliance with a plastic case is doubly insulated and therefore has no earth wire connection. These appliances carry the double insulation symbol:

Plugs, sockets, and cables

The outer casings of plugs, sockets, and cables of all mains circuits and appliances are made of hard-wearing electrical insulators. That's because plugs, sockets, and cables contain live wires. Most mains appliances are connected via a wall socket to the mains using a cable and a **three-pin plug**.

Sockets are made of stiff plastic materials with the wires inside. Figure 1 shows part of a wall socket circuit. It has an earth wire as well as a live wire and a neutral wire. The sockets are in parallel with each other so that appliances can be switched on or off without affecting other appliances.

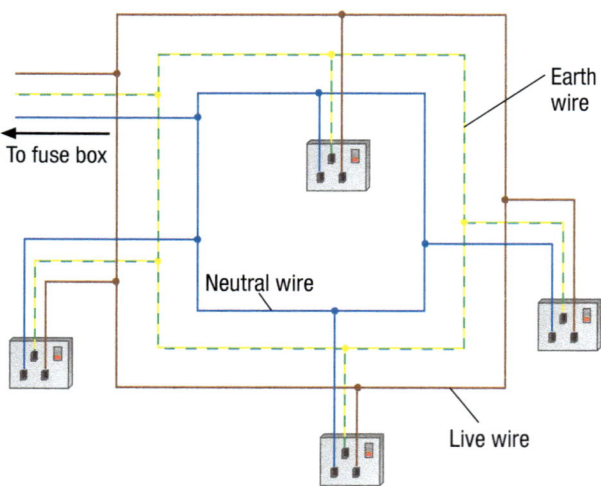

Figure 1 *A wall socket circuit*

- The earth wire of this circuit is connected to the ground at your home.
- The longest pin of a three-pin plug is designed to make contact with the earth wire of a wall socket circuit. So, when you plug an appliance with a metal case to a wall socket, the case is automatically earthed. This is explained in more detail in Topic 16.3.

Plugs have cases made of stiff plastic materials. The live pin, the neutral pin, and the earth pin stick out through the plug case.

- The pins are made of brass because brass is a good conductor and doesn't rust or oxidise. Copper isn't as hard as brass even though it conducts better.
- The case material is an electrical insulator. The inside of the case is shaped so the wires and the pins cannot touch each other when the plug is sealed.
- The plug contains a fuse between the live pin and the live wire. If too much current passes through the wire in the fuse, it melts and cuts the live wire off.

Cables used for mains appliances and circuits are made up of two or three insulated copper wires surrounded by an outer layer of rubber or flexible plastic material. The wires are colour-coded to distinguish which one is live, which one is neutral, and which one is earthed. The colour coding of electrical wires in countries outside the European Union may be different from the colour coding of electrical wires in the EU.

- Copper is used for the wires because it is a good electrical conductor and it bends easily.
- Plastic is a good electrical insulator and therefore prevents anyone touching the cable from receiving an electric shock.
- Two-core cables are used for appliances that have plastic cases (e.g., hairdryers, radios).
- Cables of different thicknesses are used for different purposes. For example, the cables joining the wall sockets in a house must be much thicker than the cables joining the light fittings. This is because more current passes along wall socket cables than along lighting circuits, so the wires in them must be much thicker. This stops the heating effect of the current making the wires too hot.

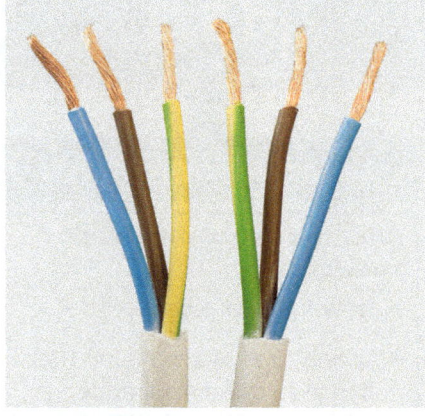

Figure 2 *In the EU, the earth wire is green and yellow, the live wire is brown, and the neutral wire is blue*

Summary questions

1 a Why are sockets wired in parallel with each other?
b Why is brass, an alloy of copper and zinc, better than copper for the pins of a three-pin plug?
c Why is it dangerous to use cables that are worn away or damaged?

2 a Match parts **1** to **4** in a three-pin plug with materials **A** to **D**.
    1 cable insulation   **A** brass
    2 case   **B** copper
    3 pin   **C** rubber
    4 wire   **D** stiff plastic
b Explain your choice of material for each part in **a**.

3 a Why is each of the three wires in a three-core mains cable insulated?
b How is the metal case of an electrical appliance connected to earth?
c Why do the cables joining the wall sockets in a house need to be thicker than the cables joining the light fittings?

Key points

- Sockets and plug cases are made of stiff plastic materials that enclose the electrical connections. Plastic is used because it is a good electrical insulator.

- A mains cable is made up of two or three insulated copper wires surrounded by an outer layer of flexible plastic material.

- The earth wire is connected to the longest pin in a plug and is used to earth the metal case of a mains appliance

- A mains appliance with a plastic case does not need to be earthed because plastic is an insulator and cannot become live.

16.3 Fuses

If you need to buy a fuse for a mains appliance, make sure you know the fuse rating. Otherwise, the new fuse might blow as soon as it is used. Worse still, it might let too much current through and cause a fire.

- A **fuse** contains a thin wire that heats up and melts if too much current passes through it. The fuse blows.
- The rating of a fuse is the maximum current that can pass through it without melting the fuse wire.
- The fuse is in series with the live wire, between the live wire and the appliance. If the fuse blows, the appliance is then cut off from the live wire.

A fuse in a mains plug must have the correct current rating for the appliance. If the current rating is too large, the fuse will not blow when it should. The heating effect of the current could set the appliance or its connecting cable on fire. The connecting cable must be thick enough (so its resistance is small enough) to make the heating effect of the current in the cable insignificant. The higher the current rating of the appliance (and fuse), the thicker the cable needs to be.

Figure 1 *A cartridge fuse*

The importance of earthing

Figure 2 shows why an electric heater is made safer by earthing its metal case.

In Figure 2a, the heater works normally and its case is earthed. The case is safe to touch, even if there was a fault and the live wire was in contact with the metal case.

In Figure 2b, the earth wire is broken. The case would become live if the live wire touched it.

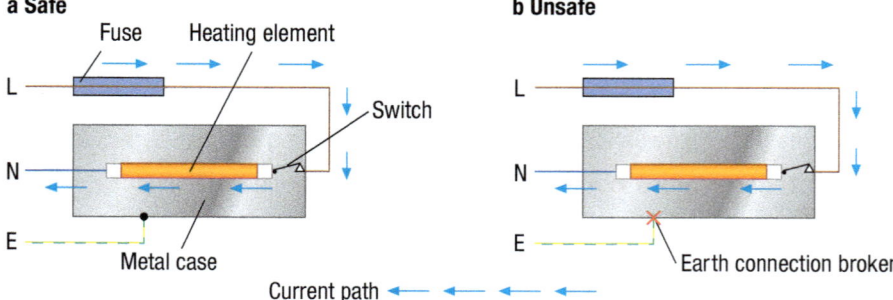

Figure 2 *Earthing an electric heater* **a** *Earthed with no fault* **b** *Not earthed*

In Figure 2c, the heater element has touched the unearthed frame, so the frame is live. Anyone touching it would be electrocuted. The fuse provides no protection to the user, because a current of just 20 mA can be lethal.

In Figure 2d, the earth wire has been repaired but the heater element still touches the frame. The current is greater than in **a** or **b** because it only passes through part of the heater element. Because the frame is earthed, anyone touching it would not be electrocuted. But Figure 2d is still dangerous if the fuse rating is incorrect. This is because, although the current might not be enough to blow the fuse, it might cause the wires of the appliance to overheat.

Circuit breakers

A **circuit breaker** is an electromagnet in series with a switch that opens (switches off or trips) when too much current passes through it. This can happen if there is a fault in an appliance that is in series with the circuit breaker. If the current is too large, the magnetic field of the electromagnet is strong enough to pull the switch contacts apart so that the current becomes zero. Once the switch is open, it stays open. It can then be reset after the fault that made it trip has been repaired. See Topic 15.2 'Electromagnets' for more about electromagnets.

Circuit breakers are often used instead of fuses. They work faster than fuses and can be reset quickly, and don't need to be replaced like fuses.

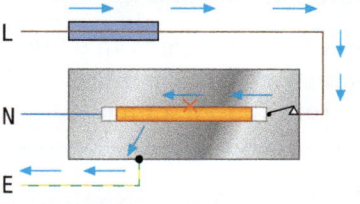

c Deadly Heating element touches the metal case, making it live

Earth connection broken

Victim touches the metal case, and because the earth wire is broken, conducts current to earth

d Still dangerous if the fuse rating is incorrect as the appliance may overheat

Figure 2 c *Not earthed with live case*
d *Earthed with live case*

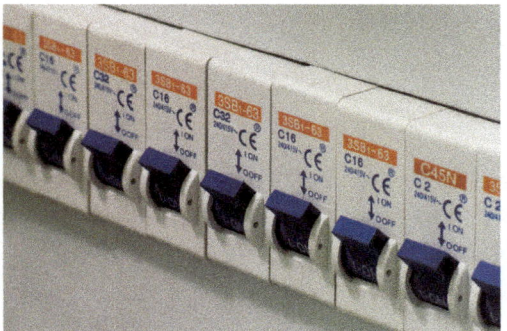

Figure 3 *A set of circuit breakers*

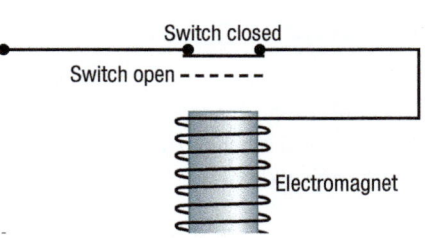

Figure 4 *Inside a circuit breaker*

Summary questions

1 **a** What is the purpose of a fuse in a mains circuit?
   b Why is the fuse of an appliance always on the live side?
   c What advantages does a circuit breaker have compared with a fuse?

2 The figure opposite shows the circuit of an electric heater that has been wired incorrectly.
   a Does the heater work when the switch is closed?
   b When the switch is open, why is it dangerous to touch the element?
   c Redraw the circuit correctly wired.

3 **a** What is the difference between a circuit breaker and a fuse?
   b What are the advantages of a circuit breaker compared with a fuse?

Key points

- A fuse contains a thin wire that heats up, melts, and cuts off the current (breaks the circuit) if the current is larger than the rating of the fuse.

- A fuse is fitted in series with the live wire. This cuts the appliance off from the live wire if the fuse blows.

- A circuit breaker is an electromagnetic switch that opens (trips) and breaks the circuit if too much current passes through it.

16.4

Electrical power and potential difference

Learning objectives

After this topic, you should know:

- the relationship between power and energy
- how to calculate electrical power and its unit
- how to calculate, from power and potential difference, the correct current for a fuse.

??? Did you know ... ?

An artificial heart is fitted into a patient during an operation. The battery needs to last a long time to avoid another operation to replace it. Most batteries have to be replaced every few years.

Figure 1 *An artificial heart*

Study tip

Be careful with units. Sometimes the power is given in kilowatts, but in the equation $P = V \times I$ you need to use watts.

When you use an electrical appliance, it transfers electrical energy into other forms of energy. The **power** of the appliance, in watts, is the energy it transfers, in joules per second. The following equation shows this:

$$\text{power (watts, W)} = \frac{\text{energy transferred (joules, J)}}{\text{time (seconds, s)}}$$

You can write the equation for the power of an appliance as:

$$P = \frac{E}{t}$$

where:
P = power in W
E = energy transferred in J
t = time taken in s.

Worked example

A light bulb transfers 30 000 J of electrical energy when it is on for 300 s. Calculate its power.

Solution

$$\text{Power} = \frac{\text{energy transferred}}{\text{time}} = \frac{30\,000\,\text{J}}{300\,\text{s}} = \textbf{100 W}$$

Calculating power

Millions of electrons pass through the circuit of an artificial heart every second. Each electron transfers a small amount of energy to the heart from the battery. So the total energy transferred to it each second is large enough to enable the appliance to work.

For any electrical appliance:

- the current through it is the charge that flows through it each second
- the potential difference across it is the energy transferred to the appliance by each coulomb of charge that passes through it
- the power supplied to it is the energy transferred to it each second. This is the electrical energy it transfers every second.

Therefore:

$$\begin{array}{rcl}\text{energy transfer to the} & = & \text{charge flow} \times \text{energy transfer} \\ \text{appliance each second} & & \text{per second} \quad \text{per unit charge}\end{array}$$

In other words:

$$\begin{array}{rcl}\textbf{power supplied} & = & \textbf{current} \quad \times \textbf{potential difference} \\ \text{(watts, W)} & & \text{(amperes, A)} \qquad \text{(volts, V)}\end{array}$$

The equation can be written as:

$$P = I \times V$$

where:
P = electrical power in W
I = current in A
V = potential difference in V.

For example, the power supplied to:
- a 4 A, 12 V electric motor is 48 W (= 4 A × 12 V)
- a 0.1 A, 3 V torch lamp is 0.3 W (= 0.1 A × 3.0 V).

Rearranging the equation $P = I \times V$ gives:

$$\text{potential difference, } V = \frac{P}{I} \text{ or}$$

$$\text{current, } I = \frac{P}{V}$$

Choosing a fuse

Domestic appliances are often fitted with a 3 A, 5 A, or 13 A fuse. If you don't know which one to use for an appliance, you can use the power rating of the appliance and its potential difference (voltage) to calculate the current through it. You then choose a fuse whose rating is just above the current calculated. The next time you change a fuse, do a quick calculation to make sure its rating is correct for the appliance (see the worked example below).

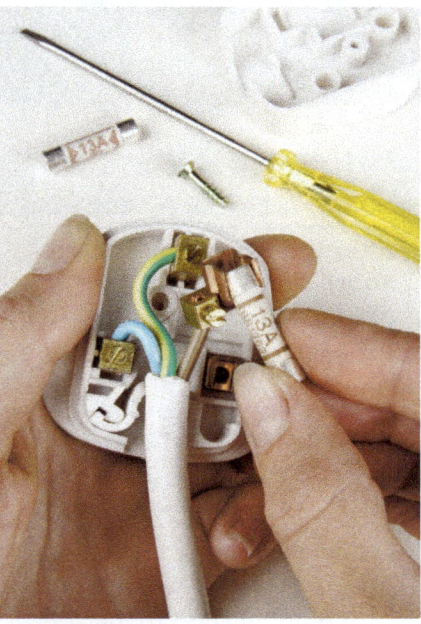

Figure 2 *Changing a fuse*

Worked example

a Calculate the normal current through a 500 W, 230 V heater.
b Which fuse, 3 A, 5 A, or 13 A, would you use for the appliance?

Solution

a Current = $\frac{500\,W}{230\,V}$ = **2.2 A**

b You would use a **3 A fuse**.

Summary questions

1 a The human heart transfers about 30 000 J of energy in about 8 hours. Calculate an estimate of the power of the human heart.
   b Calculate the power supplied to a 5 A, 230 V electric heater.
   c Why would a 13 A fuse be unsuitable for a 230 V, 100 W table lamp?

2 a Calculate the power supplied to each of the following devices in normal use:
   i a 12 V, 3 A light bulb
   ii a 230 V, 2 A heater.
   b Which type of fuse, 3 A, 5 A, or 13 A, would you select for:
   i a 24 W, 12 V heater?
   ii a 230 V, 800 W microwave oven?

3 a Why would a 3 A fuse be unsuitable for a 230 V, 800 W microwave oven?
   b The heating element of a 12 V heater has a resistance of 4.0 Ω. When the heating element is connected to a 12 V power supply, calculate:
   i the current through it
   ii the electrical power supplied to it.
   c A 6.0 kW electric oven is connected to a fuse box by a cable of resistance 0.25 Ω. When the cooker is switched on at full power, a current of 26 A passes through it.
   i Calculate the potential difference between the two ends of the cable and the power wasted in it because of the heating effect of the current.
   ii What percentage of the power supplied to the cable is wasted?

Study tip

You should be able to calculate the electric current that flows through an appliance and decide on the correct fuse to use.

Key points

- The power supplied to a device is the energy transferred to it each second.

- Power (W) = energy (J) transferred per second:
  $P = \dfrac{E}{t}$

- Electrical power supplied (W) = current (A) × potential difference (V)

- Correct rating (in A) for a fuse:
  $$= \frac{\text{electrical power (W)}}{\text{potential difference (V)}}$$
  and choose the fuse rating just higher than the current.

<table>
<tr><td>16.5</td><td># Electrical energy and charge</td></tr>
</table>

Learning objectives

After this topic, you should know:

- how to calculate the flow of electric charge from the current and time

- what energy transfers take place when charge flows through a resistor

- how the energy transferred by a flow of charge is related to potential difference

- about the electrical energy supplied by the battery in a circuit and the electrical energy transferred to the components.

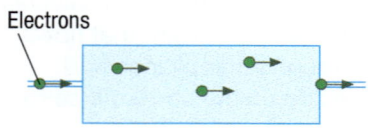

Electrons

Charge flow = current × time

Figure 1 *Charge and current*

Worked example

Calculate the charge flow when a current of 8 A passes for 80 s.

Solution

Charge flow = current × time
$$= 8\,A \times 80\,s$$
$$= 640\,C$$

∞ **links**

You calculated the energy supplied to an electrical device in Topic 12.2 'Specific heat capacity'.

Calculating charge

When an electrical appliance is on, electrons are forced through the appliance by the potential difference of the power supply unit. The potential difference causes a flow of charge through the appliance, carried by electrons.

As explained in Topic 14.2 'Electric circuits', the electric current is the rate of flow of charge through the appliance. The unit of charge, the **coulomb (C)**, is the amount of charge flowing through a wire or a component in 1 second when the current is 1 A.

The charge passing along a wire or through a component in a certain time depends on the current and the time.

We can calculate the charge using the equation:

charge   =   current   ×   time
(coulombs, C)   (amperes, A)   (seconds, s)

The equation can be written as:

$$Q = I \times t$$

where:
Q = charge in C
I = current in A
t = time in s.

Energy and potential difference

When a resistor is connected to a battery, electrons are made to pass through the resistor by the battery. Each electron repeatedly collides with the vibrating metal ions of the resistor, transferring energy to them. The ions of the resistor therefore gain kinetic energy and vibrate even more. The resistor becomes hotter.

When charge flows through a resistor, energy is transferred to the resistor, so the resistor becomes hotter.

As you saw in Topic 14.3, the energy transferred in a certain time in a resistor depends on:

- the amount of charge that passes through it

- the potential difference across the resistor.

Because energy = power × time = potential difference × current × time, you can calculate the energy transferred using the equation:

energy transferred = potential difference × charge
(joules, J)   (volts, V)   (coulombs, C)

The equation can be written as:

$$E = V \times Q$$

where:
E = energy transferred in J
V = potential difference in V
Q = charge in C.

Substituting $Q = I \times t$ into $E = V \times Q$ gives $E = V \times I \times t$. This equation can be used to calculate the energy supplied to an electrical device in a certain time if you know the current and potential difference.

Energy transfer in a circuit

The circuit in Figure 2 shows a 12 V battery in series with a torch bulb and a variable resistor. When the voltmeter reads 10 V, the potential difference across the variable resistor is 2 V.

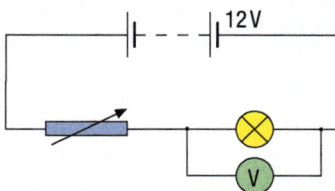

Figure 2 *Energy transfer in a circuit*

Each coulomb of charge:

- leaves the battery with 12 J of energy (because energy from the battery = charge × battery potential difference)
- transfers 10 J of energy to the torch bulb (because energy transfer to bulb = charge × potential difference across bulb)
- transfers 2 J of energy to the variable resistor.

The energy transferred to the bulb makes the bulb hot and emit light. The energy transferred to the variable resistor makes the resistor warm, so energy is therefore transferred to the surroundings by both the bulb and the resistor.

The energy from the battery is equal to the sum of the energy transferred to the bulb and to the variable resistor.

Worked example

Calculate the energy transferred in a component when the charge passing through it is 30 C and the potential difference is 20 V.

Solution

Energy transferred = 20 V × 30 C
= **600 J**

Summary questions

1 **a** Calculate the charge flowing in 50 s when the current is 3 A.
   b Calculate the energy transferred when the charge flow is 30 C and the potential difference is 4 V.
   c Calculate the energy transferred in 60 s when a current of 0.5 A passes through a 12 Ω resistor.

2 **a** Calculate the charge flow for:
     i a current of 4 A passing for 20 s
     ii a current of 0.2 A passing for 60 minutes.
   b Calculate the energy transfer:
     i for a charge flow of 20 C when the potential difference is 6.0 V
     ii for a current of 3 A that passes through a resistor for 20 s, when the potential difference is 5 V.
   c In Figure 2, an ammeter is connected in the circuit in series with the battery. The variable resistor is then adjusted until the ammeter reading is 2.5 A. The voltmeter reading is then 8.0 V.
     i   Calculate the charge that passes through the battery in 60 s.
     ii Calculate the energy transferred to or from each coulomb of charge when it passes through each component including the battery.
     iii Show that the energy transferred from the battery in 60 s is equal to the sum of the energy transferred to the lamp and to the variable resistor in this time.

3 In the figure opposite, a 4.0 Ω resistor and an 8.0 Ω resistor in series with each other are connected to a 6.0 V battery.
   Calculate:
   a the resistance of the two resistors in series
   b the current through the resistors
   c the charge flow through each resistor in 60 seconds
   d the potential difference across each resistor
   e the energy transferred to each resistor in 60 seconds
   f the energy supplied by the battery in 60 seconds.

Study tip

Make sure you know and can use the relationship between charge, current, and time.

Key points

- Charge (C) = current (A) × time (s).

- When an electrical charge flows through a resistor, energy transferred to the resistor makes it hot.

- Energy transferred (J) = potential difference (V) × charge flow (C).

- When charge flows around a circuit for a certain time, the electrical energy supplied by the battery is equal to the electrical energy transferred to all the components in the circuit.

16.6 Using electrical energy

Learning objectives

After this topic, you should know:

- what a kilowatt-hour is
- how to work out the energy used by a mains appliance
- how to work out the cost of mains electricity.

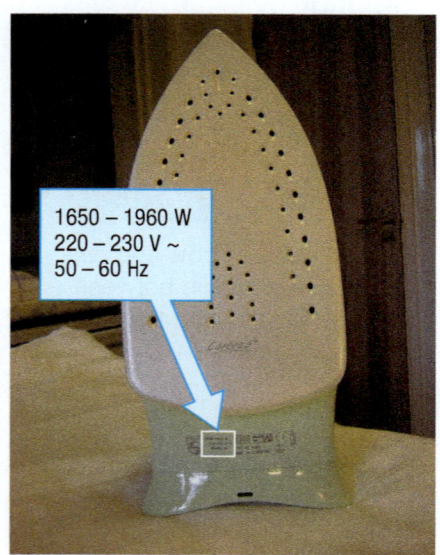

1650 – 1960 W
220 – 230 V ~
50 – 60 Hz

Figure 1 *Mains power*

When you use an electric heater, how much electrical energy is transferred from the mains? You can work this out if you know its power and how long you use it for.

For any appliance, the energy supplied to it depends on:

- how long it is switched on
- the power supplied to it.

A 1 kilowatt heater uses the same amount of electrical energy in 1 hour as a 2 kilowatt heater would use in half an hour.

The energy supplied to a 1 kW appliance in 1 hour is 1 **kilowatt-hour** (kWh).

The kilowatt-hour is the unit of energy supplied by mains electricity. You can use this equation to work out the energy, in kilowatt-hours, transferred by a mains appliance in a certain time:

energy transferred = power × time
(kilowatt-hours, kWh)   (kilowatts, kW)   (seconds, s)

$$E = P \times t$$

where:

E = energy transferred from the mains in kWh
P = power in kW
t = time taken for the energy to be transferred in h.

Worked example

You have used this equation before in Topic 5.2 'Power' to calculate the power of an appliance. It is the same equation, just rearranged and with different units.

$$E = P \times t$$

Divide both sides by *t*: $\dfrac{E}{t} = P$

This is the same as: $P = \dfrac{E}{t}$

For example:

- a 1 kW heater switched on for 1 hour uses 1 kWh of electrical energy (= 1 kW × 1 hour)
- a 1 kW heater switched on for 10 hours uses 10 kWh of electrical energy (= 10 kW × 1 hour)
- a 0.5 kW or 500 W heater switched on for 6 hours uses 3 kWh of electrical energy (= 0.5 kW × 6 hours).

If you want to calculate the energy transferred in joules, you can use the same equation:

$$E = P \times t$$

where:

E = energy transferred from the mains in *joules*, J
P = power in *watts*, W
t = time taken for the energy to be transferred in *seconds*, s.

Paying for electrical energy

The **electricity meter** in your home measures how much electrical energy your family uses. It records the total energy supplied to the appliances you all use. It gives a reading of the number of kilowatt-hours (kWh) of energy supplied by the mains. The units that appear on an electricity bill are kWh. Look at the electricity bill in Figure 3.

The difference between the two readings is the number of kilowatt-hours or 'units' supplied since the last bill. This is used to work out the cost of the electricity. For example, a cost of 12.79p per unit means that each kilowatt-hour of electrical energy costs 12.79p. Therefore:

cost = number of kWh used × cost per kWh

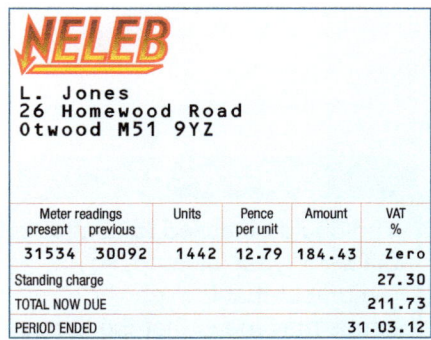

Figure 2 *An electricity meter*

NELEB

L. Jones
26 Homewood Road
Otwood M51 9YZ

Meter readings present	previous	Units	Pence per unit	Amount	VAT %
31534	30092	1442	12.79	184.43	Zero
Standing charge					27.30
TOTAL NOW DUE					211.73
PERIOD ENDED					31.03.12

Figure 3 *Checking your bill*

Summary questions

1 a How many kWh of energy are used by a 100 W lamp in 24 hours?

b How many joules of energy are used by a 5 W torch lamp in 50 minutes (3000 seconds)?

c An electricity bill showed that 1270 kWh of electricity was used. Work out the cost of this amount of electricity at 14p per kWh.

2 a Work out the number of kWh transferred in each case below.

  i A 3 kilowatt electric kettle is used six times for 5 minutes each time.

  ii A 1000 watt microwave oven is used for 30 minutes.

  iii A 100 watt electric light is used for 8 hours.

b Calculate the cost of the electricity used in part **a** if the cost of electricity is 12p per kWh.

3 a An electric heater is left on for 3 hours. During this time it uses 12 kWh of electrical energy.

  i What is the power of the heater?

  ii How many joules are supplied?

b The mains power supply of a computer provides a current of 1.7 A at 230 V.

  i Calculate the power supplied to the computer.

  ii In one month, the computer is used for 130 hours. How many kilowatt-hours of electrical energy are supplied to the computer in this time?

  iii Calculate the cost of this electrical energy if the unit cost of electricity is 12p per kWh.

16.7

Electrical issues

An electrical fault is dangerous. It could give someone a nasty shock or even electrocute them, resulting in death. Also, a fault can cause a fire. This happens when too much current passes through a wire or an appliance and heats it up.

Fault prevention

Electrical faults can happen if sockets, plugs, cables, or appliances are damaged. Users need to check for loose fittings, cracked plugs and sockets, and worn cables. Any of these damaged items need to be repaired or replaced by a qualified electrician.

- If a fuse blows or a circuit breaker trips when a mains appliance is in use, switch the appliance off. Then don't use it until it has been checked by a qualified electrician.
- If an appliance (or its cable or plug or socket) overheats and/or you get a distinctive burning smell from it, switch it off. Again, don't use it until it has been checked.

If too many appliances are connected to the same socket, this may cause the socket to overheat. If this happens, switch the appliances and the socket off and disconnect the appliances from the socket.

Smoke alarms and infrared sensors connected to an alarm system are activated if a fire breaks out. An electrical fault could cause an appliance or a cable to become hot and could set fire to curtains or other material in a room. Smoke alarms and sensors should be checked regularly to make sure they work properly.

An electrician selecting a cable for an appliance needs to use:

- a two-core cable if the appliance is **double-insulated** and no Earth wire is needed
- a three-core cable if an Earth wire is needed because the appliance has a metal case
- a cable with conductors of suitable thickness so the heating effect of the current in the cable is insignificant.

New bulbs for old

When choosing an electrical appliance, most people compare several different appliances. The cost of the appliance is just one factor that may need to be considered. Other factors might include the power of the appliance and its efficiency.

A **filament bulb** is very inefficient. The energy from the hot bulb gradually makes the plastic parts of the bulb socket brittle and they crack. If you want to replace a bulb, a visit to an electrical shop can present you with a bewildering range of bulbs.

Low-energy bulbs are much more efficient as they don't become hot like filament bulbs do. Different types of low-energy bulb are available:

- **Low-energy compact fluorescent lamp bulbs (CFLs)** are often used for room lighting instead of filament bulbs.
- **Low-energy light-emitting diodes (LEDs)** operate at low voltage and low power. They are much more efficient than filament bulbs or halogen bulbs and they last much longer.

Table 1 gives more information about these different bulbs.

Table 1 *Characteristics of different types of bulb*

Type	Power	Efficiency	Lifetime in hours	Cost of bulb	Typical use
Filament bulb	100 W	20%	1000	50p	room lighting
Halogen bulb	100 W	25%	2500	£2.00	spotlight
Low-energy compact fluorescent bulb (CFL)	25 W	80%	15 000	£2.50	room lighting
Low-energy light-emitting diode (LED)	2 W	90%	30 000	£5.00	spotlight

Summary questions

1 A socket that includes a residual current circuit breaker (RCCB) instead of a fuse should be used for mains appliances such as lawnmowers where there is a possible hazard when the appliance is used. This type of circuit breaker switches the current off if the live current and the neutral current differ by more than 30 mA. This can happen, for example, if the blades of a lawnmower cut into the cable. Create a table to show a possible electrical hazard for each of these appliances: lawnmower, electric drill, electric saw, hairdryer, vacuum cleaner. The first entry has been done for you.

Appliance	Hazard
Lawnmower	The blades might cut the cable.

2 a i If a mains appliance causes a fuse to melt, why is it a mistake to replace the fuse straight away?

  ii Should the cable of an electric iron be a two-core or a three-core cable?

 b A householder wants to replace a 100 W room light with a row of low-energy LEDs with the same light output. Use the information in Table 1 to answer the following questions.

  i How many times would the filament bulb need to be replaced in the lifetime of an LED?

  ii How many LEDs would be needed to give the same light output as a 100 W filament bulb?

  iii The householder reckons the cost of the electricity for each LED at 10p per kWh over its lifetime of 30 000 hours would be £6. Show that the cost of the electricity for a 100 W bulb over this time would be £300.

  iv Use your answers above to calculate how much the householder would save by replacing the filament bulb with LEDs.

3 An electrician needs to fit a mains cable and a mains plug to connect a 230 V, 1100 W microwave oven to a mains socket.

 a Calculate the current that passes through the appliance when it is switched on at full power.

 b The appliance has a metal case. Explain why a three-core cable rather than a two-core cable is necessary.

 c i Which type of fuse, 3 A, 5 A, or 13 A, should be fitted in the mains plug?

  ii Why is it dangerous to use a fuse with a higher current rating?

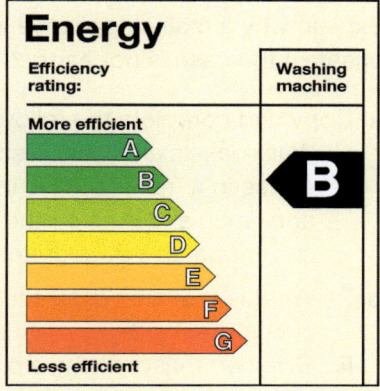
Key points

- Electrical faults are dangerous because they can cause electric shocks and fires.

- Never touch a mains appliance, plug, or socket with wet hands. Never touch a bare wire or a terminal at a potential of more than 30 V.

- Check cables, plugs, and sockets for damage regularly. Check smoke alarms and infrared sensors regularly.

- When choosing an electrical appliance, the power and efficiency rating of the appliance need to be considered.

- Filament bulbs and halogen bulbs are much less efficient than low-energy bulbs such as CFLs and LEDs.

Chapter summary questions

1 a In a mains circuit, which wire:
   i is earthed at the local substation?
   ii alternates in potential?

b An oscilloscope is used to display the potential difference of an alternating voltage supply unit. State and explain how the trace would change if:
   i the potential difference is increased
   ii the frequency is increased.

2 Explain why a mains appliance with a metal case is unsafe if the case is not earthed.

3 a Copy and complete the following sentences:
   i Wall sockets are connected in .... with each other.
   ii A fuse in a mains plug is in .... with the appliance and cuts off the .... wire if too much current passes through the appliance.

b i What is the main difference between a fuse and a circuit breaker?
   ii Give two reasons why a circuit breaker is safer than a fuse.

4 a i Calculate the current in a 230V, 2.5kW electric kettle.
   ii Which fuse, 3A, 5A, or 13A, would you fit in the kettle plug?
   iii If the kettle is used on average six times a day for 5 minutes each time, calculate the energy in kWh it uses in 28 days.

b A student uses a 4.0A, 230V microwave oven for 10 minutes every day and a 2500W electric kettle three times a day for 4 minutes each time.
   i Which appliance uses more energy in one day?
   ii Calculate the total cost of using these appliances for 7 days if the unit cost of electricity is 14p per kWh.

5 A 5Ω resistor is in series with a bulb, a switch, and a 12V battery.

a Draw the circuit diagram.

b When the switch is closed for 60 seconds, a direct current of 0.6A passes through the resistor. Calculate:
   i the energy supplied by the battery
   ii the energy transferred to the resistor
   iii the energy transferred to the bulb.

c The bulb is replaced by a 25Ω resistor.
   i Calculate the total resistance of the two resistors.
   ii Calculate the battery current.
   iii Calculate the power supplied by the battery and the power delivered to each resistor.

6 A 12V, 36W bulb is connected to a 12V supply.

a Calculate:
   i the current through the bulb
   ii the charge flow through the bulb in 200s.

b i Show that 7200J of electrical energy is delivered to the bulb in 200s.
   ii Calculate the energy delivered to the bulb by each coulomb of charge that passes through it.

c A second 12V, 36W bulb is connected to the power supply in parallel with the first bulb.
   i Calculate the current through each bulb and through the battery.
   ii Show that the energy delivered per second to the two bulbs is equal to the energy supplied per second by the battery.

7 An electrician has the job of connecting a 6.6kW electric oven to the 230V mains supply in a house.

a Calculate the current needed to supply 6.6kW of electrical power at 230V.

b The table below shows the maximum current that can pass safely through five different mains cables. For each cable the cross-sectional area of each conductor is given in square millimetres (mm²).

	Cross-sectional area of conductor (mm²)	Maximum safe current (A)
A	1.0	14
B	1.5	18
C	2.5	28
D	4.0	36
E	6.0	46

   i To connect the oven to the mains supply, which cable should the electrician choose? Give a reason for your answer.
   ii State and explain what would happen if she chose a cable with thinner conductors.

8 a i What are transformers used for in the National Grid?
   ii What type of transformer is connected between the generators in the power station and the cables of the grid system?

b i What can you say about the voltage of the cables in the grid system compared with the voltages at the power station generator and at the mains cables into your home?
   ii What can you say about the current through the grid cables compared with the current from the power station generator?
   iii What is the reason for making the grid voltage different from the generator voltage?

Practice questions

1 A hairdryer has a label on it.

Electrical supply	230 V 50 Hz
Maximum power	1200 W

a What can you deduce from the label? (5)

b The diagrams show the readings on a domestic electricity meter in February and May.

Electricity costs 15p per kilowatt-hour.

February

May

i Calculate the cost of the electricity used between the two meter readings. (3)

ii A 2000 W electric fire is used for 3 hours. Calculate the cost of using the electric fire for 3 hours. (3)

c A 60 W lamp is connected to the 230 V mains supply.
i Calculate the current flowing through the lamp. (2)
ii Calculate the amount of electrical charge that flows through the lamp in 30 hours. (3)

2 a i Calculate the current in a 230 V 1000 W heater when it operates normally. (1)
ii Which one of the following fuses is most suitable for this heater? (1)

3 A   5 A   13 A

iii Explain why each of the other fuses are unsuitable. (2)

b If your hands are wet it is dangerous to plug in an electrical appliance.

Why is this dangerous? (2)

c The diagrams show an electric kettle and an electric hairdryer.

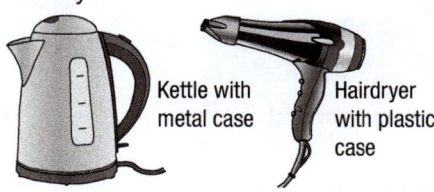

Kettle with metal case

Hairdryer with plastic case

The kettle is earthed, but the hairdryer is not earthed. Explain why it is not necessary to earth the hairdryer and how earthing the kettle protects the user. (6)

3 The diagrams give information about four types of electric lamp.

Each lamp produces the same amount of light energy in the same time.

 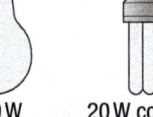 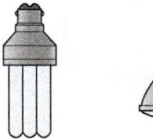  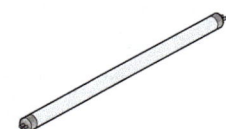

100 W filament lamp

20 W compact fluorescent lamp (CFL)

10 W LED spotlight

15 W fluorescent tube

a Which lamp is most efficient? (1)
b Which lamp would become hottest when it is working? (1)
c Which lamp would be the cheapest to run for 1000 hours? (1)

4 A power supply is connected to a cathode ray oscilloscope.

The trace on the oscilloscope screen is shown below.

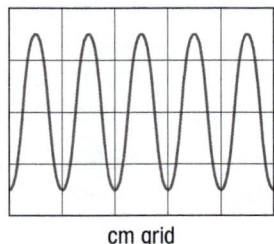

cm grid

The oscilloscope settings are:

X-axis 0.02 s per cm

Y-axis 2 V per cm

What is the:

a frequency of the supply? (3)

b peak potential difference? (2)

17.1

Atoms and radiation

Learning objectives

After this topic, you should know:

- what a radioactive substance is

- the types of radiation given out from a radioactive substance

- when a radioactive source will give out radiation (radioactive decay)

- the origins of background radiation.

Figure 2 *Marie Curie 1867–1934*

??? Did you know ... ?

Becquerel and the Curies were awarded the Nobel Prize for the discovery of radioactivity. After Pierre died in a road accident, Marie continued their work. She was awarded a second Nobel Prize in 1911 for the discovery of polonium and radium. She died in 1934 from leukaemia, a disease of the blood cells. It was probably caused by the radiation from the radioactive materials she worked with.

A key discovery

Figure 1 *Becquerel's key*

If you took a photo that showed a mysterious image, what would you think? In 1896, a French physicist, **Henri Becquerel**, discovered the image of a key on a film he developed. He remembered the film had been in a drawer under a key (Figure 1). On top of that there had been a packet of uranium salts. The uranium salts must have sent out some form of radiation that passed through paper (the film wrapper) but not through metal (the key).

Becquerel asked a young research worker, **Marie Curie**, to investigate. She found that the salts gave out radiation all the time. It happened no matter what was done to them. She used the word **radioactivity** to describe this strange new property shown by uranium.

She and her husband, Pierre, did more research into this new branch of science. They discovered new radioactive elements. They named one of the elements **polonium**, after Marie's native country, Poland.

Practical

Investigating radioactivity

You can use a **Geiger counter** to detect radioactivity. Look at Figure 3. The counter clicks each time a particle of radiation from a radioactive substance enters the Geiger tube.

Safety: Avoid touching or inhaling radioactive material.

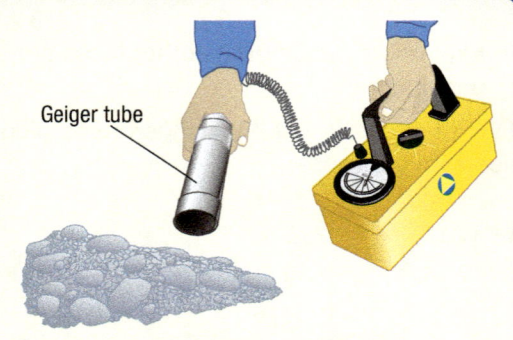

Figure 3 *Using a Geiger counter*

Inside the atom

What materials can prevent radiation from passing through? Ernest Rutherford carried out tests to answer this question about a century ago. He placed different materials between various radioactive substances and a detector.

He discovered two types of radiation:

- One type (**alpha radiation**, symbol α) was stopped by paper.
- The other type (**beta radiation**, symbol β) went through the paper.

Scientists later discovered a third type, **gamma radiation** (symbol γ), even more penetrating than β radiation, and also **neutron radiation** (See Topic 17.7 'Nuclear fission').

Rutherford carried out further investigations and discovered that α radiation is made up of positively charged particles. He realised that these particles could be used to probe the atom. His research students included Hans Geiger, who invented what was later called the Geiger counter. They carried out investigations in which a narrow beam of α particles was directed at a thin metal foil. Rutherford was astonished that some of the α particles rebounded from the foil. He proved that this happens because every atom has at its centre a positively charged nucleus containing most of the mass of the atom. He went on to propose that the nucleus contains two types of particle – protons and neutrons.

A radioactive puzzle

Why are some substances radioactive? Every atom has a nucleus made up of protons and neutrons. Electrons move around within energy levels (or shells) surrounding the nucleus.

Most atoms each have a stable nucleus that doesn't change. But the atoms of a radioactive substance have a nucleus that is unstable. An unstable nucleus becomes stable by emitting α, β, or γ radiation. An unstable nucleus **decays** when it emits radiation (Figure 4).

It is not possible to tell when an unstable nucleus will decay. It is a **random** event that happens without anything being done to the nucleus.

The origins of background radiation

A Geiger counter clicks even when it is not near a radioactive source. This is caused by **background radiation** (Figure 5). This is ionising radiation from radioactive substances:

- in the environment (e.g., in the air, the ground, or building materials)
- from space (cosmic rays)
- from man-made sources such as fallout (leftover radioactive substances in the air) from nuclear weapons testing and nuclear accidents at some power stations.

Most background radiation is from naturally occurring substances in the Earth. For example, radon gas is radioactive and is a product of the decay of uranium found in the rocks in certain areas. Medical sources of background radiation include X-ray tubes and radioactive substances used in hospitals.

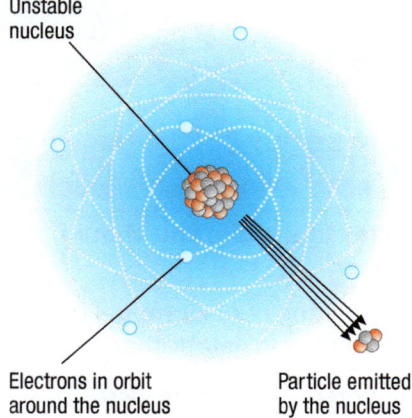

Unstable nucleus

Electrons in orbit around the nucleus

Particle emitted by the nucleus

Figure 4 *Radioactive decay*

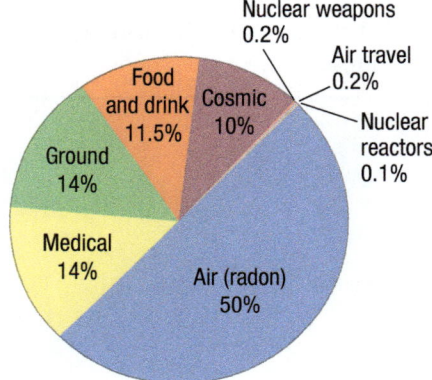

Nuclear weapons 0.2%
Air travel 0.2%
Food and drink 11.5%
Cosmic 10%
Nuclear reactors 0.1%
Ground 14%
Medical 14%
Air (radon) 50%

Figure 5 *The origins of background radiation*

∞ links
Look back at Topic 8.5 'X-rays in medicine' for more about X-rays.

Key points

- A radioactive substance contains unstable nuclei that become stable by emitting radiation.

- There are three main types of radiation from radioactive substances – alpha (α), beta (β), and gamma (γ) radiation.

- Radioactive decay is a random event – you cannot predict or influence when it will happen.

- Most background radiation results from radioactive substances that occur naturally in rocks or space or from man-made sources such as nuclear fallout.

Summary questions

1 **a** State two differences between the radiation from uranium and the radiation from a lamp.
  b State two differences between radioactive atoms and the atoms in a lamp filament.

2 **a** The radiation from a radioactive source is stopped by paper. What type of radiation does the source emit?
  b The radiation from a different source goes through paper. What can you say about this radiation?

3 **a** Explain why some substances are radioactive.
  b State two sources of background radioactivity.
  c A Geiger counter clicks very rapidly when a certain substance is brought near it. After the substance is taken away, the Geiger counter still clicks, but much less often.
    i What can you say about the substance that made the Geiger counter click?
    ii Why did the counter click after the source had been moved away?

17.2 The discovery of the nucleus

After this topic, you should know:

- how the nuclear model of the atom was established

- what conclusions were made about the atom from experimental evidence

- why the nuclear model was accepted.

The physicist Ernest Rutherford discovered that alpha and beta radiation is made up of different types of particles. He realised that alpha (α) particles could be used to probe the atom. He asked two of his research workers, Hans Geiger and Ernest Marsden, to investigate how a thin metal foil scatters a beam of alpha particles. Figure 1 shows the arrangement they used.

The apparatus was in a vacuum chamber to prevent air molecules absorbing the alpha α particles. The detector was moved to different positions. At each position, the number of spots of light observed in a certain time was counted. The detector consisted of a microscope focused on a small glass plate. Each time an alpha particle hit the plate, a spot of light was observed.

Their results showed that:

- most of the alpha particles passed straight through the metal foil

- the number of alpha particles deflected per minute decreased as the angle of deflection increased

- about 1 in 10 000 alpha particles were deflected by more than 90°.

Rutherford was astonished by the results. He said it was like firing naval shells at tissue paper and discovering that a small number of the shells rebound. He knew that α particles are positively charged and that the radius of an atom is about 10^{-10} m. He deduced from the results that there is a nucleus at the centre of every atom that is:

- much smaller than the atom because most α particles pass through without deflection

- where most of the mass of the atom is located.

Rutherford's nuclear model of the atom was quickly accepted because it:

- agreed exactly with the measurements Geiger and Marsden made in their experiments

- explains radioactivity in terms of changes that happen to an unstable nucleus when it emits radiation

- predicted the existence of the neutron, which was later discovered.

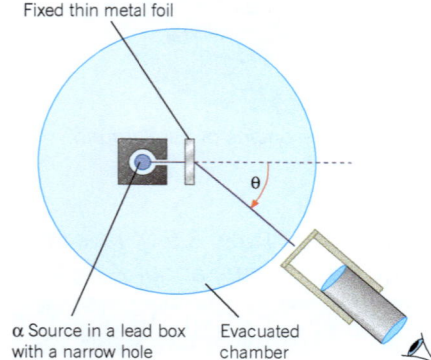

Figure 1 *Alpha particle scattering*

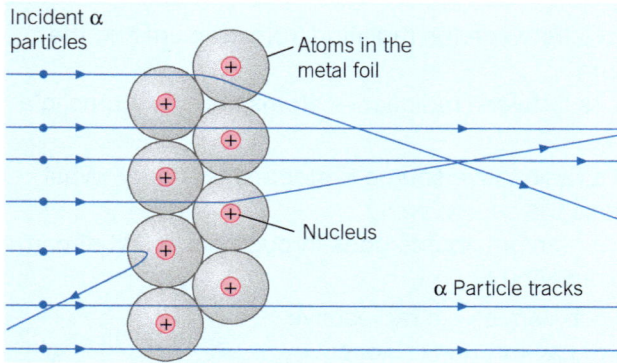

Figure 2 *Alpha (α) particle tracks*

Bohr's model of the atom

After Rutherford's discovery, scientists knew that every atom has a positively charged nucleus that negatively charged electrons move around. The physicist Niels Bohr put forward the theory that the electrons in an atom orbit the nucleus at specific distances and specific energy values or **energy levels**. His model of the atom showed that the electrons in an orbit can move to another orbit by absorbing electromagnetic radiation to move away from the nucleus or by emitting electromagnetic radiation to move closer to the nucleus. His calculations based on his atomic model agreed with experimental observations of the light emitted by atoms.

A nuclear puzzle

More α-scattering experiments showed that:

- the hydrogen nucleus has the least amount of charge
- the charge of any nucleus is shared equally between a whole number of smaller particles, each with the same amount of positive charge.

The name proton was given to the hydrogen nucleus because scientists reckoned that every other nucleus contained hydrogen nuclei. But they also knew that the mass of every nucleus except for the hydrogen nucleus is bigger than the total mass of its protons. So there must be an uncharged type of particle with about the same mass as a proton in every nucleus except the hydrogen nucleus. They called this uncharged particle the neutron. The proton–neutron model of the nucleus explains all the mass and charge values of every nucleus. Direct experimental evidence for its existence was found by the physicist James Chadwick about 20 years after Rutherford's discovery of the nucleus.

Go further!

An atom emits electromagnetic radiation when an electron moving around the nucleus jumps from one energy level to a lower energy level. The radiation is emitted as a **photon**, which is a packet of waves emitted in a short burst. The energy of the emitted photon is equal to the energy change of the electron. Einstein put forward the photon theory and Bohr used it in his calculations.

Summary questions

1   Write four features of every nucleus of every atom.

2 a Figure 3 shows four possible paths, labelled A, B, C, and D, of an alpha particle deflected by a nucleus. Write the path the alpha particle would travel along.
  b Explain why each of the other paths in part **a** is not possible.
  c Explain why the alpha-scattering experiment led to the acceptance of the nuclear model of the atom.

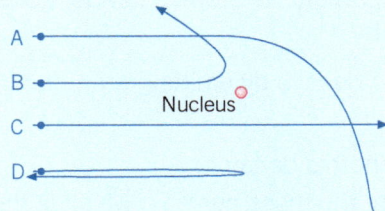

Figure 3

3 a Write one difference and one similarity between a proton and a neutron.
  b Explain why the mass of a helium nucleus is four times the mass of a hydrogen nucleus and its charge is only twice as much as the charge of a hydrogen nucleus.

Key points

- Rutherford used α particles to probe inside atoms. He found that some of the α particles were scattered through large angles.

- An atom has a small positively charged central nucleus where most of the atom's mass is locate

- The nuclear model of the atom correctly explained why some α particles scattered through large angles.

17.3 Nuclear reactions

Nuclear physics

Learning objectives

After this topic, you should know:

- what an isotope is

- how the nucleus of an atom changes when it emits an α particle or a β particle

- how to represent the emission of an α particle or a β particle from a nucleus.

Table 1 *Relative masses and charges*

	Relative mass	Relative charge
Proton	1	+1
Neutron	1	0
Electron	$\approx \frac{1}{2000}$	−1

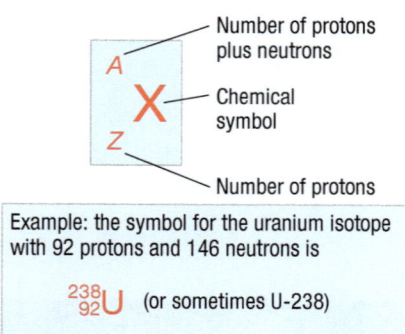

Number of protons plus neutrons

Chemical symbol

Number of protons

Example: the symbol for the uranium isotope with 92 protons and 146 neutrons is

$^{238}_{92}U$ (or sometimes U-238)

Figure 1 *Representing an isotope*

The nucleus emits an α particle and forms a new nucleus

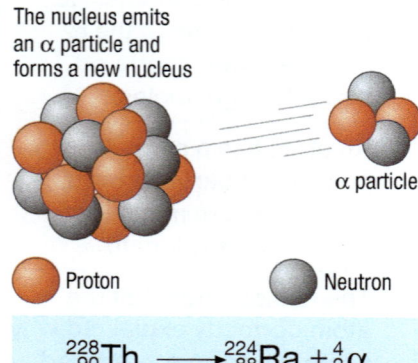

α particle

○ Proton   ○ Neutron

$^{228}_{90}Th \longrightarrow ^{224}_{88}Ra + ^{4}_{2}\alpha$

Figure 2 *α emission*

In α (alpha) or β (beta) decay, the number of protons in a nucleus changes. In α decay, the total number of neutrons and protons also changes. This topic looks at the changes that happen in α and β decay and how these changes are represented.

Table 1 gives the relative masses and the relative electric charges of a proton, a neutron, and an electron.

Atoms are uncharged. They have equal numbers of protons (+) and electrons (–). A charged particle, called an **ion**, is formed when an atom gains or loses one or more electrons. Then there are unequal numbers of protons and electrons in the ion.

Atoms of the same element each have the same number of protons. The number of protons in a nucleus is given the symbol Z. It is called the **atomic number** (or **proton number**).

Isotopes are atoms of the same element with different numbers of neutrons. Isotopes of an element have nuclei with the same number of protons but a different number of neutrons.

The number of protons plus neutrons in a nucleus is called its **mass number**. It has the symbol A.

Figure 1 shows how to represent an isotope of an element X, which has Z protons and A protons plus neutrons. For example, the uranium isotope $^{238}_{92}U$ contains 92 protons and 146 neutrons (= 238 – 92) in each nucleus. So, its mass number is 238 and the relative charge of the nucleus is +92.

Radioactive decay

An unstable nucleus becomes more stable by emitting an α particle or a β particle or by emitting γ radiation.

α emission

An α particle consists of two protons plus two neutrons. Its relative mass is 4 and its relative charge is +2. So it is represented by the symbol $^{4}_{2}\alpha$.

When an unstable nucleus emits an α particle, its atomic number goes down by 2 and its mass number goes down by 4.

For example, the thorium isotope $^{228}_{90}Th$ decays by emitting an α particle. So it forms the radium isotope $^{224}_{88}Ra$.

Figure 2 shows an equation to represent this decay.

- The numbers along the top represent the mass number, which is the number of protons and neutrons in each nucleus and in the α particle.

- The equation shows that the total number of protons and neutrons after the change (= 224 + 4) is equal to the total number of protons and neutrons before the change (= 228).

- The numbers along the bottom represent the atomic number, which is the number of protons in each nucleus and in the α particle.

- The equation shows that the total number of protons after the change (= 88 + 2) is equal to the total number of protons before the change (= 90).

β emission

A β particle is an electron created and emitted by a nucleus which has too many neutrons compared with its number of protons. A neutron in the nucleus changes into a proton and a β particle. The β particle is instantly emitted at high speed by the nucleus.

The relative mass of a β particle is effectively zero and its relative charge is –1. So a β particle is represented by the symbol $_{-1}^{0}\beta$.

When an unstable nucleus emits a β particle, the atomic number of the nucleus goes up by 1 but its mass number stays the same (because a neutron changes into a proton).

For example, the potassium isotope $_{19}^{40}$K decays by emitting a β particle. So it forms a nucleus of the calcium isotope $_{20}^{40}$Ca (Figure 3).

- The numbers along the top represent the mass number, which is the number of protons and neutrons in each nucleus and zero for the β particle, as explained above.
- The equation shows that the total number of protons and neutrons after the change (= 40 + 0) is equal to the total number of protons and neutrons before the change (= 40).
- The numbers along the bottom represent the atomic number. This is the number of protons in each nucleus and –1 for the β particle. (Note that the relative charge of the β particle is –1, so its atomic number is shown as –1 in nuclear equations, even though it has no protons.)
- The equation shows that the total charge (in relative units) after the change (= 20 – 1) is equal to the total charge before the change (= 19).

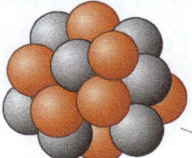

A β particle is created in the nucleus and instantly emitted

A neutron in the nucleus changes into a proton

$$_{19}^{40}K \longrightarrow\ _{20}^{40}Ca + _{-1}^{0}\beta$$

Figure 3 β emission

γ emission

γ radiation is emitted by some unstable nuclei after an α particle or a β particle has been emitted. γ radiation is uncharged and has no mass. So it does not change the number of protons or the number of neutrons in a nucleus.

Study tip

Make sure you know the changes to mass number and to atomic number that occur in α decay and in β decay.

Key points

- Isotopes of an element are atoms with the same number of protons but different numbers of neutrons. Therefore they have the same atomic number but different mass numbers.

α decay	β decay
Change in the nucleus	
Nucleus loses 2 protons and 2 neutrons	A neutron in the nucleus changes into a proton
Particle emitted	
2 protons and 2 neutrons emitted as an α particle	An electron is created in the nucleus and instantly emitted
Equation	
$_{Z}^{A}X \rightarrow\ _{Z-2}^{A-4}Y +\ _{2}^{4}\alpha$	$_{Z}^{A}X \rightarrow\ _{Z+1}^{A}Y +\ _{-1}^{0}\beta$

Summary questions

1 How many protons and how many neutrons are there in the nucleus of each of the following isotopes:

a $_{6}^{12}$C? b $_{27}^{60}$Co? c $_{92}^{235}$U?

d How many more protons and how many more neutrons are there in $_{92}^{238}$U than in $_{88}^{224}$Ra?

2 A substance contains the radioactive isotope $_{92}^{238}$U, which emits α radiation. The product nucleus X emits β radiation and forms a nucleus Y. How many protons and how many neutrons are present in:

a a nucleus of $_{92}^{238}$U? b a nucleus of X? c a nucleus of Y?

3 a Copy and complete the following equations for α and β decay.

i $_{92}^{235}U \rightarrow\ _{?}^{?}Th +\ _{2}^{4}\alpha$ ii $_{29}^{64}Cu \rightarrow\ _{?}^{?}Zn +\ _{-1}^{0}\beta$

b A radioactive isotope of polonium (Po) has 84 protons and 126 neutrons.

i The isotope is formed from the decay of a radioactive isotope of bismuth (Bi), which emits a β particle in the process. Copy and complete the equation below with the correct values of A and Z to represent this decay.

Bi → Po + β

ii The polonium isotope decays by emitting an α particle to form a stable isotope of lead (Pb). Copy and complete the equation below with the correct values of A and Z to represent this decay.

Po → Pb + α

More about alpha, beta, and gamma radiation

Penetrating power

α radiation cannot penetrate paper. But what materials stop β and γ radiation? And how far can each type of radiation travel through air? You can use a Geiger counter to find out, but you must take account of background radiation. To do this you should:

1 Measure the count rate (which is the number of counts per second) without the radioactive source present. This is the background count rate.

2 Measure the count rate with the source in place. Subtracting the background count rate from this gives the count rate from the source alone.

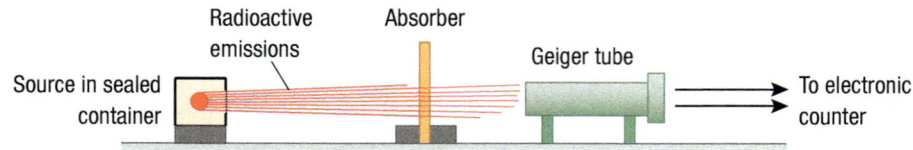

Figure 1 *Absorption tests*

You can then test absorber materials for each source. You can also test the range in air – how far away from the source you can detect a count rate due to the source.

- To test different absorber materials, place each material between the tube and the radioactive source. Then measure the count rate. You can add more layers of material until the count rate from the source is zero. The radiation from the source has then been stopped by the absorber material.

- To test the range in air, move the tube away from the source, measuring the distance each time. When the tube is beyond the range of the radiation, the count rate from the source is zero.

Table 1 shows some results of these two tests.

Table 1 *Testing the penetrating power of α, β, and γ radiation*

Type of radiation	Absorber materials	Range in air
Alpha (α)	Thin sheet of paper	About 5 cm
Beta (β)	Aluminium sheet (about 5 mm thick) Lead sheet (2–3 mm thick)	About 1 m
Gamma (γ)	Thick lead sheet (several cm thick) Concrete (more than 1 m thick)	Unlimited

γ radiation spreads out in air without being absorbed. It gets weaker as it spreads out.

Radioactivity dangers

Radioactive substances must always be kept in sealed containers to prevent **contamination** of any other materials by the unwanted presence of a radioactive substance on the materials.

The radiation from a radioactive substance can knock electrons out of atoms. The atoms become charged because they lose electrons. The process is called **ionisation**. (Remember that a charged particle is called an ion.) X-rays also cause ionisation.

Objects exposed to ionising radiation are said to be **irradiated** by the radiation. The irradiated object itself does not become radioactive, even though atoms within it may be ionised.

Ionisation in a living cell can damage or kill the cell. Damage to the genes in a cell can be passed on if the cell generates more cells. Strict safety rules must always be followed when radioactive substances are used.

α radiation is more dangerous in the body than β or γ radiation. This is because it ionises substances much more than β radiation, which is more ionising than γ radiation. In other words, the ionising power of α radiation is much greater than the ionising power of β or γ radiation.

Figure 2 *Radioactive warnings*

Summary questions

1 **a** Why is a radioactive source stored in a lead-lined box?
 b How do physicists know that γ radiation is **not** made up of charged particles?
 c Why should long-handled tongs be used to move a radioactive source?
 d What type or types of radiation from a radioactive source is stopped by a thick aluminium plate?

2 **a** Which type of radiation is:
   i uncharged
   ii positively charged
   iii negatively charged?
 b A narrow beam of α, β, and γ radiation was directed at a sheet of paper in front of a Geiger counter, as shown in Figure 1.
   i When the paper was removed, state and explain how the Geiger counter reading changed.
   ii A thick sheet of lead was placed in front of the Geiger tube where the sheet of paper was. State and explain how the Geiger counter reading changed.
   iii When the radioactive source was removed, explain why the Geiger counter continued to count.

3 **a** Explain why ionising radiation is dangerous.
 b Explain how you would use a Geiger counter to find the range of the radiation from a source of α radiation.

Key points

- **α radiation** is stopped by paper, has a range of a few centimetres in air, and consists of particles, each composed of two protons and two neutrons. It has the greatest ionising power.

- **β radiation** is stopped by a thin sheet of metal, has a range of about a metre in air, and consists of fast-moving electrons emitted from the nucleus. It is less ionising than α radiation and more ionising than γ radiation.

- **γ radiation** is stopped by thick lead, has an unlimited range in air, and consists of electromagnetic radiation.

- α, β, and γ radiation ionise substances they pass through. Ionisation in a living cell can damage or kill the cell.

17.5 Half-life

17.5

Learning objectives

After this topic, you should know:

- what is meant by the 'half-life' of a radioactive source
- what is meant by the count rate from a radioactive source
- what happens to the count rate from a radioactive isotope as it decays.

Half-lives and instability

Radioactive isotopes have a wide range of half-lives. Some radioactive isotopes have half-lives of a fraction of a second, but others have half-lives of more than a billion years. Isotopes with:

- nuclei that are the most unstable have the shortest half-lives. They emit a lot of radiation in a short amount of time.
- nuclei that are the least unstable have the longest half-lives. They emit little radiation each second, but for a long time.

For a radioactive source made up of a single isotope, the half-life tells you how quickly its activity decreases. As its activity decreases, the rate at which it gives out radiation decreases. So the hazards caused by the ionising effect of the radiation from radioactive materials decreases with time according to the half-lives of the isotopes they contain.

Study tip

Remember that the half-life of a radioactive isotope is the time taken for the number of nuclei to halve or for the count rate to halve.

Every atom of an element has the same number of protons in its nucleus. However, different atoms of the element can have different numbers of neutrons. Each different type of atom (with the same number of protons but different numbers of neutrons) is called an isotope.

The **activity** of a radioactive isotope is the number of atoms that decay per second. As the nucleus of each unstable parent atom decays, the number of undecayed parent atoms goes down. So the activity of the sample decreases.

You can use a Geiger counter to monitor the activity of a radioactive sample. You need to measure the **count rate** from the sample. This is the number of counts per second (or per minute). The graph below shows how the count rate of a sample decreases.

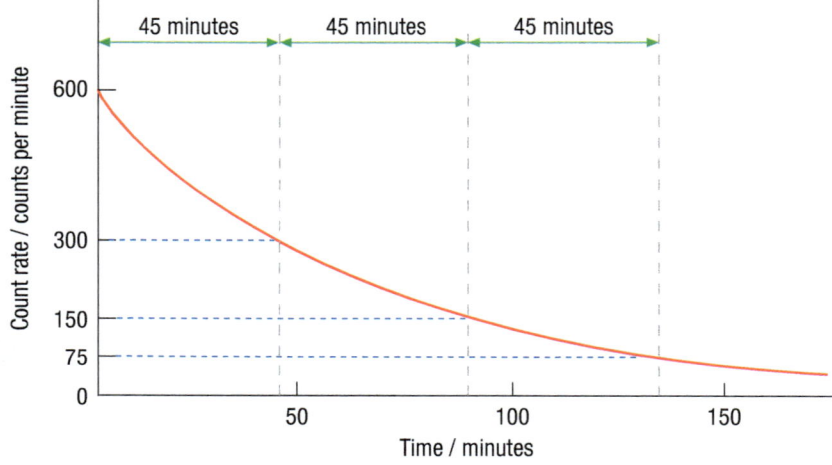

Figure 1 *Radioactive decay – a graph of count rate against time*

The graph shows that the count rate decreases with time. The count rate falls from:

- 600 counts per minute (c.p.m.) to 300 c.p.m. in the first 45 minutes
- 300 c.p.m. to 150 c.p.m. in the next 45 minutes.

The average time taken for the count rate (and therefore the number of parent atoms) to fall by half is always the same. This time is called the **half-life**. The half-life shown on the graph is 45 minutes.

The half-life of a radioactive isotope is the average time it takes:
- **for the number of nuclei of the isotope in a sample (and therefore the mass of parent atoms) to halve**
- **for the count rate from the isotope in a sample to fall to half its initial value.**

??? Did you know …?

The nitrogen isotope N-12 has a half-life of 0.0125 seconds. The uranium isotope U-238 has a half-life of 4.5 billion years.

The random nature of radioactive decay

Radioactive decay is a random process. It is not possible to predict *when* an individual atom will suddenly decay. But it *is* possible to predict how many atoms will decay in a certain time – because there are so many of them. This is a bit like throwing dice. You can't predict what number you will get with a single throw. But if you threw a dice 1000 times, you would expect one-sixth to come up with a particular number.

Suppose there are 1000 unstable atoms to start with. Look at Figure 2.

If 10% of the atoms decay every hour:

- 100 atoms will decay in the first hour, leaving 900
- 90 atoms (= 10% of 900) will decay in the second hour, leaving 810.

Table 1 shows the result if you continue the calculations. The results are plotted in Figure 2.

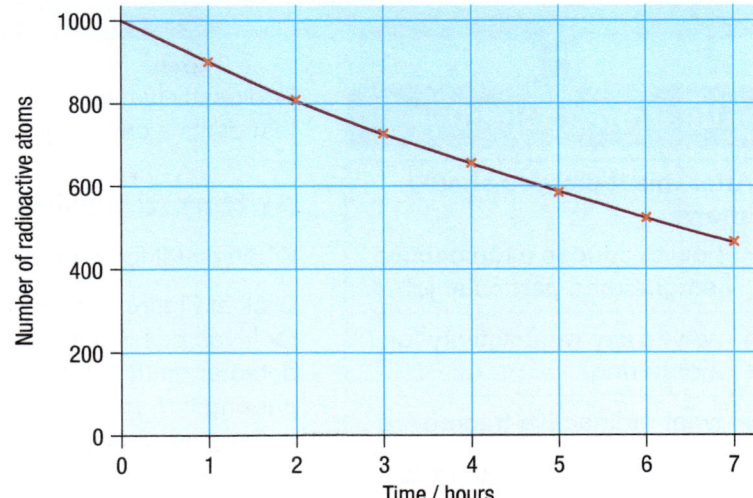

Figure 2 *Half-life*

Table 1 *Radioactive decay*

Time from start (hours)	0	1	2	3	4	5	6	7
No. of unstable atoms present	1000	900	810	729	656	590	531	478
No. of unstable atoms that decay in the next hour	100	90	81	73	66	59	53	48

Summary questions

1 a What is meant by the half-life of a radioactive isotope?
  b What will the count rate in Figure 1 be after 135 minutes from the start?
  c Use the graph in Figure 2 to work out the half-life of the radioactive isotope.

2 A radioactive isotope has a half-life of 15 hours. A sealed tube contains 8 milligrams of this isotope.
  a What mass of the isotope is in the tube:
    i 15 hours later?
    ii 45 hours later?
  b Estimate how long it would take for the mass of the isotope to decrease to less than 5% of the initial mass.

3 a A sample of a radioactive isotope contains 320 million atoms of the isotope. How many atoms of the isotope are present after:
    i one half-life?   ii five half-lives?
  b In Figure 1, estimate how long it would take for the count rate to decrease to less than 40 counts per minute.

Key points

- The half-life of a radioactive isotope is the average time it takes for the number of nuclei of the isotope in a sample to halve.

- The count rate of a Geiger counter due to a radioactive source decreases as the activity of the source decreases.

- The number of atoms of a radioactive isotope and the count rate both decrease by half every half-life.

17.6

Radioactivity at work

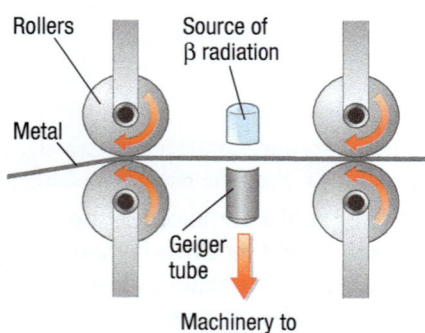

Figure 1 *Thickness monitoring using a radioactive source*

Radioactivity has many uses. For each use, a radioactive isotope is chosen that emits a certain type of radiation and has a suitable half-life.

Automatic thickness monitoring

When metal foil is made, the manufacturer monitors its thickness.

Look at Figure 1. The radioactive source emits β radiation. The amount of radiation passing through the foil depends on the thickness of the foil. A detector on the other side of the metal foil measures the amount of radiation passing through it.

- If the thickness of the foil increases too much, the detector reading drops.
- The detector sends a signal to the rollers to increase the pressure on the metal sheet.
- This makes the foil thinner again.

Radioactive tracers

Radioactive tracers are used to trace the flow of a substance through a system. Doctors use radioactive iodine to find out if a patient's kidney is working properly.

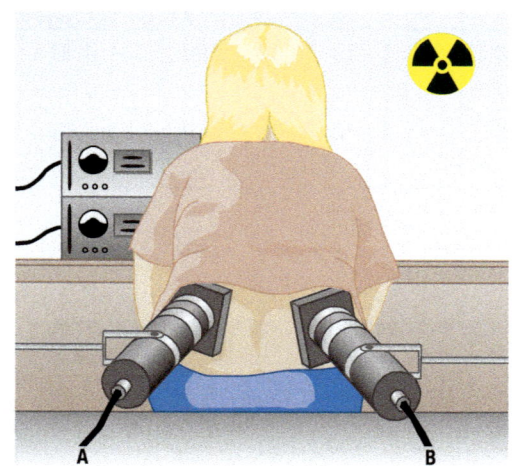

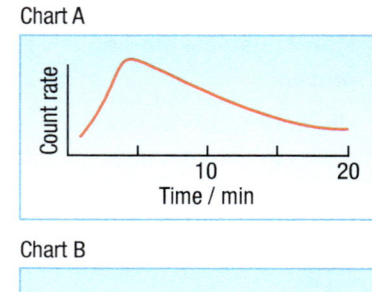

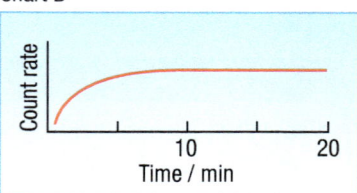

Figure 2 *Using a tracer to monitor a patient's kidneys*

Before the test, the patient drinks water containing a tiny amount of the radioactive substance. A detector is then placed against each kidney. Each detector is connected to a chart recorder.

- The radioactive substance flows into a normal kidney which removes it from the body. So the detector reading goes up and then down (see Chart A).
- In a damaged kidney, the reading goes up and stays up. This is because the radioactive substance goes into the kidney and is not removed (see Chart B).

Radioactive iodine is used for this test because:

- Its half-life is 8 days, so it lasts long enough for the test to be done but decays almost completely after a few weeks.
- It emits γ radiation, so it can be detected outside the body.
- It decays into a stable product.

Radioactive dating

Radioactive dating is used to find the age of ancient material. Two techniques are used:

- **Carbon dating** is used to find the age of ancient wood and other organic material. Living wood contains a tiny proportion of radioactive carbon. This has a half-life of 5600 years. When a tree dies, it no longer absorbs any carbon. So the amount of radioactive carbon in it decreases. To find the age of a sample, scientists need to measure the count rate from the wood. This is compared with the count rate from the same mass of living wood. For example, suppose the count rate in a sample of wood is half the count rate of an equal mass of living wood. Then the sample must be 5600 years old.

- **Uranium dating** is used to find the age of igneous rocks. These rocks contain radioactive uranium, which has a half-life of 4500 million years. Each uranium atom decays into an atom of lead. Scientists work out the age of a sample by measuring the number of atoms of uranium and lead. For example, if a sample contains one atom of lead for every atom of radioactive uranium, the age of the sample must be 4500 million years. This is because there must have *originally* been two atoms of uranium for each atom of uranium now present, and it would take one half-life for this to fall to one.

> ### Did you know … ?
>
> Smoke alarms save lives. A radioactive source inside the alarm sends out α particles into a gap in a circuit in the alarm. The α particles ionise the air in the gap so it conducts a current across the gap. In a fire, smoke absorbs the ions created by the α particles so they don't cross the gap. The current across the gap falls and the alarm sounds.
>
> The battery in a smoke alarm needs to be checked regularly to make sure it is still working.

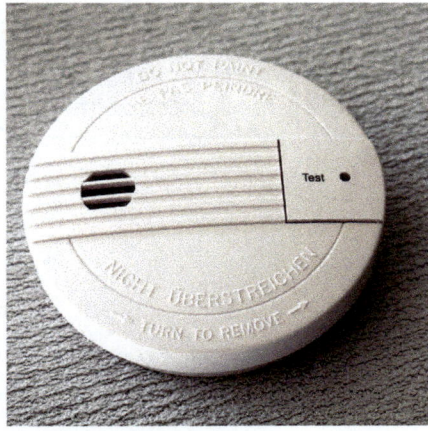

Figure 3 *A smoke alarm*

Study tip

Make sure you can choose an appropriate radioactive isotope for a particular task.

Summary questions

1 Radiation from radioactive sources is used for different purposes. Which type of radiation, α, β, or γ, is used in the examples **a** to **c** below? Give a reason for your choice in each example.
   a monitoring the continuous production of thin metal sheets
   b finding out whether a kidney in a patient is working
   c monitoring a leak in an underground pipeline

2 a Explain why γ radiation is not suitable for monitoring the thickness of metal foil.
   b When a radioactive tracer is used, why is it best to use a radioactive isotope that decays into a stable isotope?

3 a What are the ideal properties of a radioactive isotope used as a medical tracer?
   b i A sample of old wood was carbon dated and found to have 25% of the count rate measured in an equal mass of living wood. The half-life of the radioactive carbon is 5600 years. How old is the sample of wood?
     ii When the count rate measurements were made, they needed to be corrected to take account of background radiation. Explain why this correction is necessary, and describe how it is made.

Key points

- The suitable uses of a radioactive isotope depend on:
  a its half-life
  b the type of radiation it gives out.

- For monitoring uses, an isotope should have a long half-life.

- Radioactive tracers should be β or γ emitters that last long enough to be monitored, but not too long to endanger the people using them.

- For radioactive dating of a sample, a radioactive isotope is needed that is present in the sample and has a half-life about the same as the age of the sample.

17.7 Nuclear fission

Chain reactions

Energy is released in a nuclear reactor as a result of **nuclear fission**. The amount of energy released during nuclear fission is much greater than that released in a chemical reaction involving a similar mass of material. In nuclear fission, the nucleus of an atom of a fissionable substance splits into two smaller 'fragment' nuclei as a result of absorbing a neutron. This event can release several more neutrons, which can cause other fissionable nuclei to split.

This then produces a **chain reaction** of fission events.

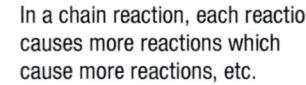

In a chain reaction, each reaction causes more reactions which cause more reactions, etc.

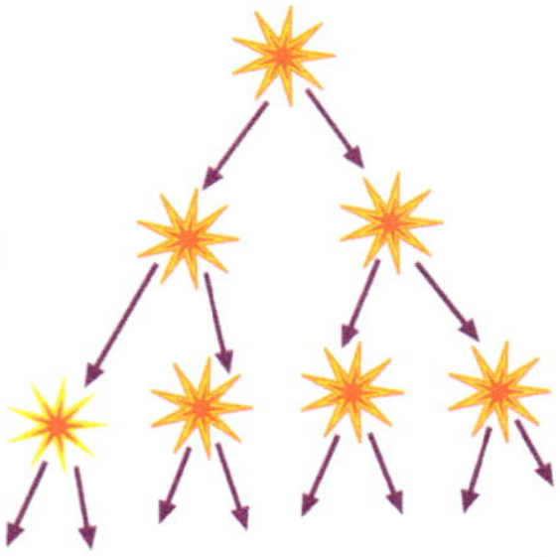

Figure 1 *A chain reaction*

Fission neutrons

When a nucleus undergoes fission, it releases:

- two or three neutrons (called fission neutrons) at high speeds
- energy, in the form of radiation, plus kinetic energy of the fission neutrons and the fragment nuclei.

The fission neutrons may cause further fission resulting in a chain reaction. In a **nuclear fission reactor**, the reaction is controlled and on average only one fission neutron from each fission event goes on to produce further fission. This ensures energy is released at a steady rate in the reactor.

Fissionable isotopes

The fuel in a nuclear reactor must contain fissionable isotopes.

- Most reactors at the present time are designed to use enriched uranium as the fuel. This consists mostly of the non-fissionable uranium isotope $^{238}_{92}$U (U-238) and about 2–3% of the uranium isotope $^{235}_{92}$U (U-235), which *is* fissionable. In comparison, natural uranium is more than 99% U-238.
- The U-238 nuclei in a nuclear reactor do not undergo fission but they change into other heavy nuclei, including plutonium-239. The isotope $^{239}_{94}$Pu is fissionable. It can be used in a different type of reactor, but not in a uranium-235 reactor, which is the most common type of reactor.

Inside a nuclear reactor

A nuclear reactor consists of uranium fuel rods spaced evenly in the reactor core. Figure 4 shows a cross-section of a pressurised water reactor.

- The reactor core contains the fuel rods, control rods, and water at high pressure. The fission neutrons are slowed down by collisions with the atoms in the water molecules – fast neutrons do not cause further fission of U-235. The water acts as a **moderator** because it slows down the fission neutrons.

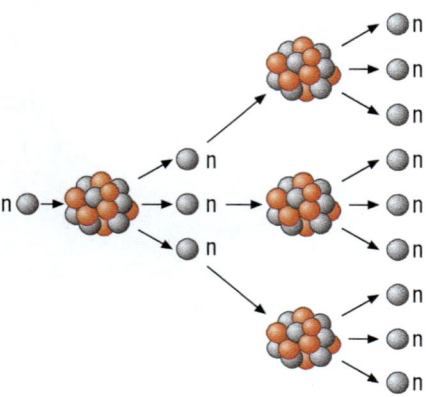

Figure 2 *A chain reaction of fission events*

Figure 3 *Ernest Rutherford*

- **Control rods** in the core absorb surplus neutrons. This keeps the chain reaction under control. The depth of the rods in the core is adjusted to maintain a steady chain reaction.
- The water acts as a **coolant**. Its molecules gain kinetic energy from the neutrons and the fuel rods. The water is pumped through the core. Then it goes through sealed pipes to and from a heat exchanger outside the core. The water transfers energy from the core to the heat exchanger.
- The **reactor core** is made of thick steel to withstand the very high temperature and pressure in the core. The core is enclosed by thick concrete walls. These absorb radiation that escapes through the walls of the steel vessel.

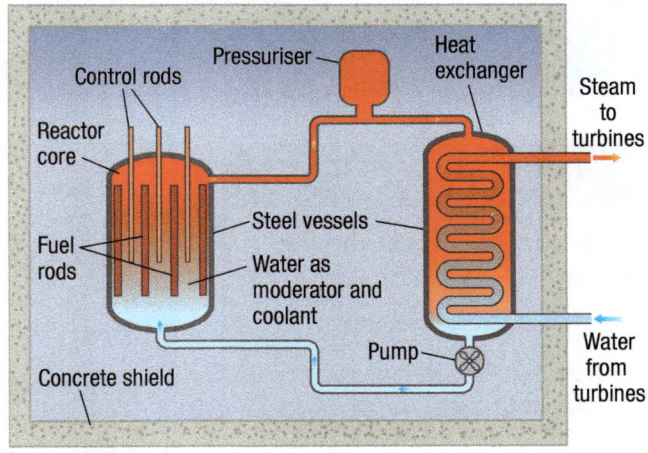

Figure 4 *A nuclear reactor*

Summary questions

1 Natural uranium consists mainly of uranium-238. Uranium fuel is produced from natural uranium by increasing the proportion of uranium-235 in it.

  a What happens to a uranium-235 nucleus when a neutron collides with it and causes it to undergo fission?

  b What happens to a uranium-238 nucleus when a neutron collides with it?

2 **a** List statements **A** to **D** below in the correct sequence, starting with B, to describe a steady chain reaction in a nuclear reactor.

    A a U-235 nucleus splits   **C** neutrons are released

    B a neutron hits a U-235 nucleus   **D** energy is released

  b **i** In a nuclear reactor, what is the purpose of the control rods?

    ii If the control rods in a nuclear reactor are pushed further into the reactor core, state and explain what would happen to the number of fission neutrons in the reactor.

3 **a** Look at the chain reaction shown in the figure.

    i Which of the nuclei A to F have been hit by a neutron?

    ii What has happened to these nuclei?

    iii Which two of the other nuclei A to F could undergo fission from a fission neutron shown?

  b In a chain reaction, a neutron causes a fission event X that releases three neutrons, two of which go on to produce further fission events Y and Z that each cause two further fission events.

    i Copy and complete the figure by adding neutron arrows to represent this chain reaction.

    ii Give one reason why one of the three neutrons released by event X did not cause further fission.

Study tip

'Fission' means splitting. Don't confuse nuclear fission and nuclear fusion.

Key points

- Nuclear fission is the splitting of an atomic nucleus into two smaller nuclei and the release of two or three neutrons and energy.

- **Nuclear fission** occurs when a neutron is absorbed by a uranium-235 nucleus or a plutonium-239 nucleus and the nucleus splits.

- A **chain reaction** occurs in a nuclear reactor when each fission event causes further fission events.

- In a **nuclear reactor**, control rods absorb fission neutrons to ensure that, on average, only one neutron per fission goes on to produce further fission.

17.8

Nuclear fusion

Learning objectives

After this topic, you should know:

- what nuclear fusion is
- how nuclei can be made to fuse together
- where the Sun's energy comes from
- why it is difficult to make a nuclear fusion reactor.

Stars release energy as a result of fusing small nuclei such as hydrogen to form larger nuclei. Water contains lots of hydrogen atoms. Imagine if scientists could obtain energy from water. A glass of water could provide the same amount of energy as a tanker full of petrol, if a fusion reactor could be made on Earth.

Fusion reactions

Two small nuclei release energy when they are fused together to form a single larger nucleus. This process is called **nuclear fusion**. It releases energy only if the relative mass of the nucleus formed is no more than about 56 (about the same as an iron nucleus). Energy must be supplied to create bigger nuclei.

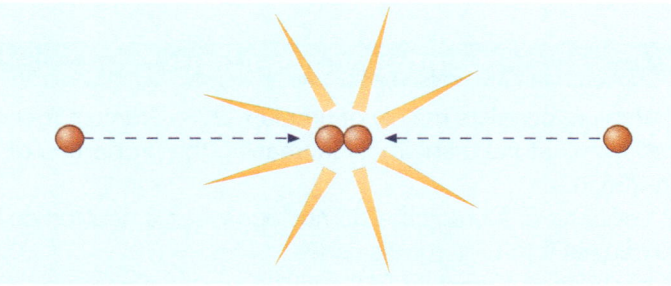

Figure 1 *A nuclear fusion reaction*

Study tip

Fusion means joining together. Don't confuse fusion with fission.

The Sun is about 75 per cent hydrogen and 25 per cent helium. The core is so hot that it consists of a plasma of bare nuclei with no electrons moving round them and unattached electrons. These nuclei move around, and if they collide they fuse. This fusion releases energy. Figure 2 shows how protons fuse together to form a $^4_2$He nucleus. Energy is released at each stage.

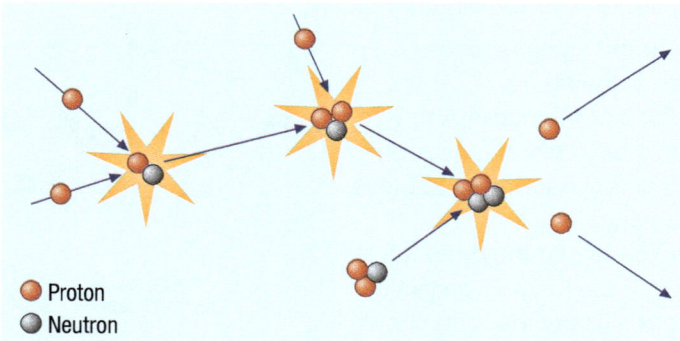

○ Proton
○ Neutron

Figure 2 *Fusion reactions in the Sun*

1 When two protons (i.e., hydrogen nuclei) fuse, they form a 'heavy hydrogen' nucleus, $^2_1$H as a result of one of the protons changing into a neutron. Other particles are created and emitted at the same time.
2 Two more protons collide separately with two $^2_1$H nuclei and turn them into heavier nuclei.
3 The two heavier nuclei collide to form the helium nucleus $^4_2$He.
4 The energy released at each stage is carried away as kinetic energy of the product nucleus and other particles emitted.

Did you know ...?

A hydrogen bomb contains uranium surrounded by the hydrogen isotope, $^2_1$H. When detonated, the uranium explodes, making the surrounding hydrogen nuclei fuse and release even more energy. A single hydrogen bomb could completely destroy 'a large city'.

Fusion reactors

Could fusion provide useful energy in a power station? Engineers are working to overcome technical difficulties with this process. The plasma of light nuclei must be heated to very high temperatures before the nuclei will fuse. This is because two nuclei approaching each other will repel each other due to their positive charges. If the nuclei are moving fast enough, they can overcome the force of repulsion and fuse together.

In a fusion reactor:
- plasma is heated by passing a very large electric current through it
- the plasma is contained by a magnetic field so it cannot touch the reactor walls. If it did, it would go cold and fusion would stop.

Scientists have been working on these problems since the 1950s. A successful fusion reactor would release more energy than it uses to heat the plasma. At the present time, scientists working on experimental fusion reactors are able to fuse heavy hydrogen nuclei to form helium nuclei – but so far the process only works for a few minutes at a time.

Figure 3 *An experimental fusion reactor*

A promising future

Practical fusion reactors could meet all our energy needs.
- The fuel for fusion reactors is heavy hydrogen, which is naturally present in sea water.
- The reaction product, helium, is a non-radioactive inert gas, so is harmless.
- The energy released could be used to generate electricity.

In comparison, fission reactors mostly use uranium, which is only found in certain parts of the world. Also, they produce nuclear waste that has to be stored securely for many years. However, fission reactors have been in operation for over 50 years, unlike fusion reactors, which are still under development.

Summary questions

1 a What is meant by nuclear fusion?
  b Look at Figure 2 and work out what is formed when a proton collides with a $^2_1$H nucleus.

2 a i Why does the plasma of light nuclei in a fusion reactor need to be very hot?
    ii Why would a fusion reactor that needs more energy than it produces not be much use?
  b State one advantage and one disadvantage a fusion reactor has compared with a fission reactor.

3 a How many protons and how many neutrons are present in a $^2_1$H nucleus?
  b Write an equation to represent the fusion of a proton and a $^2_1$H nucleus when they form a helium $^3_2$He nucleus. The symbol for a proton is $^1_1$p – its proton number (the lower number) is 1 and its mass number (the top number) is also 1.
  c Copy and complete the equation below to show the reaction in Figure 2 that takes place when two $^3_2$He nuclei fuse together to form a $^4_2$He nucleus.

$$^3_2\text{He} + {}^3_2\text{He} \rightarrow {}^4_2\text{He} + \text{_____} + \text{_____}$$

Key points

- Nuclear fusion is the process of forcing two atomic nuclei close enough together so they form a single larger nucleus.

- Nuclear fusion can be brought about by making two light nuclei collide at very high speed.

- Energy is released when two light nuclei are fused together. Nuclear fusion in the Sun's core releases energy.

- A fusion reactor needs to be at a very high temperature before nuclear fusion can take place. The nuclei to be fused are difficult to contain.

17.9 Nuclear issues

Learning objectives

After this topic, you should know:

- what radon gas is and why it is dangerous

- how safe nuclear reactors are

- what happens to nuclear waste.

 links

For more information on ionising radiation, look back at Topic 17.4 'More about alpha, beta *and* gamma *radiation'.*

??? Did you know ...?

Nuclear waste

Used fuel rods are very hot and very radioactive.

- After removal from a reactor, they are stored in large tanks of water for up to a year. The water cools the rods down.

- Remote-control machines are then used to open the fuel rods. The unused uranium and plutonium are removed chemically from the used fuel. These are stored in sealed containers so they can be used again.

- The remaining material contains many radioactive isotopes with long half-lives. This radioactive waste must be stored in secure conditions for many years.

Figure 2 *Storage of nuclear waste*

Radioactivity all around us

When you use a Geiger counter, it clicks even without a radioactive source near it. This is due to background **radiation**. Radioactive substances are found naturally all around us.

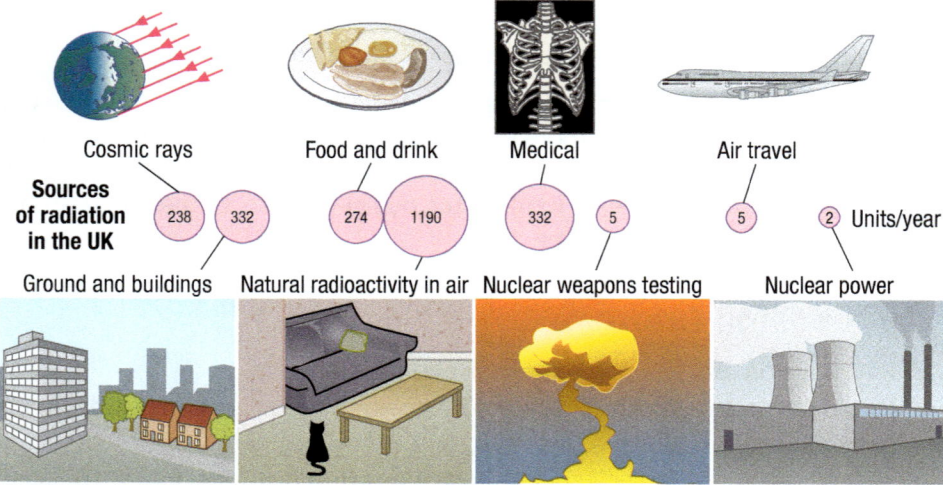

Figure 1 *Radioactivity*

Figure 1 shows some sources of background radiation. The radiation from radioactive substances is hazardous because it ionises substances it passes through. The numbers in Figure 1 tell you the **radiation dose**, which is how much radiation on average each person experiences in a year from each source.

- Medical sources include X-rays as well as radioactive substances, because X-rays have an ionising effect. People whose jobs involve the use of ionising radiation have to wear personal radiation monitors to ensure they are not exposed to too much ionising radiation.

- Background radiation in the air is caused mostly by radon gas that rises through the ground from radioactive substances in rocks deep underground. Radon gas emits α particles, so radon is a health hazard if it is breathed in. It can find its way into homes and other buildings in certain locations. In homes and other buildings where people spend long periods, measures need to be taken to reduce exposure to radon gas. For example, pipes under the building can be installed and fitted to a suction pump to draw the gas out of the ground before it enters the building.

Chernobyl and Fukushima

In 1986, a nuclear reactor in Ukraine exploded. Emergency workers and scientists struggled for days to contain the fire. A cloud of radioactive material from the fire drifted over many parts of Europe, including Britain. More than 100 000 people were evacuated from Chernobyl and the surrounding area. Over 30 people died in the accident. More have developed leukaemia or cancer since then. It was the world's worst nuclear accident. Could it happen again?

- Most nuclear reactors are of a different design. There are hundreds of nuclear reactors in the world that have been working safely for many years.
- The Chernobyl accident did not have a high-speed shutdown system like most reactors have. The operators at Chernobyl ignored safety instructions.
- The lessons learnt from Chernobyl were put into practice at Fukushima in Japan after three nuclear reactors were crippled in March 2011 by an earthquake and a tsunami. All the people living within 20 kilometres of the reactors were evacuated from their homes, and they are unlikely to be allowed to return for many years. Radiation levels and health effects over a much wider area are being monitored. Food and milk production in the entire area will also be monitored and controlled for many years. Nearby reactors with greater protection from tsunamis were much less affected than the three crippled reactors. Major lessons from Fukushima will need to be learnt about how to minimise the effect of natural disasters on nuclear reactors and the local people.

Radioactive risks

The effect of radiation on living cells depends on:

- the type and the amount of radiation received (the dose)
- whether the source of the radiation is inside or outside the body
- how long the living cells are exposed to the radiation.

The larger the dose of radiation someone receives, the greater the risk of cancer. High doses kill living cells.

Table 1 *Risks from radiation*

	α radiation	β radiation	γ radiation
Source inside the body	**Very dangerous** – affects all the surrounding tissue	**Dangerous** – reaches cells throughout the body	
Source outside the body	**Some danger** – absorbed by skin and damages skin cells and retinal cells		

The smaller the dose, the lower the risk. There is a very low level of risk to each of us because of background radiation.

??? Did you know ... ?

New improved nuclear reactors

Most of the world's nuclear reactors in use now will need to be replaced in the next 20 years. New improved 'third generation' nuclear reactors will replace them. The new types of reactors have:

- a standard design to cut down costs and construction time
- a longer operating life – typically 60 years
- more safety features, such as convection of outside air through cooling panels along the reactor walls
- much less effect on the environment.

Figure 3 *Chernobyl*

Key points

- Radon gas is an α-emitting isotope that enters houses in certain areas through the ground.

- There are hundreds of fission reactors safely in use in the world. None of them is of the same type as the Chernobyl reactors that exploded.

- Nuclear waste is stored in safe and secure conditions for many years after unused uranium and plutonium have been removed from it to be used in the future.

Summary questions

1 a Why does radioactive waste need to be stored:
      i securely?   ii for many years?
   b Why is a source of α radiation very dangerous inside the human body but not as dangerous outside it?

2 In some locations, the biggest radiation hazard comes from radon gas that rises through the ground and into buildings. The dangers of radon gas can be minimised by building new houses that are slightly raised on brick pillars and modifying existing houses. Radon gas is an α-emitting isotope.
   a Why is radon gas dangerous in a house?
   b Describe one way of making an existing house safe from radon gas.

3 Should the UK government replace our existing nuclear reactors with new reactors, either fission or fusion or both? Answer this question by discussing the benefits and drawbacks of new fission and fusion reactors.

Chapter summary questions

1 a How many protons and how many neutrons are in a nucleus of each of the following isotopes:

   i $^{14}_{6}C$?   **ii** $^{228}_{90}Th$?

b $^{14}_{6}C$ emits a β particle and becomes an isotope of nitrogen (N).

   i Write how many protons and how many neutrons are in this nitrogen isotope.

   ii Write the symbol for this isotope.

c $^{228}_{90}Th$ emits an α particle and becomes an isotope of radium (Ra).

   i Write how many protons and how many neutrons are in this isotope of radium.

   ii Write the symbol for this isotope.

2 Copy and complete the following table about the properties of alpha (α), beta (β), and gamma (γ) radiation.

	α	β	γ
Identity		electrons	
Stopped by			thick lead
Range in air		about 1 m	
Relative ionisation			weak

3 The following measurements were made of the count rate from a radioactive source.

Time (hours)	0	0.5	1.0	1.5	2.0	2.5
Count rate due to the source (counts per minute)	510	414	337	276	227	188

a Plot a graph of the count rate (on the vertical axis) against time.

b Use your graph to find the half-life of the source.

4 In a radioactive carbon dating experiment of ancient wood, a sample of the wood had an activity of 40 Bq. The same mass of living wood had an activity of 320 Bq.

a i Explain what is meant by the activity of a radioactive source.

   ii Work out how many half-lives the activity took to decrease from 320 to 40 Bq.

b The half-life of the radioactive carbon in the wood is 5600 years. Work out the age of the sample.

5 In an investigation to find out what type of radiation was emitted from a given source, the following measurements were made with a Geiger counter.

Source	Average count rate (counts per minute)
No source present	29
Source at 20 mm from tube with no absorber between	385
Source at 20 mm from tube with a sheet of metal foil between	384
Source at 20 mm from tube with a 10 mm thick aluminium plate between	32

a What caused the count rate when no source was present.

b What is the count rate from the source with no absorbers present.

c What type of radiation was emitted by the source. Explain how you arrive at your answer.

6 Figure 1 shows the path of two α particles labelled A and B that are deflected by the nucleus of an atom.

a Explain why they are deflected by the nucleus.

b Explain why B is deflected less than A.

c Explain why most α particles directed at a thin metal foil pass straight through it.

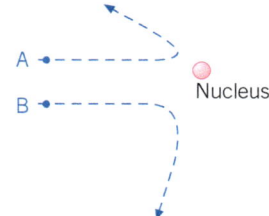

Figure 1

7 a Explain what is meant by a nuclear chain reaction.

b Explain what would happen in a nuclear reactor if:

   i the coolant fluid leaked out of the core

   ii the control rods were pushed further into the reactor core.

8 a i Explain what is meant by nuclear fusion.

   ii Explain why two nuclei repel each other when they get close.

   iii Explain why they need to collide at high speed to fuse together.

b Give two reasons why nuclear fusion is difficult to achieve in a reactor.

Practice questions

1 A group of students investigated the nature of radioactive decay. They used eighty 1p penny coins and a stopwatch to perform the following investigation:

  1 Place the eighty 1p coins in a container with lid.

  2 Shake container for 10 s and tip out coins onto bench.

  3 Count the number of 'heads' showing and record result.

  4 Replace only the 'tails' into container and repeat the tests.

Table of results

Time in seconds	Number of 'heads' shown
0	80
10	39
20	20
30	11
40	5
50	3
60	2
70	1

a Draw a graph of number of 'heads' (y-axis) against time (x-axis). (3)

b Give some conclusions about the results. (2)

c Give a reason why only 1p coins were used and not a mix of coins. (1)

d Suggest one reason for a possible human error in the investigation. (1)

e State what the 'head' coins are meant to represent in nuclear decay. (1)

2 Background radiation depends on many sources. The pie chart shows some of the most common sources of radiation.

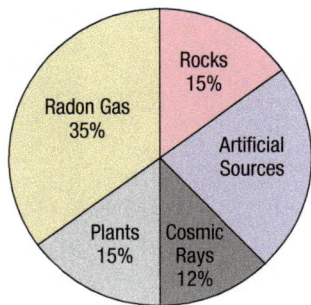

a Calculate the percentage of radiation from artificial sources. (1)

b Give the names of two artificial sources of radiation. (2)

c In Australia adults receive on average 1.5–2.0 millisieverts each year of radiation. The maximum recommended dose is 20.0 millisieverts each year.

Passengers on each long haul flight from Sidney to London will receive an additional 0.3 millisieverts of radiation.

Explain why airline pilots are only allowed to make a certain number of long haul flights. Your answer should include a calculation. (3)

3 a Copy and complete the table of information for an atom of Radon (1).

  Number of protons 86
  Number of electrons ?
  Number of neutrons 132

b Calculate the mass number of this Radon atom (1)

c Radon can exist in the form of many different isotopes. Give the difference in the nucleus of these isotopes. (1)

d Radon Spa Therapy is used in the treatment of Rheumatoid Arthritis. A 3g minimum amount of Radon-222 needs to be dissolved in the water for the treatment to be effective. The half-life of Radon-222 is 3.8 days.

  Calculate how long a 48g sample of Radon will last before it stops being effective. (2)

4 a Copy the diagram and draw a line from each type of radiation to the correct description. (3)

  Alpha radiation   Electromagnetic radiation
  Beta radiation   Same as a helium nucleus
  Gamma radiation   High speed electron

b When nuclear isotopes decay nuclear radiation is emitted. Copy and complete the nuclear decay diagram. (3)

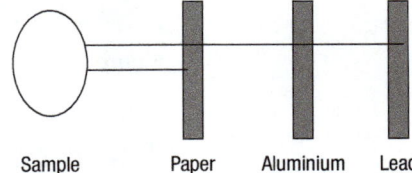

c A sample of radioactive material is tested to find out whether it passes through different materials. The results are shown in the diagram.

  Sample   Paper   Aluminium   Lead

  Name the types of radiation in the sample. (1)

5 Evaluate the advantages and disadvantage of nuclear fusion reactors compared to the traditional nuclear fission reactors. (6)

Chapter 18 Space

The Solar System

After this topic, you should know:

- how the Solar System formed
- what is meant by a protostar
- how energy is released inside the Sun
- why the Sun is stable.

How the Solar System formed

When you look at the night sky you sometimes see unexpected objects such as comets or meteors. The Solar System contains lots of objects as well as the Sun, the planets, and their moons. Comets are frozen rocks that move around the Sun in orbits that are elliptical in shape (like squashed circles). These elliptical orbits take them far away from the Sun. You only see them when they return near the Sun because then they heat up so much that they emit light. Meteors or shooting stars are small bits of rocks that burn up when they enter the Earth's atmosphere.

The Solar System also includes minor (dwarf) planets and asteroids hundreds of kilometres in size orbiting the Sun, mostly between the orbits of Mars and Jupiter. Figure 1 shows how the Sun formed from clouds of dust and gas pulled together by gravitational attraction billions of years ago.

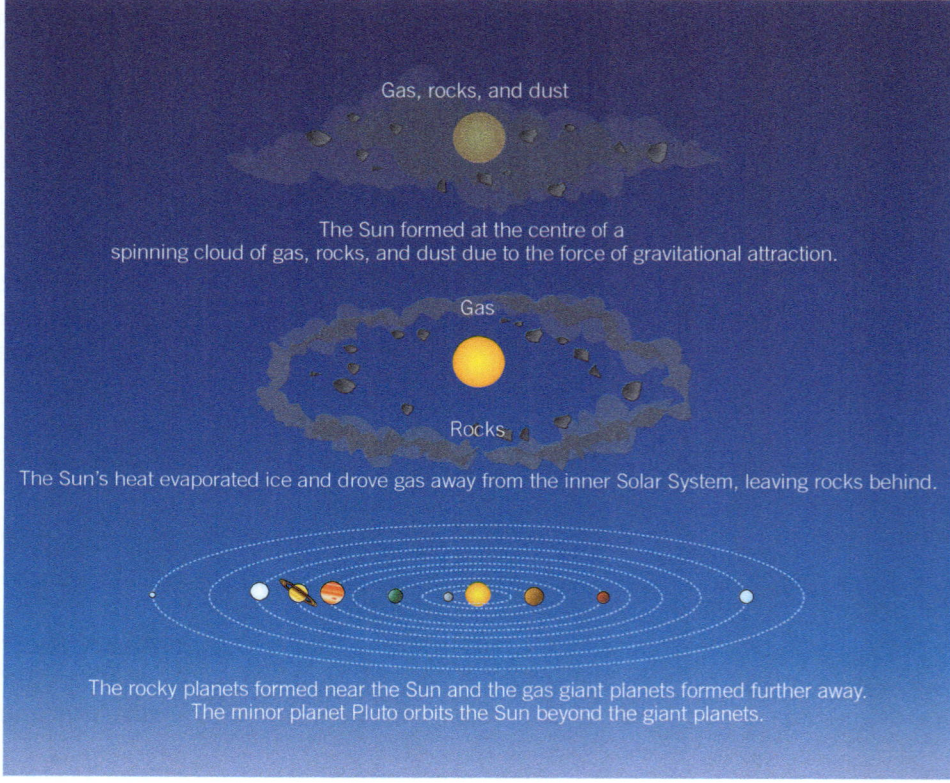

Gas, rocks, and dust

The Sun formed at the centre of a spinning cloud of gas, rocks, and dust due to the force of gravitational attraction.

Gas

Rocks

The Sun's heat evaporated ice and drove gas away from the inner Solar System, leaving rocks behind.

The rocky planets formed near the Sun and the gas giant planets formed further away. The minor planet Pluto orbits the Sun beyond the giant planets.

Figure 1 *The formation of the Solar System*

The birth of a star

All stars including the Sun form out of clouds of dust and gas.

- The particles in the clouds are pulled together by their own gravitational attraction so the particles speed up. The clouds merge together and become more and more concentrated to form a **protostar**, which is a star-to-be.

- As a protostar becomes denser, its particles speed up more and collide more, so its temperature increases and it gets hotter. The process transfers energy from the protostar's gravitational store to its thermal store. If the protostar becomes hot enough, the nuclei of hydrogen atoms fuse together, forming helium nuclei. Energy is released in this fusion, so the protostar gets hotter and brighter and starts to shine. A star is born!
- Objects can form that are too small to become stars. These kinds of objects can be attracted by a protostar to become **planets** orbiting the star.

Shining stars

Stars such as the Sun radiate energy because of hydrogen fusion in the core. They are called **main sequence** stars because this is the main stage in the life of a star. Such stars can maintain their energy output for millions of years until there are no more hydrogen nuclei left to fuse together.

- Energy released in the core (the central part of the star) keeps the core hot, so the process of fusion continues. Radiation such as γ radiation flows out steadily from the core in all directions.
- The star is stable because the forces within it are balanced. The force of gravity acts inwards trying to make the star contract. This is balanced by the outward force of the radiation from nuclear fusion in its core trying to make the star expand. These forces stay in equilibrium until most of the hydrogen nuclei in the core have been fused together.

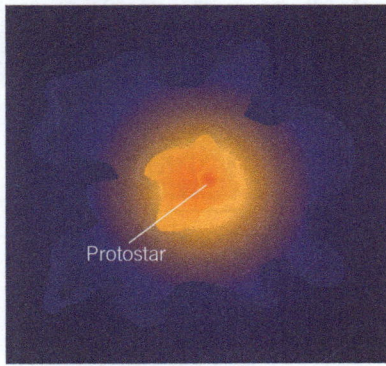

Protostar

Figure 2 *The birth of a star*

Summary questions

1 a Explain why a comet cannot be seen when it is far away from the Sun.
   b Write one difference and one similarity between a comet and an asteroid.

2 a Which planet in the Solar System is:
   i the largest?   ii nearest the Sun?
   b Explain why the Earth is likely to be the only planet in the Solar System where liquid water is always present on the surface.

3 The Earth's orbit is almost circular. State and explain two ways in which conditions on the Earth would differ if the Earth's orbit was like the orbit of a comet.

4 a Describe how the Sun formed from dust and gas clouds in space.
   b Explain what a main sequence star is and why the Sun is a main sequence star.

Key points

- The Solar System formed from gas and dust clouds that gradually became more and more concentrated because of gravitational attraction.

- A protostar is a concentration of gas and dust that becomes hot enough to cause nuclear fusion.

- Energy is released inside a star because of hydrogen nuclei fusing together to form helium nuclei.

- The Sun is stable because gravitational forces acting inwards balance the forces of nuclear fusion energy in the core acting outwards.

18.2 The life history of a star

Learning objectives

After this topic, you should know:

- why stars eventually become unstable
- the stages in the life of a star
- what will eventually happen to the Sun
- what a supernova is.

Study tip

Make sure you can recall the life cycle of star.

When a star runs out of hydrogen nuclei to fuse together in its core, it reaches the end of its main sequence stage. Its core collapses, and its outer layers swell out.

Stars about the same size as the Sun (or smaller) swell out, cool down, and turn red.

- These stars are now **red giants**. At this stage, helium and other light elements in the core fuse to form heavier elements.
- When there are no more light elements in the core, fusion stops, and no more radiation is released. Because of its own gravity, the star collapses in on itself. As it collapses, it heats up and turns from red to yellow to white. It becomes a **white dwarf**. This is a hot, dense, white star much smaller in diameter than it was before. Stars such as the Sun then fade out, go cold, and become **black dwarfs**.

Stars much bigger than the Sun end their lives after the main-sequence stage much more dramatically.

- These stars swell out to become **red supergiants**. They then collapse.
- In the collapse, the matter surrounding the star's core compresses the core more and more. Then the compression suddenly reverses in a cataclysmic explosion called a **supernova**. This event can outshine an entire galaxy for several weeks.

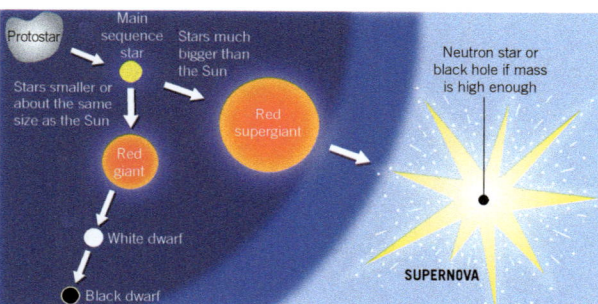

Figure 1 *The life cycle of a star*

The birthplace of the elements

Light elements are formed from fusion in stars. Stars such as the Sun fuse hydrogen nuclei (i.e., protons) into helium and similar small nuclei, including carbon nuclei. When the star becomes a red giant, it fuses helium and the other small nuclei into larger nuclei.

Nuclei larger than iron nuclei cannot be formed by this process because too much energy is needed.

Heavy elements are formed when a massive star collapses then explodes as a supernova. The enormous force of the collapse fuses small nuclei into nuclei bigger than iron nuclei. The explosion scatters the elements throughout the universe.

The debris from a supernova contains all the known elements, from the lightest to the heaviest. Eventually, new stars form as gravity pulls the debris together. Planets form from debris surrounding a new star. Because of this, these planets will be made up of all the known elements too.

Figure 2 *The Crab Nebula is the remnants of a supernova explosion that was observed in the 11th century. In 1987, a star in the southern hemisphere exploded and became the biggest supernova to be seen for four centuries. Astronomers realised that it was Sandaluk II, a star in the Andromeda galaxy millions of light years from Earth. If a star near the Sun exploded, the Earth would probably be blasted out of its orbit. You would see the explosion before the shock wave hit Earth*

The future of the Sun

The Sun is about 5000 million years old and will probably continue to shine for another 5000 million years.

The Sun will turn into a red giant bigger than the orbit of Mercury. By then, the human race will probably have long passed into history.

Planet Earth

The heaviest known natural element is uranium. It has a half-life of 4500 million years. The presence of uranium on the Earth is evidence that the Solar System must have formed from the remnants of a supernova.

Elements such as plutonium are heavier than uranium. Scientists can make these elements by bombarding heavy elements such as uranium with high-speed neutrons. They would have been present in the debris that formed the Solar System. Elements heavier than uranium formed then have long since decayed.

What's left after a supernova?

The explosion compresses the core of the star into a **neutron star**. This is an extremely dense object made up only of neutrons. If the star is massive enough, it becomes a **black hole** instead of a neutron star. The gravitational field of a black hole is so strong that nothing can escape from it. Not even light, or any other form of electromagnetic radiation, can escape a black hole.

Figure 3 *M87 is a galaxy that spins so fast at its centre that it is thought to contain a black hole with a billion times more mass than the Sun*

∞ links
To remind yourself about the heavier elements and half-lives, look back at Topic 7.3 and Topic 7.5.

Study tip

Make sure you know how the heavier elements were formed and that they were not formed during the Big Bang.

Summary questions

1 **a** The list below shows some of the stages in the life of a star such as the Sun. Put the stages in the correct sequence.
    A white dwarf **B** protostar **C** red giant **D** main sequence.
  b i Name the stage in the above list that the Sun is at now.
    ii Describe what will happen to the Sun after it has gone through the above stages.

2 **a i** What is the force that makes a red supergiant collapse.
    ii What is the force that prevents a main sequence star from collapsing.
  b Explain why a white dwarf eventually becomes a black dwarf.

3 **a** Match each statement below with an element or elements in the list.

   helium   hydrogen   iron   uranium

    i Helium nuclei are formed when nuclei of this element are fused.
    ii This element is formed in a supernova explosion.
    iii Stars form nuclei of these two elements (and others not listed) by fusing smaller nuclei.
  b Describe two differences between a red giant star and a neutron star.

4 Uranium-238 is a radioactive isotope found naturally in the Earth. It has a half-life of about 4500 million years. It was formed from lighter elements.
  a i Name the physical process in which this isotope is formed.
    ii Name the astronomical event in which the above process takes place.
  b i Explain why all the uranium in the Earth has not decayed by now.
    ii Plutonium-239 has a half-life of 24 000 years. It is formed in a nuclear reactor from uranium-238. Explain why plutonium-239 is not found naturally.

Key points

- Stars become unstable when they have no more hydrogen nuclei that they can fuse together.

- Sequence of stars with about the same mass as the Sun: protostar → main sequence star → red giant → white dwarf → black dwarf.

- Sequence of stars much more massive than the Sun: protostar → main sequence star → red supergiant → supernova → neutron star (or black hole if enough mass).

- The Sun will eventually become a black dwarf.

- A supernova is the explosion of a red supergiant after it collapses.

18.3

Planets, satellites, and orbits

Learning objectives

After this topic, you should know:

- what force keeps planets and satellites moving along their orbits

- the direction of the force on an orbiting body in a circular orbit

- how the velocity of a body in a circular orbit changes as the body moves around the orbit

- why an orbiting body needs to move at a particular speed for it to stay in a circular orbit.

The Earth orbits the Sun in an orbit that is almost circular. Most of the other planets orbit the Sun in orbits that are ellipses or slightly squashed circles. The Moon is a natural satellite that orbits the Earth in a circular orbit. Artificial satellites also orbit the Earth. In each case, a body orbits a much bigger body. The force on the orbiting body is the force of gravitational attraction between it and the larger body.

Circular orbits

Figure 1 shows the force of gravity acting on a planet in a circular orbit around the Sun.

- The force of gravity on the planet from the Sun acts towards the centre of the Sun. This force is the resultant force on the planet because no other forces act on it. The force is an example of a **centripetal force** because it acts towards the centre of the circle.

- The direction of the planet's velocity (i.e., its direction of motion) is changed by this force. So it continues to orbit the Sun. The direction of motion of any planet (i.e., the direction of its velocity) in a circular orbit is *at right angles* to the direction of the force of gravity on it.

- A planet in a circular orbit experiences acceleration towards the centre of the circle because the resultant force on it acts towards the centre of the circle. The acceleration is its change of velocity per second, and its change of velocity is directed towards the centre.

- The speed of a planet in a circular orbit does not change, even though its velocity changes its direction. This is because the force on it is at right angles to its direction of motion. So no work is done by the force on the planet. So the kinetic energy and the speed of the planet do not change.

Into orbit

Satellites in orbits too close to the Earth gradually lose speed. This happens because of atmospheric drag when a satellite's orbit is in the Earth's upper atmosphere. If a satellite loses speed, it gradually spirals inwards until it hits the Earth's surface.

A satellite in a circular orbit above the Earth's atmosphere moves around the Earth at a constant height above the surface. The satellite is in a stable orbit. To stay in an orbit of a particular radius, the satellite has to move at a particular speed around the Earth. The same is true for a planet moving in a circular orbit around the Sun.

Imagine a satellite is launched from the top of a very tall mountain. Figure 2 shows what would happen if the satellite was launched too fast or too slow. If the launch speed is too slow, the satellite falls to the surface. If the launch speed is too high, the satellite flies off into space. At the correct speed, the satellite moves around the Earth in a circular orbit at a constant height and a constant speed.

For any small body to stay in a circular orbit around another bigger body, the smaller body must move at a particular speed around the bigger body.

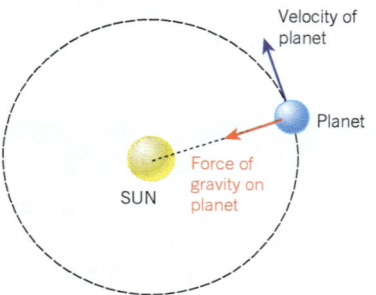

Figure 1 *A circular orbit*

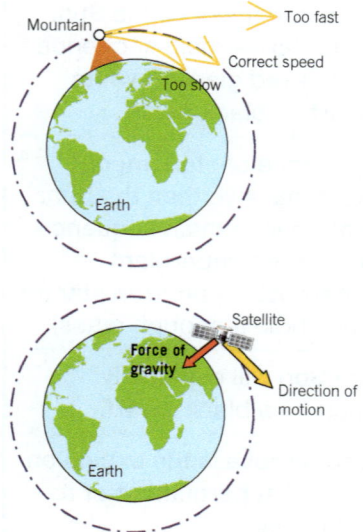

Figure 2 *Launching a satellite*

Using satellites

The further a satellite is from the Earth or a planet is from the Sun:

- the less the particular speed needed for it to stay in a circular orbit. The force of gravity on a satellite is weaker further from the Earth. The same is true for a planet moving around the Sun. If the speed of an orbiting body was too great or small, it would move out of its circular orbit.
- the longer the orbiting body takes to move around the orbit once. This is because the circumference of the orbit is bigger, and the orbiting body moves slower in order to remain in orbit. So the time for each complete orbit (= circumference ÷ speed) is longer.

If the speed of a satellite in a stable orbit changes, then the radius of the orbit has to change, for example, suppose a space vehicle above the Earth is in a circular orbit and its engines are used briefly to increase its speed. The vehicle moves out of its orbit and gains height. Because of this, its potential energy increases and its kinetic energy decreases. So its speed decreases and it can move into a higher orbit.

For a stable orbit, the radius must change if the speed of the orbiting object changes.

Figure 3 *A satellite in orbit*

Summary questions

1. A satellite is in a circular orbit around the Earth.
   a. Write the direction of:
      i the force of gravity on the satellite
      ii its acceleration.
   b. Explain why its velocity continually changes even though its speed is constant.

2. a GPS satellites orbit the Earth about once every 12 hours. Write and explain whether or not a GPS satellite orbits the Earth above or below:
      i a geostationary satellite
      ii a weather satellite that has a period of two hours.
   b. Explain why GPS satellites are easier to launch than geostationary statellites.

	Radius of orbit (A.U.)	Time for each orbit (years)
Mercury	0.39	0.24
Venus	0.72	0.61
Earth	1.00	1.00
Mars	1.52	1.88
Jupiter	5.20	11.9
Saturn	9.53	29.5

1 astronomical unit (A.U.) = the mean distance from the sun to the Earth

3. Use the information in the table above to determine which of the three planets, Venus, Earth, or Jupiter, travels **a** slowest, **b** fastest on its orbit. Explain your reasoning.

4. The Earth moves on its orbit at a speed of 30 km/s. Use the information in the table to estimate the speed of Mercury in km/s in its orbit.

Key points

The force of gravity between:

- a planet and the Sun keeps the planet moving along its orbit
- a satellite and the Earth keeps the satellite moving along its orbit.

- The force of gravity on an orbiting body in a circular orbit is towards the centre of the circle.
- As a body in a circular orbit moves around the orbit:
  - the magnitude of its velocity (its speed) does not change
  - the direction of its velocity continually changes and is always at right angles to the direction of the force
  - so it experiences an acceleration towards the centre of the circle.
- To stay in orbit at a particular distance, a small body must move at a particular speed around a larger body.

18.4 The expanding universe

Learning objectives

After this topic, you should know:

- what is meant by the red-shift of a light source

- how red-shift depends on speed

- how people know that the distant galaxies are moving away from Earth

- why people think the universe is expanding.

Red-shift

The Earth is the third planet from the Sun. The Sun is a star on the outskirts of the Milky Way galaxy. A galaxy is an enormous collection of stars that stay together because of the force of gravity between them. The Milky Way galaxy contains about 100 000 million stars. Its size is about 100 000 light years across. This means that light takes 100 000 years to travel across it. But it is just one of billions of galaxies in the universe. The furthest known galaxies are about 13 000 million light years away.

Figure 1 *In places where there is little light pollution you can see the Milky Way galaxy*

People can find out lots of things about stars and galaxies by studying the light from them. You can use a prism to split the light into a spectrum. The wavelength of light increases across the spectrum from blue to red. You can tell from its spectrum if a star or galaxy is moving towards Earth or away from Earth. This is because:

- the light waves are stretched out if the star or galaxy is moving away from you. The wavelength of the waves is increased. This is called a **red-shift** because the spectrum of light is shifted towards the red part of the spectrum.
- the light waves are squashed together if the star or galaxy is moving towards you. The wavelength of the waves is reduced. This is called a **blue-shift** because the spectrum of light is shifted towards the blue part of the spectrum.

The dark spectral lines shown in Figure 2 are caused by absorption of light by specific atoms such as hydrogen that make up a star or galaxy. The position of these lines tells you if there is a shift, and if there is a shift, whether it is a red-shift or a blue-shift.

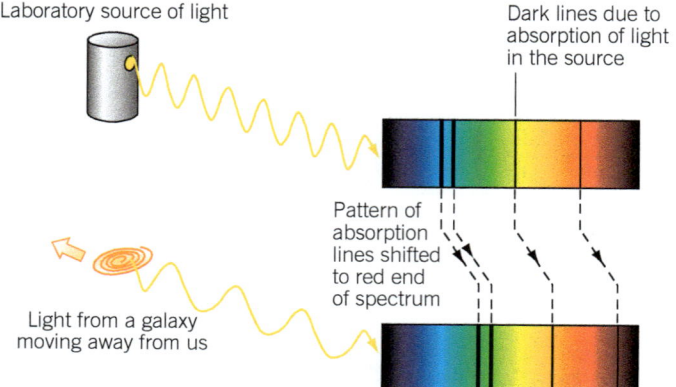

Laboratory source of light

Dark lines due to absorption of light in the source

Pattern of absorption lines shifted to red end of spectrum

Light from a galaxy moving away from us

Figure 2 *Red-shift*

The bigger the shift, the more the waves are squashed together or stretched out. So the faster the star or galaxy must be moving towards or away from you. In other words:

The faster a star or galaxy is moving (relative to you), the bigger the shift is.

Expanding universe

In 1929, the astronomer Edwin Hubble discovered that:

1 the light from distant galaxies was red-shifted

2 the further a galaxy is from Earth, the bigger its red-shift is.

He concluded that:

- the distant galaxies are moving away from Earth (i.e., receding)
- the greater the distance a galaxy is from Earth, the greater the speed at which it is moving away from Earth (its speed of recession).

Why should the distant galaxies be moving away from Earth? Humans have no special place in the universe, so all the distant galaxies must be moving away from each other. In other words, *the whole universe is expanding*.

Summary questions

1 a State whether each of the following is approaching the Earth or receding from the Earth:
   i a distant galaxy
   ii a galaxy that shows a blue-shift in its light.

 b The Sun is in the Milky Way galaxy. Astronomers think that the Andromeda Galaxy will eventually collide with the Milky Way galaxy. Write the evidence that astronomers have to support this prediction.

2 a Put these objects in order of increasing size:

   Andromeda galaxy   Earth   Sun   universe

 b Quasars are astronomical objects much smaller in appearance than galaxies, and they have red-shifts of the same order of magnitude as the distant galaxies.
   i Which part of the above statement makes you conclude that quasars are much further away than nearby galaxies.
   ii Quasars can be as bright as a distant galaxy even though they are much smaller. Use this information to describe the power of the radiation emitted by a quasar.

3 Galaxy X has a larger red-shift than galaxy Y.

 a Explain what is meant by a red-shift.
 b Write which galaxy, X or Y, is:
   i nearer to Earth   ii moving away faster.

4 Some of the nearest galaxies to Earth have different red-shifts, and some have different blue-shifts. Use this information to describe these galaxies.

Key points

- The red-shift of a distant galaxy is the shift to longer wavelengths (and lower frequencies) of the light from the galaxy because it is moving away from you.

- The faster a distant galaxy is moving away from you, the greater its red-shift is.

- All the distant galaxies show a red-shift. The further away a distant galaxy is from you, the greater its red-shift is.

- The distant galaxies are all moving away from you because the universe is expanding.

18.5 The Big Bang

Learning objectives

After this topic, you should know:
- what the Big Bang theory of the universe is
- why the universe is expanding
- what cosmic microwave background radiation is
- what evidence there is that the universe was created in a Big Bang.

The universe is expanding, but what is making it expand? The **Big Bang theory** was put forward as a model to explain the expansion. This says that:

- the universe is expanding after exploding suddenly (the Big Bang) from a very small and extremely hot and dense region
- space, time, and matter were created in the Big Bang.

Many scientists disagreed with the Big Bang theory. They put forward an alternative theory called the Steady State theory. These scientists said that the galaxies are being pushed apart. They thought that this is caused by matter entering the universe through 'white holes' (the opposite of black holes).

Which theory is weirder – everything starting from a Big Bang, or matter leaking into the universe from outside? Until 1965, most people supported the Steady State theory.

Evidence for the Big Bang

Scientists had two conflicting theories about the evolution of the universe: it was in a Steady State or it began at some point in the past with a Big Bang. Both theories could explain why *distant* galaxies are moving apart, so scientists needed to find some way of deciding which theory was correct. They worked out that if the universe began in a Big Bang, then high-energy electromagnetic radiation should have been produced very soon after the universe began. This radiation would have stretched as the universe expanded and become lower-energy radiation. Scientists thought up experiments to look for this trace energy as extra evidence for the Big Bang model.

It was in 1965 that scientists first detected microwaves coming from every direction in space. The existence of this **cosmic microwave background radiation (CMBR)** can be explained only by the Big Bang theory.

Cosmic microwave background radiation

Cosmic microwave background radiation was created as high-energy gamma radiation just after the Big Bang. It has been travelling through space since then. As the universe has expanded, it has stretched out to longer and longer wavelengths and is now microwave radiation. It has been mapped out using microwave detectors on the Earth and on satellites.

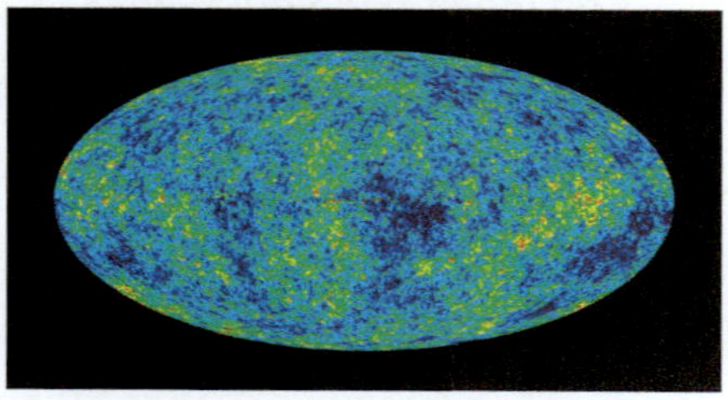

Figure 1 *A microwave image of the universe from the Cosmic Background Explorer satellite*

The future of the universe

Will the universe expand forever? Or will the force of gravity between the distant galaxies stop them moving away from each other? The answer to this question depends on the total mass of the galaxies, how much matter is between them, and how much space they take up – in other words, the density of the universe.

Astronomers think that the stars in a galaxy account for only a small percentage of the total mass of a galaxy. They know that galaxies would spin much faster if their stars were the only matter in galaxies. The missing mass is called **dark matter** because it can't be seen. Its presence means that the average density of the universe is much bigger than if dark matter didn't exist.

- If the density of the universe is less than a particular amount, it will expand forever. The stars will die out, and so will everything else as the universe heads for a Big Yawn!

- If the density of the universe is more than a particular amount, it will stop expanding and go into reverse. Everything will head for a Big Crunch!

Observations since 1998 of supernovae in distant galaxies suggest that the distant galaxies are accelerating away from each other. These observations have been checked and confirmed by other astronomers. So astronomers have concluded that the expansion of the universe is accelerating. It could be that the universe is in for a Big Ride followed by a Big Yawn.

The discovery that the distant galaxies are accelerating is puzzling astronomers. Scientists think some unknown source of energy, called dark energy, must be causing this accelerating motion. The only known force on the distant galaxies, the force of gravity, can't be used to explain dark energy, because it is an attractive force and so it acts against the outward motion of the distant galaxies away from each other.

There is still a lot about the universe, for example dark mass and dark energy, that astronomers don't understand. New telescopes and technologies will help to improve humans' understanding of the universe – and will certainly create more questions for scientists to investigate.

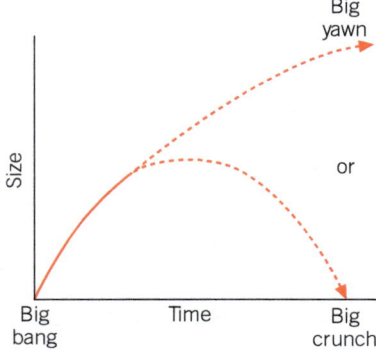

Figure 2 *The future of the universe?*

Study tip

Make sure you know more about the Big Bang theory than just 'the universe started with a big bang'.

Summary questions

1 **a** Describe the Big Bang theory of the universe.

   b Explain why many astronomers did not support the Big Bang theory when it was first proposed.

   c Explain the significance of the discovery of cosmic microwave background radiation.

2 Put the following events **A–D** in the correct time sequence:

   A the distant galaxies were created
   B cosmic microwave background radiation was first detected
   C the Big Bang happened
   D the expansion of the universe began.

3 **a** Explain why astronomers think that the expansion of the universe is accelerating.

   b Describe what would have been the effect on the expansion of the universe if its density had been greater than a particular value.

4 **a** Hubble estimated that the speed of recession of a distant galaxy increases by about 22 km/s for every extra distance of one million light years. A distant galaxy is receding away at a speed of 150 000 km/s. Estimate how far away the galaxy is in light years.

   b The Milky Way galaxy is about 100 000 light years across. Make an order of magnitude estimate of the ratio of the distance to the galaxy in part **a** to the distance across the Milky Way.

Key points

- The universe started with the Big Bang, which was a massive explosion from a very small point.

- The universe has been expanding ever since the Big Bang.

- Cosmic microwave background radiation (CMBR) is electromagnetic radiation that was created just after the Big Bang.

- The red-shifts of the distant galaxies provide evidence that the universe is expanding. CMBR can be explained only by the Big Bang theory.

Chapter summary questions

1 a The stages in the development of the Sun are listed below. Put the stages in the correct sequence.

   A dust and gas   **D** red giant
   B main sequence   **E** white dwarf
   C protostar

b i Describe what will happen to the Sun after its present stage.
  ii Describe what will happen to a star that has much more mass than the Sun.

c The Earth moves around the Sun in a circular orbit at a constant speed. Explain why the velocity of the Earth changes and why it accelerates as it moves around the Sun.

2 a i Define a supernova.
  ii Explain how you could tell the difference between a supernova and a distant star like the Sun at present.

b i Define a black hole.
  ii Describe what would happen to stars and planets near a black hole.
  iii Define a neutron star, and describe how it is formed.

3 a i Which element that, as well as hydrogen, was formed in the early universe?
  ii Which of the two elements in part **i** is formed from the other one in a star?

b i Write which *two* of the elements listed below are *not* formed in a star that gives out radiation at a steady rate.
    carbon   *iron*   *lead*   *uranium*
  ii Explain how the two elements given in your answer to part **i** would have been formed.
  iii Explain how you know that the Sun formed from the debris of a supernova.

4 a Put these events in the correct sequence with the earliest event first:
  1 cosmic microwave background radiation was released
  2 hydrogen nuclei were first fused to form helium nuclei
  3 the Big Bang took place
  4 neutrons and protons formed.

b The stars were formed from clouds of dust and gas.
  i What force can cause dust and gas particles to attract each other?
  ii Where did the energy that heated the stars come from?

iii The stars in a galaxy revolve about the centre of the galaxy. Explain why the stars in a galaxy do not pull each other into a large single massive object at the centre.

5 Light from a distant galaxy has a change of wavelength because of the motion of the galaxy.

a i Is this change of wavelength an increase or a decrease?
  ii Explain why this effect is called a red-shift.
  iii Explain what the change of wavelength tells you about the motion of the galaxy.

b Light from a particular nearby galaxy is found to have undergone a blue-shift because of the motion of the galaxy. Explain what this tells you about the motion of this galaxy.

c i Edwin Hubble discovered that the further a distant galaxy is from Earth, the greater the red-shift of the light from it. Explain what this tells you about the universe.
  ii Give the crucial observational evidence that led scientists to accept the Big Bang theory of the universe.

6 a Galaxy **A** is further from us than galaxy **B**.
  i Which galaxy, **A** or **B**, produces light with a greater red-shift?
  ii Galaxy **C** gives a bigger red-shift than galaxy **A**. Describe the distance to galaxy **C** compared with galaxy **A**.

b All the distant galaxies are moving away from each other.
  i Explain what this tells you about the universe.
  ii Explain what it tells you about your place in the universe?

7 In a demonstration of the expansion of a one-dimensional universe, 11 students representing galaxies stand along a straight line spaced 1 metre apart. The students are told to increase their spacing by 0.5 metres every 10 seconds.

Figure 1

Work out the average speed of separation of two students who are:

a next to each other

b at opposite ends of the line.

Practice questions

1 One theory for the origin of the universe is that it began from a single point.

  a One of the boxes gives the correct name of the theory.

    Tick (✓) the correct box. (1)

Red Bang	
Big Bang	
Loud Bang	

  b Observation of distant galaxies shows that the universe is expanding. Evidence for this is the light from the galaxies appears stretched. Give the name of this evidence. (1)

  c Complete the sentence using the correct words from the list.

    the Sun   all directions   the centre of the Earth

    Cosmic background radiation has been found in the universe. It appears to come from _____. (1)

2 In 1929 Edwin Hubble investigated the light from distant galaxies. He stated that distant galaxies were moving away from us. Figure 1 shows a graph of the Hubble Data.

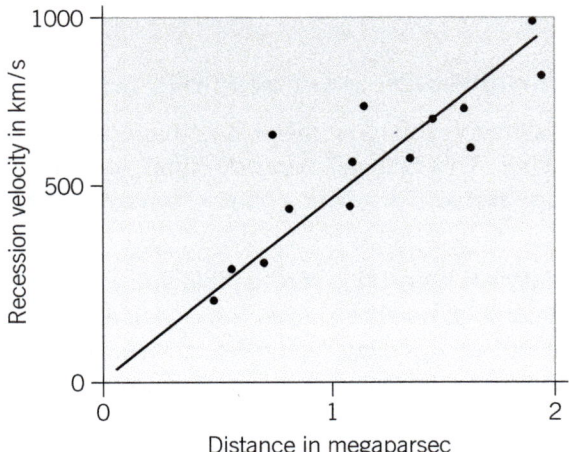

Figure 1

  a State what is meant by a distance of 1 light year. (1)

  b Describe the relationship between the recession velocity of a galaxy and the distance from the Earth. (2)

  c In 1929 Hubble used a large reflective telescope and calculated the Hubble Constant to determine the distance of galaxies from the Earth.

In 2013 NASA researchers surveyed more than 125 000 galaxies and claim to have measured the Hubble Constant with an uncertainty of less than 5%. Suggest two reasons why modern researchers can claim to be more accurate in measuring the Hubble Constant. (2)

3 a Describe how the Sun was formed and became a star. You should include details of how the Sun releases energy. (3)

  b Give two reasons why the Sun is a main sequence star and is stable. (2)

  c A star many times bigger than the Sun will eventually complete its life cycle and become a black hole. Explain what a black hole in space is. (2)

  d Complete the following sentence.

    Elements heavier than iron are formed in a _____. The explosion of this massive star distributes _____ throughout the _____. (3)

4 Table 1 shows some information about the Earth, the Moon, and satellites.

Table 1

Name	Mass (kg)	Altitude (km)	Orbit time (hours)	Speed (km/h)
Earth	5.98×10^{24}	0	0	0
Moon	7.45×10^{22}	384 000	672	3600
GPS satellite	2.2×10^2	20 000	12	13 900
Weather satellite	2.4×10^2	36 000	24	11 100
Spy satellite	2.8×10^2	705	1.45	27 500

Using the information given in Table 1, evaluate the differences and similarities between planets, moons, and artificial satellites. (6)

Investigations

After this topic, you should know:

- what continuous and categoric variables are

- what is meant by repeatable evidence, reproducible evidence, and valid evidence

- what the link is between the independent and dependent variable

- what a hypothesis and a prediction are

- how to reduce risks in hazardous situations.

Science is relevant to people's lives every day. If you work as a scientist you will use your knowledge of the world around you, and particularly about the subject you are working with. You will observe the world around you and ask questions about what you have observed.

Science usually moves forward by slow steady steps. Each small step is important in its own way as it builds on the existing body of knowledge.

Thinking scientifically

Deciding what to measure

Variables can be one of two different types:

- A **categoric variable** is one that is best described by a label (usually a word). The type of material is a categoric variable, such as metal or plastic.

- A **continuous variable** is one that is measured, so its value could be any number. Temperature (as measured by a thermometer or temperature sensor) is a continuous variable, such as 37.6 °C, 45.2 °C. Continuous variables can have values that are measured (e.g., light intensity, voltage, etc.). A particular measured variable may be called a quantity.

When designing an investigation you should always try to measure continuous data whenever you can. If this is not possible, you should try to use the data in alphabetic or numerical order. If you cannot measure your variable then it is a categoric variable that you will label rather than measure.

Making your investigation repeatable, reproducible, and valid

When you are designing an investigation you must make sure that others can carry it out and obtain similar results. This makes it **reproducible**. You should also plan to make each result **repeatable**. This means getting consistent sets of repeat measurements.

You must also make sure you really are measuring the variable you want to measure. If you don't, your data can't be used to answer your original question. This seems very obvious but it is not always easy. You need to make sure that you have controlled as many other variables as you can, so that no-one can say that your investigation is not **valid**.

How might an independent variable be linked to a dependent variable?

The **independent variable** is the one you choose to vary in your investigation.

The **dependent variable** is what you measure to judge the effect of varying the independent variable.

These variables may be linked. If there is a pattern (for example, as the value of one variable gets bigger, so does the value of the other variable), it may be that:

- changing one has caused the other to change

- the two are related, but one is not necessarily the cause of the other.

Starting an investigation

Observation

As a scientist you make observations and then ask questions about them. You can only ask useful questions if you know something about the observed event. You will not have all of the answers, but you should know enough to start asking sensible questions.

When you are designing an investigation you have to observe carefully which variables are likely to have an effect.

What is a hypothesis?

A hypothesis is an idea based on an observation that uses scientific ideas to try to explain the observation.

When making hypotheses you can be very imaginative with your ideas. However, you should have some scientific reasoning behind those ideas so that they are not totally bizarre.

Remember, your suggested hypothesis might not be correct. To check your hypothesis, you make a prediction from it, and then test it by carrying out an investigation.

observation + knowledge → hypothesis → prediction → investigation

Starting to design an investigation

An investigation starts with a question, followed by a prediction. You may well predict that there is a relationship between two variables.

You should start by planning a preliminary investigation to find the most suitable range and interval for the independent variable.

Making your investigation safe

Remember that when you design your investigation, you must:
- look for any potential hazards
- decide how you will reduce any risk.

In the practical questions in the examination, you may need to:
- write down your plan
- make a risk assessment
- make a prediction
- draw a blank table ready for the results.

Study tip

Observations, backed up by creative thinking and good scientific knowledge can lead to a hypothesis.

Key points

- Continuous variables are measurable variables whereas categoric variables are labels.

- Your evidence is:
  - repeatable if you get consistent sets of repeat measurements
  - reproducible if others can repeat your results
  - valid if you have controlled all the variables that need to be controlled.

- A pattern of change observed by two variables does not mean that change of one is the cause of the change of the other.

- A hypothesis is a scientific idea based on observation. A hypothesis can lead to predictions and investigations.

- You must make a risk assessment, make a prediction, and write a plan.

Setting up investigations

Learning objectives

After this topic, you should know:

- what a fair test is
- how a survey is set up
- what a control group is
- how to decide on the conditions, range, and intervals
- how to ensure accuracy and precision
- the causes of error and anomalies.

Study tip

Trial runs will tell you a lot about how your investigation might work out. Trial runs help you establish:

- Do you have the correct conditions?
- Have you chosen a sensible range?
- Have you got enough readings that are close together?
- Will you need to repeat your readings?

Study tip

Precision is not the same as accuracy.

Imagine you carry out an investigation into the energy value of a type of fuel. Your results for the amount of energy released are all about the same. Your data has precision, but this doesn't necessarily mean that it is accurate.

Fair testing

A fair test is one in which only one independent variable is changed, so only this variable can affect the dependent variable. All other variables are controlled.

This is usually straightforward to set up in the laboratory, but almost impossible in fieldwork. Investigations in the environment are not easy to control. There are complex variables that are changing constantly.

How do you set up fieldwork investigations? The best approach is to make sure that all of the many variables change in much the same way, except for the one you are investigating.

You might carry out an investigation in a large population using a survey. You would need to survey a smaller sample of the population rather than the whole population. You need to control as many variables within your sample as possible. Imagine scientists were investigating the effect of mobile phone usage on health. They would try to choose people of the same age and same family history to test. The larger the sample size tested, the more valid the results will be.

Control groups are used in investigations like this, to try to compare the effect of the variable being investigated. This would involve monitoring the health of similar groups of people who do not use mobile phones.

Designing an investigation

Accuracy

Your investigation must provide **accurate** data. Accurate results are very close to the true value. This is essential if your results are going to have any meaning.

How do you know if you have accurate data?

It is very difficult to be certain whether your data is accurate. It is not always possible to know what that true value is.

- Sometimes you can calculate a theoretical value and check it against the experimental evidence. Close agreement between these two values could indicate accurate data.
- You can draw a graph of your results and see how close each result is to the line of best fit.
- Try repeating your measurements with a different instrument and see if you get the same readings.

How do you obtain accurate data?

- Using instruments that measure accurately will help.
- The more carefully you use the measuring instruments, the more accuracy you will achieve.

Precision

Your investigation must provide data with sufficient **precision**. This means that repeated results under the same conditions are similar. Without precise data, you will not be able to make a valid conclusion.

How do you obtain precise and repeatable data?

- Repeat your tests as often as necessary to achieve repeatability.
- Repeat your tests in exactly the same way each time.
- Use measuring instruments that have appropriate scale divisions for a particular investigation.

Making measurements

Using instruments

You cannot expect perfect results, but you should choose an instrument that will give you as much accuracy as possible, that is, it will give you a true reading.

Some instruments have smaller scale divisions than others. Instruments that measure the same quantity can have different sensitivities. The **resolution** of an instrument refers to the smallest change in a value the instrument can detect. Choosing the wrong scale could lead you to miss important data, or make inappropriate conclusions.

You also need to be able to use an instrument properly.

Errors

Repeated results may show differences because of errors in taking measurements. Values for a particular result may differ from the rest because of a **random error**. This is most likely to be due to a measurement being made or recorded inaccurately. It could be due to not carrying out the method consistently, for example forgetting to stir a beaker of hot water before using a thermometer to measure the water temperature.

There may be a **systematic error** in an investigation. This means that the method was carried out consistently, but the same error was being repeated each time, for example using an ammeter that has a 'zero error' which means it does not read zero when there is no current in it.

Anomalies

Anomalies are results that are clearly out of line with the rest. They are outside the range of natural variation that is present in any measurement. Anomalies should be looked at carefully. There might be a very interesting reason why one result is very different. If an anomaly is simply due to a random error then it should be ignored.

If you can recognise an anomaly whilst you are doing an investigation, then you can repeat that part of the investigation to obtain a more accurate and precise result. If you do not find anomalies until after you have finished collecting data, then those results must be discarded.

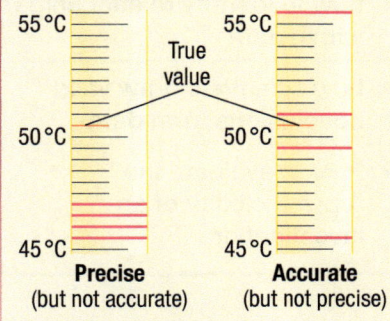

Using data

After this topic, you should know:

- what is meant by the range and the mean of a set of data
- how data should be displayed
- how to identify relationships within data
- how scientists draw valid conclusions from data
- how to evaluate the reproducibility of an investigation.

Presenting data

Tables

Tables are the best method of recording results quickly and clearly. You should design your results table before you start your investigation.

The range of the data

The range is the spread between the maximum and the minimum values of your results. For example, if your results for temperature measurements vary between 5 °C and 15 °C, the range is 5–15 °C. Don't forget to include the units!

The mean of the data

To find the mean, add up the values of all your results and divide by how many results there are.

Bar charts

If you have a categoric independent variable and a continuous dependent variable then you should display your results as a bar chart.

Line graphs

If you have a continuous independent and a continuous dependent variable then you should display your results as a line graph.

Using data to draw conclusions

Identifying patterns and relationships

Once you have a bar chart or a line graph of your results, you can begin looking for patterns in your results. You must keep an open mind at this point.

Firstly, there could still be some anomalous results. You might not have picked these out earlier. How do you spot an anomaly? It will vary significantly from the pattern of the rest of the results, outside the normal variation. Drawing a line of best fit will help to identify any anomalies. Ask yourself – do the anomalies represent something important or were they just a mistake?

Secondly, remember a line of best fit can be a straight line or it can be a curve. You need to decide from your results which is best.

A line of best fit will also lead you into thinking what the relationship is between your two variables. You need to consider whether your graph shows a linear relationship. Can you be confident about drawing a straight line of best fit on your graph? If the answer is yes, then does the gradient of this line slope upwards or downwards (does it have a positive or negative gradient)?

A directly proportional relationship is shown by a straight line with a positive gradient that passes through the origin (0, 0).

Your results might show a curved line of best fit. These can be quite straightforward, or might show a more complex relationship.

Scatter graphs

A scatter graph is plotted in the same way as a line graph, but you cannot draw a line of best fit. For example, if you investigated whether people's shoe size is related to their age, you might draw a scatter graph of your results.

Note that an **inversely proportional relationship** gives a graph like the one in Figure 1. This shows how the current I through a resistor varies according to the resistance R when there is a fixed p.d. V across the resistor. As explained in Topic 14.3 'Potential difference and resistance', the relationship between I and R is given by the equation $I = \dfrac{V}{R}$. The graph in Figure 1 shows this relationship because $I \times R$ always has the same value. For example, I would be 10 times smaller if resistance R was 10 times bigger.

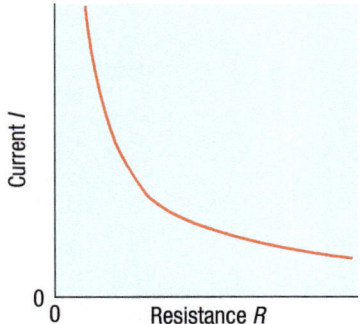

Figure 1 *Current and resistance are inversely proportional*

Drawing conclusions

Your graphs will hopefully show a relationship between your two chosen variables. You then need to consider what this relationship means to draw your conclusion. You must also take into account the repeatability and validity of the data you are considering when drawing a conclusion. You should continue to have an open mind when writing your conclusion.

At the start of your investigation you will have made a prediction. This might be supported by your results, it might not be supported, or it might be partly supported. Your results might suggest some other hypothesis.

You must be willing to think carefully about your results. Remember it is quite rare for a set of results to be perfectly repeatable and to clearly support a prediction.

Consider the possible links between variables. It may be that:
- changing one has caused the other to change
- the two are related, but one is not necessarily the cause of the other.

You must decide which relationship is the most likely. Remember that a positive relationship correlation does not always mean a causal link between the two variables.

Your conclusion must go no further than the evidence that you have. Any patterns you spot are only strictly valid in the range of values you tested. Further tests would be needed to check whether the pattern continues beyond this range.

The purpose of your prediction was to test a hypothesis. In your conclusion, your hypothesis may be supported or refuted, or it can lead to another hypothesis.

You have to decide which is most likely based on the evidence of your results.

Evaluation

If you cannot draw a clear conclusion, this may be because of a lack of repeatability, reproducibility, or validity of the results. You could check reproducibility by:
- looking for other similar work on the Internet or from others in your class
- asking somebody else to redo your investigation
- trying an alternative method to see if you get the same results.

To check validity, you could revise and improve your practical procedure to ensure the control variables stay the same. Then you could repeat your tests and obtain extra measurements that may help you to draw conclusions.

Key points

- The range states the maximum and minimum values.
- The mean is the sum of the values divided by how many values there are.
- Tables are used to record results during an investigation.
- Bar charts are used for a categoric independent variable and a continuous dependent variable.
- Line graphs are used to display data that are continuous.
- Drawing a line of best fit helps to show the relationship between variables. Possible relationships are: linear – positive or negative; directly proportional; predictable and complex curves.
- Conclusions must go no further than the data available.
- The reproducibility of data can be checked against published data, by repeating the test, or by using a different method.

1 Some students were asked to investigate the factors that determine the resistance of a piece of wire.

One group of students decided to find out how the resistance depends on the length of the wire.

1 (a) Which of the following variables should be controlled?

Tick **three** boxes.

colour of wire	
diameter of wire	✓
length of wire	
temperature of wire	✓
type of wire	✓

(3)

> The correct control variables have been ticked.

1 (b) The students set up the circuit below.

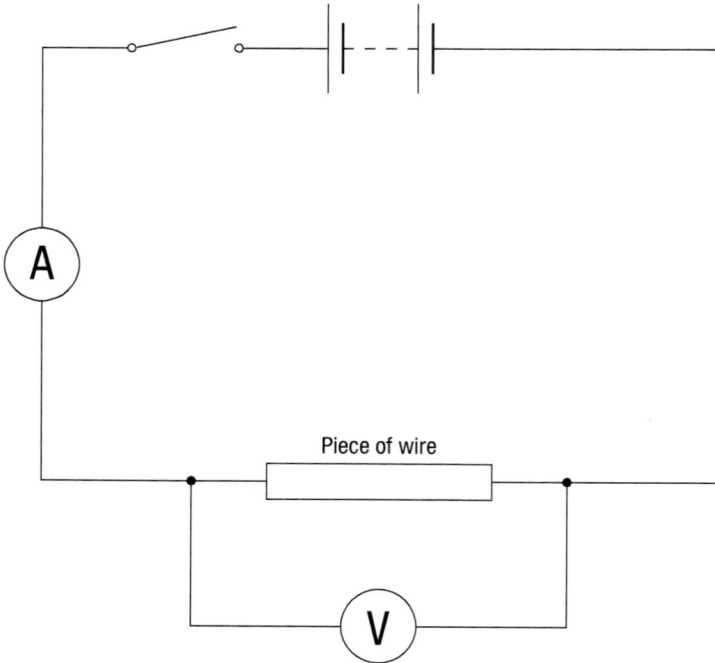

Piece of wire

One of the meters had a zero error.

What is a *zero error* and how would you correct readings taken from the meter with the zero error?

A zero error occurs when the pointer on a meter does not return to 0 when there is no current. If the meter has a reading when it should read zero, this value should be subtracted from every reading.

(2)

> The candidate has correctly stated what a zero error is and given the way of correcting the meter reading if the zero error is positive. They could have said that it is necessary to add the zero error to each reading if the meter shows a negative reading when it should read 0.

1 (c) The students cut six different lengths from a reel of wire. They measured the length of each piece of wire.

Then they used crocodile clips to connect a length of wire in the circuit.

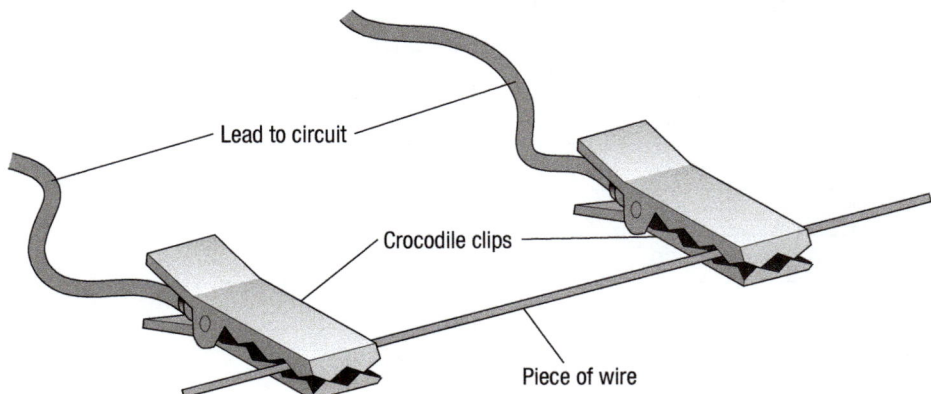

Lead to circuit

Crocodile clips

Piece of wire

1 (c) **(i)** Identify a fault in their technique and suggest how the measurement of length could be improved.

> The length of wire in the circuit is less than the measured length. The measurement could be improved by measuring the distance between the points of contact of the crocodile clips with the wire.

(2)

The candidate has given the correct answer – this is a case where a diagram, marking exactly what should be measured, would help.

1 (c) **(ii)** Suggest reasons why they should not use pieces of wire less than 5 cm long or pieces of wire more than 1 m long.

> If the wire is too short, its resistance will be low and the current may be high enough to melt some wires. If the wire is long it will be cumbersome and in fact it could short out.

(5)

The candidate would be awarded five marks, since they have made five relevant points, but a better answer would explain what 'short out' means, especially since there is a different meaning of the word 'short' already in the answer. Other points could gain credit: if the wire gets hot, one of the control variables has not been kept constant and the results will not be valid; or the percentage error in measuring length will be higher for shorter wires: 1 mm in 5 cm = 2%; 1 mm in 1 m = 0.1%.

1 (d) The students obtained the data they needed and repeated the procedure with the other pieces of wire. They then calculated the resistance of each piece of wire.

Their results are given in the table.

Length of wire in cm	Resistance in Ω
40	0.8
50.00	1.0
60.0	1.2
70	0.4
80.0	1.6
90	1.8

> Some candidates may put $V = I \times R$ or put the equation in words rather than letters and still score the mark.

1 (d) **(i)** What equation should the students use to calculate resistance?

$R = V \div I$

(1)

> The answer could have been improved by pointing out the inconsistencies – some lengths to the nearest cm, others to the nearest mm and one to 0.01 cm, but both marks would be scored.

1 (d) **(ii)** Comment on the recorded readings of *length of wire* in the table.

The readings are inconsistent. A normal ruler can measure to the nearest mm so all results should be recorded to the first decimal place.

(2)

1 (d) **(iii)** The results have been plotted on the grid below. Draw a line of best fit.

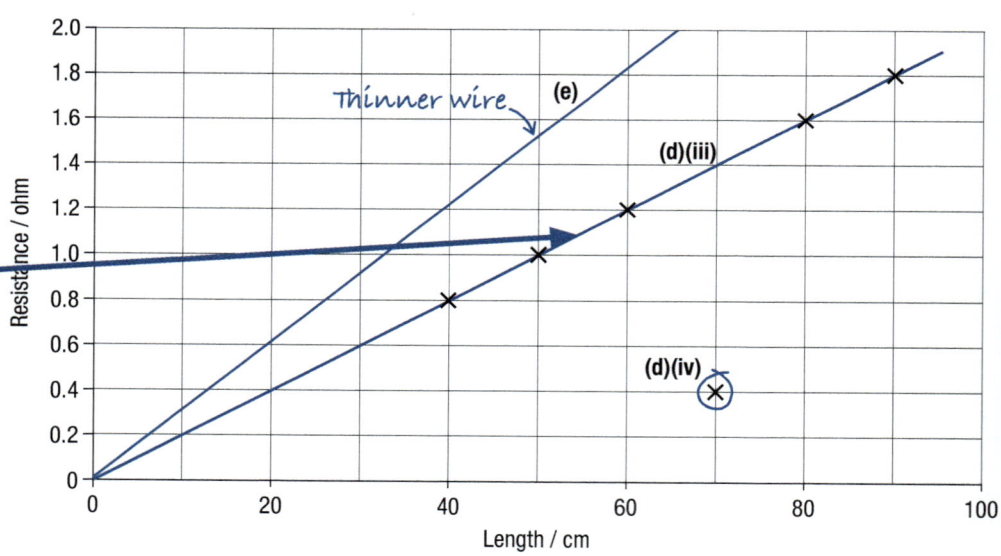

> A straight line, through (0, 0), has been drawn with the aid of a ruler and the anomalous result has been ignored.

(2)

1 (d) **(iv)** One of the results is anomalous.
Put a ring around the anomalous result on the graph.

Ring around (70, 0.4)

(1)

> The anomalous point has been correctly identified.

1 (d) **(v)** Describe the relationship between the length of a wire and its resistance.

As the length is doubled, the resistance is doubled.

(2)

> The candidate could have said that the two quantities are directly proportional. This is indicated by the straight line through the origin of the graph.

1 (e) Another group of students carried out the same investigation. They used a reel of thinner wire.

Draw a line on the graph to show how their results were different to those of the first group of students. Label this line '**Thinner wire**'.

Graph line with greater gradient

(1)

> Thinner wires have a higher resistance per unit length so the line has a greater slope.

Glossary

A

Acceleration – Change of velocity per second (in metres per second per second, m/s^2).

Acid rain – Rain that has absorbed sulfur dioxide gas released from burning fossil fuels.

Alpha (α) radiation – α particles, each composed of two protons and two neutrons, emitted by unstable nuclei.

Alternator – An alternating current generator.

Alternating current – Electric current in a circuit that repeatedly reverses its direction.

Ampere – The unit of electric current (unit symbol A).

Amplitude – The height of a wave crest or a wave trough of a transverse wave from the rest position. Of oscillating motion, is the maximum distance moved by an oscillating object from its equilibrium position.

Angle of incidence – The angle between the incident ray and the normal.

Angle of reflection – The angle between the reflected ray and the normal.

Anomalous result – A result that does not match the pattern seen in the other data collected or is well outside the range of other repeat readings. It should be retested and if necessary discarded.

Atomic number – The number of protons (which equals the number of electrons) in an atom. It is sometimes called the proton number.

B

Bar chart – A chart with rectangular bars with lengths proportional to the values that they represent. The bars should be of equal width and are usually plotted horizontally or vertically. Also called a bar graph.

Beta (β) radiation – β particles that are high-energy electrons created in and emitted from unstable nuclei.

Big Bang theory – The theory that the universe was created in a massive explosion (the Big Bang) and that the universe has been expanding ever since.

Biofuel – Any fuel taken from living or recently living materials such as animal waste.

Black dwarf – A star that has faded and cooled.

Black body radiation – The radiation emitted by a perfect black body which is a body that absorbs all the radiation that hits it

Black hole – An object in space that has so much mass that nothing, not even light, can escape from its gravitational field.

Blue-shift – A decrease in the wavelength of electromagnetic waves emitted by a star or galaxy due to its motion towards us. The faster the speed of the star or galaxy, the greater the blue-shift is.

Boiling point – The temperature at which a pure substance boils or condenses.

Braking distance – The distance travelled by a vehicle during the time its brakes act.

C

Cable – Two or three insulated wires surrounded by an outer layer of rubber or flexible plastic.

Camera – An instrument for photographing an object by using a converging lens to form a real image of the object on a film (or on electronic pixels) in a lightproof box.

Carbon-neutral – A biofuel from a living organism that takes in from the atmosphere as much carbon dioxide as is released when the fuel is burnt.

Centre of mass – The point where an object's mass may be thought to be concentrated.

Centripetal force – The resultant force towards the centre of a circle acting on an object moving in a circular path.

Chain reaction – A series of reactions in which one reaction causes further reactions, which in turn cause further reactions, etc. A nuclear chain reaction occurs when fission neutrons cause further fission, so more fission neutrons are released. These go on to produce further fission.

Circuit breaker – An electromagnetic switch that opens and cuts the current off if too much current passes through it.

Compression – Squeezing together.

Condense – Turn from vapour into liquid.

Conservation of energy – Energy cannot be created or destroyed.

Conservation of momentum – In a closed system, the total momentum before an event is equal to the total momentum after the event. Momentum is conserved in any collision or explosion provided no external forces act on the objects that collide or explode.

Contrast medium – An X-ray absorbing substance used to show up a body organ so the organ can be seen on a radiograph.

Control group – In an experiment to determine the effect of changing a single variable, a control is set up in which the independent variable is not changed therefore enabling a comparison to be made. In a survey, a control group is usually established to serve the same purpose.

Convection – Circulation of a liquid or gas caused by increasing its thermal energy.

Convex (converging) lens – A lens that makes light rays parallel to the principal axis converge to (i.e., meet at) a point; also referred to as a convex lens.

Cosmic microwave background radiation – Electromagnetic radiation that has been travelling through space ever since it was created, shortly after the Big Bang.

Coulomb (C) – The unit of electrical charge, equal to the charge passing a point in a (direct current) circuit in 1 second when the current is 1 A.

Count rate – The number of counts per second detected by a Geiger counter.

Critical angle – The angle of incidence of a light ray in a transparent medium that produces refraction along the boundary.

CT scanner – A medical scanner that uses X-rays to produce a digital image of any cross-section through the body or a three-dimensional image of an organ.

D

Dark matter – Matter in a galaxy that cannot be seen and its presence is deduced because the galaxies would spin much faster if their stars were their only matter.

Data – Pieces of information, either qualitative or quantitative, that have been collected.

Deceleration – Change of velocity per second when an object slows down.

Density – The mass per unit volume of a substance.

Diffraction – The spreading of waves when they pass through a gap or around the edges of an obstacle which has a similar size as the wavelength of the waves.

Diffusion – The spreading out of particles away from each other.

Diode – A non-ohmic conductor that has a much higher resistance in one direction (its reverse direction) than in the other direction (its forward direction).

Direct current – Electric current in a circuit that is in one direction only.

Directly proportional – A graph shows this relationship if the line of best fit is a straight line through the origin.

Dispersion – The splitting of white light into the colours of the spectrum.

Displacement – Distance in a given direction.

Diverging (concave) lens – A lens that makes light rays parallel to the axis diverge (i.e., spread out) as if from a single point; also referred to as a concave lens.

Doppler effect – The change of wavelength (and frequency) of the waves from a moving source due to the motion of the source towards or away from the observer.

Driving force – Force on a vehicle that makes it move; sometimes referred to as motive force.

Dynamo – A direct-current generator.

E

Echo – Reflection of sound, that can be heard.

Efficiency – Useful energy transferred by a device ÷ total energy supplied to the device.

Effort – The force applied to a device used to raise a weight or shift an object.

Elastic – A material is elastic if it is able to regain its shape after it has been squashed or stretched.

Elastic potential energy – Energy stored in an elastic object when work is done to change its shape.

Electric current – Flow of electric charge. The size of an electric current (in amperes, A) is the rate of flow of charge.

Electrical energy – Energy transferred by the movement of electrical charge.

Electricity meter – Meter in a home that measures the amount of electrical energy supplied.

Electromagnet – An insulated wire wrapped around an iron bar that becomes magnetic when there is a current in the wire.

Electromagnetic induction – The process of inducing a potential difference in a wire by moving the wire so it cuts across the lines of force of a magnetic field.

Electromagnetic spectrum – The continuous spectrum of electromagnetic waves.

Electromagnetic waves – Electric and magnetic disturbances that transfer energy from one place to another. The spectrum of electromagnetic waves, in order of increasing wavelength, is as follows: γ and X-rays, ultraviolet radiation, visible light, infrared radiation, microwaves, radio waves.

Electron – A tiny negatively charged particle that moves around the nucleus of an atom.

Endoscope – A medical instrument that uses optical fibres to see inside the body.

Error – A mistake or uncertainty.

Error – human – Often present in the collection of data, and may be random or systematic, for example, the effect of human reaction time when recording short time intervals with a stopwatch.

Error – random – Causes readings to be spread about the true value, due to results varying in an unpredictable way from one measurement to the next. Random errors are present when any measurement is made, and cannot be corrected. The effect of random errors can be reduced by making more measurements and calculating a new mean.

Error – systematic – Causes readings to be spread about some value other than the true value, due to results differing from the true value by a consistent amount each time a measurement is made. If a systematic error is suspected, the data collection should be repeated using a different technique or a different set of equipment, and the results compared.

Evaporate – Turn from liquid into vapour.

Evidence – Data which has been shown to be valid.

Extension – An increase in length of a spring (or a strip of material) from its original length.

F

Fair test – A test in which only the independent variable has been allowed to affect the dependent variable.

Far point – The furthest point from an eye at which an object can be seen in focus by the eye. The far point of a normal eye is at infinity.

Field line – See line of force.

Fleming's left-hand rule – A rule that gives the direction of the force on a current-carrying wire in a magnetic field according to the directions of the current and the field.

Fluid – A liquid or a gas.

Focal length – The distance from the centre of a lens to the point where light rays parallel to the principal axis are focused (or, in the case of a diverging lens, appear to diverge from).

Force – A force can change the motion of an object (in newtons, N).

Force diagram – A diagram showing the forces on an object.

Force multiplier – A lever used so that a weight or force can be moved by a smaller force.

Free electron – An electron that moves around freely within the structure of a metal and is not held inside an atom.

Frequency of an alternating current – The number of complete cycles an alternating current passes through each second. The unit of frequency is the hertz (Hz).

Frequency of oscillating motion – The number of complete cycles of oscillation per second, equal to 1/the time period. The unit of frequency is the hertz (Hz).

Frequency of a wave – The number of wave crests passing a fixed point every second. The unit of frequency is the hertz (Hz).

Friction – A force opposing the relative motion of two solid surfaces in contact.

Fuse – This contains a thin wire that melts and cuts the current off if too much current passes through it.

G

Gamma (γ) radiation – Electromagnetic radiation emitted from unstable nuclei in radioactive substances.

Generator effect – The production of a potential difference using a magnetic field.

Geothermal – Energy comes from energy released by radioactive substances deep within the Earth.

Geostationary – A satellite that stays in the same place as seen from the Earth because it takes 24 hours to move round the Earth and is in a circular orbit above the Earth's equator.

Global warming – The increase in the average temperature of the Earth because of greenhouse gases in the atmosphere. Increasing global warming is occurring due to the increase of greenhouse gases in the atmosphere.

Gradient (of a straight-line graph) – The change in the quantity plotted on the y-axis divided by the change in the quantity plotted on the x-axis.

Gravitational field strength, g – The force of gravity on an object of mass 1 kg (in newtons per kilogram, N/kg). The acceleration of free fall.

Gravitational potential energy (GPE) – The energy of an object due to its position in a gravitational field. Near the Earth's surface, change of GPE (in joules, J) = weight (in newtons, N) × vertical distance moved (in metres, m).

Greenhouse gases – Gases in the atmosphere, such as water vapour, methane, and carbon dioxide, that absorb longer wavelength infrared radiation from the Earth and prevent it escaping into space.

H

Half-life – The average time taken for the number of nuclei of the isotope (or mass of the isotope) in a sample to halve, or for its activity to drop to half the initial value.

Half-wave rectification – The use of a diode in a circuit with an alternating supply p.d. to allow current only in one direction every second half-cycle of the supply p.d.

Hazard – Something (for example, an object, a property of a substance, or an activity) that can cause harm.

Hypothesis – A proposal intended to explain certain facts or observations.

I

Infrared radiation – Electromagnetic waves between visible light and microwaves in the electromagnetic spectrum.

Input energy – The energy supplied to a device.

Interval – The spacing between readings, for example, a set of 11 readings equally spaced over a distance of 1 m would give an interval of 10 cm.

Inverse proportionality – The relationship shown when making one variable n times bigger causes the other one to become n times smaller (e.g., doubling one quantity causes the other one to halve).

Ion – A charged atom.

Ionisation – Any process in which atoms become charged.

Isotopes – Atoms with the same number of protons and different numbers of neutrons.

J

Joule (J) – The unit of energy.

K

Kilowatt (kW) – 1000 watts.

Kilowatt-hour (kWh) – The unit that measures the electrical energy supplied to a 1 kW electrical device in 1 hour.

Kinetic energy – Energy of a moving object due to its motion; kinetic energy (in joules, J) = ½ × mass (in kilograms, kg) × (speed)^2 (in m^2/s^2).

L

Light-dependent resistor (LDR) – A resistor whose resistance depends on the intensity of the light incident on it.

Light-emitting diode (LED) – A diode that emits light when an electric current passes through it.

Limit of proportionality – The limit for Hooke's law applied to the extension of a stretched spring.

Line graph – This is used to display data when both variables are continuous. A line of best fit may be straight or a smooth curve.

Line of action – The line along which a force acts.

Line of force – The line in a magnetic field along which a magnetic compass points; also called a magnetic field line.

Live wire – The mains wire that has a voltage that alternates (between +325 V and −325 V in Europe).

Load – The weight of an object raised by a device used to lift the object, or the force applied by a device when it is used to shift an object.

Long sight – Describes the vision of an eye that cannot focus on nearby objects but can focus on distant objects.

Longitudinal wave – A wave in which the vibrations are parallel to the direction of energy transfer.

M

Magnetic field – The space around a magnet or a current-carrying wire.

Magnetic field line – A line in a magnetic field along which a magnetic compass points; also called a line of force.

Magnetic poles – The ends of a bar magnet or a magnetic compass.

Magnification – The image height ÷ the object height.

Magnifying glass – A converging lens used to magnify a small object (make it appear larger). The object is be placed between the lens and its focal point.

Magnitude – The size or amount of a physical quantity.

Main sequence – The main stage in the life of a star, during which it radiates energy because of fusion of hydrogen nuclei in its core.

Mass – The quantity of matter in an object; a measure of the difficulty of changing the motion of an object (in kilograms, kg).

Mass number – The number of protons and neutrons in a nucleus.

Mean – The arithmetical average of a series of numbers.

Mechanical wave – A vibration that travels through a substance.

Medium – A substance.

Melting point – The temperature at which a pure substance melts or freezes (solidifies).

Microwaves – Electromagnetic waves between infrared radiation and radio waves in the electromagnetic spectrum.

Moment – The turning effect of a force defined by the equation: Moment of a force (in newton metres) = force (in newtons) × perpendicular distance from the pivot to the line of action of the force (in metres).

Momentum – Mass (in kg) × velocity (in m/s). The unit of momentum is the kilogram metre per second (kg m/s).

Monitor – To make observations over a period of time.

Motive force – See driving force.

Motor effect – This occurs when a current is passed along a wire in a magnetic field and the wire is not parallel to the lines of the magnetic field, a force is exerted on the wire by the magnetic field.

N

National Grid – The network of cables and transformers used to transfer electricity from power stations to consumers (i.e., homes, shops, offices, factories, etc.).

Near point – The nearest point to an eye at which an object can be seen in focus by the eye. The near point of a normal eye is 25 cm from the eye.

Neutral wire – The wire of a mains circuit that is earthed at the local substation so its potential is close to zero.

Neutron star – The highly compressed core of a massive star that remains after a supernova explosion.

Neutron – An uncharged particle of the same mass as a proton. The nucleus of an atom consists of protons and neutrons.

Newton (N) – The unit of force.

Newton's First Law of motion – If the resultant force on an object is zero, the object stays at rest if it is stationary or it keeps moving with the same speed in the same direction.

Newton's Second Law of motion – The acceleration of an object is proportional to the resultant force on the object and inversely proportional to the mass of the object.

Newton's Third Law – When two objects interact with each other, they exert equal and opposite forces on each other.

Normal – A straight line through a surface or boundary perpendicular to the surface or boundary.

North pole – The north-pointing end of a freely suspended bar magnet or of a magnetic compass.

Nuclear fission – The process in which certain nuclei (uranium-235 and plutonium-239) split into two fragments, releasing energy and two or three neutrons as a result.

Nuclear fission reactor – A reactor that release energy steadily due to the fission of a suitable isotope such as uranium-235.

Nuclear fusion – The process in which small nuclei are forced together so they fuse to form a larger nucleus.

Nucleus – The positively charged centre of an atom, composed of protons and neutrons.

O

Ohm – The unit of electrical resistance (symbol Ω).

Ohm's law – The current through a resistor at constant temperature is directly proportional to the potential difference across the resistor.

Opinion – A belief not backed up by facts or evidence.

Optical fibre – A fine glass fibre used to transmit light signals.

Oscillate – To move to and fro about a central position along a line.

Oscilloscope – A device used to display the shape of an electrical wave.

P

Parallel – Components connected in a circuit so that the potential difference is the same across each one are in parallel with each other.

Parallelogram of forces – A geometrical method used to find the resultant of two forces that do not act along the same line.

Payback time – The time taken to pay back the cost of an energy-saving device from the savings on fuel bills.

Peak potential difference – The maximum voltage of an a.c. supply measured from zero volts.

Perpendicular – At right angles.

Pitch – The pitch of a sound increases if the frequency of the sound waves increases.

Pivot – The point about which an object turns when acted on by a force that makes it turn.

Plane mirror – A flat mirror.

Planet – A large object that moves in an orbit round a star. A planet reflects light from the star and does not produce its own light.

Plug – A device with an insulated case used to connect the cable from an electrical appliance to a socket.

Potential difference – A measure of the work done or energy transferred to a device by each coulomb of charge that passes through it. The unit of potential difference is the volt (V).

Power – The energy transformed or transferred per second. The unit of power is the watt (W).

Power of a lens – 1/The focal length of the lens in metres. The unit of lens power is the dioptre, d.

Precise – A precise measurement is one in which there is very little spread about the mean value. Precision depends only on the extent of random errors – it gives no indication of how close results are to the true value.

Prediction – A forecast or statement about the way something will happen in the future. In science a prediction is based on prior knowledge or on a hypothesis.

Pressure – Force per unit cross-sectional area for a force acting on a surface at right angles to the surface. The unit of pressure is the pascal (Pa) or newton per square metre (N/m^2).

Principal focus – The point where light rays parallel to the principal axis of a lens are focused (or, in the case of a diverging lens, appear to diverge from).

Principle of moments – For an object in equilibrium, the sum of all the clockwise moments about any point = the sum of all the anticlockwise moments about that point.

Proton – A positively charged particle with an equal and opposite charge to that of an electron. The nucleus of an atom consists of protons and neutrons.

Protostar – A concentration of dust clouds and gas in space that forms a star.

R

Radiation dose – The amount of ionising radiation a person receives.

Radio waves – Electromagnetic waves of wavelengths greater than 0.10 m.

Random – Cannot be predicted and has no recognisable cause.

Range – The maximum and minimum values of a variable.

Range of vision – The distance from the near point of an eye to its far point.

Rarefaction – Stretched apart.

Reactor core – The thick steel vessel used to contain the fuel rods, the control rods, and the moderator of a nuclear fission reactor.

Real image – An image formed by a lens that can be projected on a screen.

Red giant – A star that has expanded and cooled, resulting in it becoming red.

Red supergiant – A star much more massive than the Sun that has expanded after the main sequence stage.

Red-shift – An increase in the wavelength of electromagnetic waves emitted by a star or galaxy due to its motion away from us. The faster the speed of the star or galaxy, the greater the red-shift is.

Reflection – The change of direction of a light ray or a wave at a boundary when the ray or wave stays in the incident medium.

Refraction – The change of direction of a light ray when it passes across a boundary between two transparent substances (including air).

Refractive index – The refractive index, n, of a transparent substance is a measure of how much the substance refracts a light ray.

Relationship – The link between variables that were investigated. These relationships may be: causal, i.e., changing x is the reason why y changes; by association, i.e., both x and y change at the same time, but the changes may both be caused by a third variable changing; by chance occurrence.

Relay – A switch opened or closed by an iron armature that is attracted to the relay's electromagnet when a current is in the electromagnet.

Renewable energy – Energy from natural sources that is always being replenished so it never runs out.

Repeatable – A measurement is repeatable if the original experimenter repeats the investigation using same method and equipment and obtains the same results.

Reproducible – A measurement is reproducible if the investigation is repeated by another person, or by using different equipment or techniques, and the same results are obtained.

Resistance – Resistance (in ohms, Ω) = potential difference (in volts, V) ÷ current (in amperes, A).

Resistive force – A force such as friction or air resistance that opposes the motion of an object.

Resolution – The smallest change in the quantity being measured (input) of a measuring instrument that gives a perceptible change in the reading.

Resultant force – The combined effect of the forces acting on an object.

Resultant moment – The difference between the sum of the clockwise moments and the anticlockwise moments about the same point.

Risk – The likelihood that a hazard will cause harm. You can reduce risk by identifying the hazard and doing something to protect against that hazard.

S

Sankey diagram – An energy transfer diagram.

Scalar – A physical quantity such as mass or energy that has a magnitude only, unlike a vector which has magnitude and direction.

Sensor circuit – An electrical circuit containing a component that produces a change in its output voltage when there is a change in its surroundings

Series – Components connected in a circuit so that the same current passes through them are in series with each other.

Short sight – Describes the vision of an eye that cannot focus on distant objects but can focus on near objects.

Simple pendulum – A pendulum consisting of a small spherical bob suspended by a thin string from a fixed point.

Snell's law – Refractive index of a medium $n = \sin i / \sin r$, where i is the angle of incidence and r is the angle of refraction of a light ray passing from air into the medium.

Socket – This is used to connect the plug of an electrical appliance to the mains circuit.

Solar heating panel – Sealed panel designed to use sunlight to heat water running through it.

South pole – South-pointing end of a freely suspended bar magnet or of a magnetic compass.

Specific heat capacity – Energy needed to raise the temperature of 1 kg of a substance by 1 °C.

Specific latent heat of fusion – Energy needed to melt 1 kg of a substance with no change of temperature.

Specific latent heat of vaporisation – Energy needed to boil away 1 kg of a substance with no change of temperature.

Speed – Distance moved ÷ time taken.

Split-ring commutator – Metal contacts on the coil of a direct current motor that connects the rotating coil continuously to its electrical power supply.

States of matter – The states that matter can exist in: solid, liquid, and gas.

Step-down transformer – An electrical device that is used to step down (reduce) an alternating voltage.

Step-up transformer – An electrical device that is used to step up (increase) an alternating voltage.

Stopping distance – Thinking distance + braking distance for a moving vehicle.

Supernova – The explosion of a massive star after fusion in its core ceases and the matter surrounding its core collapses onto the core and rebounds.

Switch mode transformer – A transformer that works at much higher frequencies than a traditional transformer. It has a ferrite core instead of an iron core.

T

Tangent – A straight line drawn to touch a point on curve so it has the same gradient as the curve at that point.

Terminal velocity – The velocity reached by an object when the drag force on it is equal and opposite to the force making it move.

Thermistor – A resistor whose resistance depends on its temperature.

Thinking distance – The distance travelled by a vehicle in the time it takes the driver to react.

Three-pin plug – A plug that has a live pin, a neutral pin, and an earth pin.

The earth pin is used to earth the metal case of an appliance so the case cannot become live.

Time period – The time taken for one complete cycle of oscillating motion.

Total internal reflection – The total reflection of a light ray inside a transparent substance when it reaches a boundary with air or another transparent substance. Total reflection happens only if the angle of incidence is greater than the critical angle.

Transformer – An electrical device used to change an (alternating) voltage. See also Step-up transformer and Step-down transformer.

Transverse wave – A wave in which the vibration is perpendicular to the direction of energy transfer.

Trial run – Preliminary work that is often done to establish a suitable range or interval for an investigation.

Turbine – A machine that uses steam or hot gas to turn a shaft.

U

Ultrasound wave – A sound wave at frequency greater than 20 000 Hz, which is the upper frequency limit of the human ear.

Ultraviolet radiation – Electromagnetic waves between visible light and X-rays in the electromagnetic spectrum.

Useful energy – Energy transferred to where it is wanted in the form it is wanted.

V

Valid – Suitability of the investigative procedure to answer the question being asked.

Variable – A physical, chemical, or biological quantity or characteristic.

Variable – categoric – A variable that has values that are labels, for example, names of plants or types of material.

Variable – continuous – A variable that has values that are measured (for example, light intensity, flow rate, etc.).

Variable – control – A variable separate from the independent variable that may affect the outcome of an investigation and therefore is kept constant or monitored.

Variable – dependent – A variable whose value is measured for each change in the independent variable.

Variable – independent – A variable whose values are changed or selected by the investigator.

Vector – A physical quantity such as displacement or velocity that has a magnitude and a direction, unlike a scalar which has magnitude only.

Velocity – Speed in a given direction (in metres/second, m/s).

Vibrate – To oscillate rapidly (or move to and fro rapidly) about a certain position.

Virtual image – An image, seen in a lens or a mirror, from which light rays appear to come after being refracted by the lens or reflected by the mirror.

Visible light – Electromagnetic waves that can be detected by the normal human eye. Visible light has a wavelength range from about 350 nm for violet light to about 650 nm for red light. (1 nm = 1 nanometre = 1 millionth of 1 millimetre).

Volt (V) – The unit of potential difference, equal to energy transfer per unit charge in joules per coulomb (J/C).

W

Wasted energy – Energy that is not usefully transferred.

Watt (W) – The unit of power.

Wave speed – The distance travelled per second by a wave crest or wave trough.

Wavelength – The distance from one wave crest to the next wave crest (along the waves).

Weight – The force of gravity on an object (in newtons, N).

White dwarf – A star that has collapsed from the red giant stage to become much hotter and denser.

Work – The energy transferred by a force, given by: work done (in joules, J) = force (in newtons, N) × distance moved in the direction of the force (in metres, m).

X

X-rays – Electromagnetic waves shorter in wavelength than ultraviolet radiation, produced in an X-ray tube.

Answers

1 Resultant forces

1.1

1 a The car decelerates
  b The gravitational force (i.e., your weight), the support force on you from the cushion.
  c An equal and opposite force acts on your foot due to the ball.
2 a i 50 N upwards   ii 200 N
  b i equal, opposite
     ii downwards, upwards
     iii upwards
3 a 500 N downwards
  b 500 N upwards
  c 500 N upwards

1.2

1 a The glider decelerates and stops. Without the air blower on, the glider is in contact with the track and friction acts on it so its velocity decreases to zero and it stops.
  b They are equal and opposite.
2 a It is in the opposite direction to the velocity.
  b It is zero.
3 a The force of the mud on the car is greater than the force on the car from the tractor.
  b 50 N

1.3

1 a 10 N to the left.
  b 50 N vertically upwards.
  c 500 N up the slope.
2 a 5.0 N
  b 6.1 N
  c 6.5 N
3 5400 N (to 2 s.f.)

1.4

1 The force of gravity on a 1 kg object is 10 N.
2 a 500 N   b 80 N
3 1200 N (to 2 significant figures)
4 a When the probe is near the Earth, the force due to the Earth is much larger than the force due to the Moon. As the probe move from the Earth to the Moon, the force due to the Earth decreases and the force due to the Moon increases. At the Moon, the force due to the Earth is (much) smaller than the force due to the Moon. The direction of each force does not change.
  b Because the forces are always in opposite directions, the resultant force is the difference between the two forces. As the probe moves away from the Earth, the resultant force decreases in magnitude and becomes zero where the two forces are equal and opposite. Before reaching this point, its direction is towards the Earth and after this point its direction is towards the Moon. As the probe moves further towards the Moon, the resultant force increases in magnitude.

1.5

1 a i The extension of a spring is directly proportional to the force applied, as long as its limit of proportionality is not exceeded.
     ii 2.5 N
  b i It does not return to its original length when it is released.
     ii The rubber band does return to its original length when it is released whereas the polythene strip does not.
2 a The limit beyond which the tension is no longer proportional to the extension.
  b The force per unit extension as long as the limit of proportionality is not reached.
  c The increase of its length from its unstretched length.
3 a i 80 mm   ii 54 mm   iii 10 mm
  b i 60 mm   ii 50 N/m   iii 1/the spring constant

Answers to end of chapter summary questions

1 a i 1.6 N vertically downwards.
     ii 1.6 N
  b i 0.4 N vertically downwards.
     ii 0.4 N vertically upwards.
2 a i It is zero.
     ii The upthrust acting on the balloon is equal to the sum of the gravitational force on the balloon and the downward pull of the

thread on the balloon. The upthrust is therefore greater than the gravitational force on the balloon.
  b The balloon would move upwards because there is no downward pull on it from the thread. The upthrust is greater than the gravitational force on the balloon so the resultant force on the balloon is vertically upwards.
3 a 320 N   b 80 N
4 a 79 mm, 121 mm, 160 mm, 201 mm, 239 mm
  b Suitable scales correctly plotted points best fit line
  c Suitable method shown (either graphical or numerical) = 280 N
  d i Suitable method shown (eg values inserted into $F = k L$ etc 25 N/m
     ii $(W =)$ $F = k L = 25$ N/m 0.140 m $= 3.5$ N
5 a i 7200 N
     ii 0
  b 13 300 N

Answers to end of chapter practice questions

1 a acceleration; force; velocity (3)
  b Acceleration is rate of change of velocity; velocity is a vector quantity – it has direction; train is changing direction; so although at constant speed, it is accelerating. (4)
  c Longer tow ropes would make the angle between the tow ropes smaller (for the same distance between the tugboats).(1) The parallelogram would therefore have a longer diagonal so the resultant force would be greater (1)
2 a 1600 N (1)
  b A resultant force of 2400 N acts on the car (1) in the direction of force B(1) so the car speeds up (or accelerates) (1)
3 a The parallelogram of forces should be a rectangle with each unequal side proportional to the force it represents. [2] The angle between the resultant and the longer side should be 7°[1]
  b Use your diagram to measure the length of the diagonal [1] then use your scale to show that the magnitude of the resultant force is 9.6 kN.[1]
4 a Choose a scale for the force diagram; draw a line 100 units long; construct an angle of 30° to the line; draw a second line, at this angle, 100 units long; complete the parallelogram; measure the length of the diagonal; convert to size of force using the scale; measure the angle of the diagonal from one of the force vectors. (6)
  b He would need to make the angle between his rope and the other rope smaller; so that the diagonal of the force parallelogram is still in the same direction. (2)
5 a Force B is equal in magnitude to force A (or 120 N) [1] and is opposite in direction [1]
  b i Weight = 300 kg 10 N/kg = 3000 N (1)
     ii Friction between the tyres and the road; the support force of the road on the car. [2]
     iii Zero (because the car is moving at constant velocity)
  c i Force A increases [1], the weight and support forces stay the same[1]; friction increases (due to increased air resistance) [1]

2 Forces and motion

2.1

1 a i The distance travelled each second does not change.
     ii The gradient of the graph is constant.
  b i 30 m/s
     ii 500 s
     iii 13.3 m/s
2 a 30 m/s
  b 9000 m
  c 110 s
3 $d = 7560$ m, speed $= 18$ m/s
4 a 30 m/s   b 75 minutes

2.2

1 a Speed is distance travelled ÷ time taken regardless of direction. Velocity is speed in a given direction.
  b 2400 m
2 1.25 m/s^2

3   a   i   As it left the motorway.
        ii When it travelled at constant velocity.
    b   27 m/s
4   a   3.0 m/s²   b   9.6 m/s

2.3
1   a   B
    b   A
    c   D
    d   C
2   a   i   A   ii   C
    b   B
3   a   Distance = velocity × time = 8 × 20 = 160 m
    b   C
4   a   40 m
    b   100 m

2.4
1   a   15 m/s
    b   i   The speed is zero at time = 0.
        ii The speed increases gradually.
2   a   The cyclist accelerates at a constant acceleration from rest to 8 m/s
        for 40 s then decelerates at constant deceleration for the last 20 s.
    b   i   0.20 m/s², 160 m.
        ii −0.40 m/s², 80 m.
    c   4.0 m/s
3   a   Student's graph, accurately drawn.
    b   2.0 m/s²
    c   i   400 m
        ii 400 m

2.5
1   a   640 N   b   4.0 m/s²
2   a   16 N   b   40 kg   c   12 m/s²
    d   2.4 N   e   25 000 kg
3   a   3000 N
    b   i   600 N   ii   2400 N

2.6
1   a   Braking distance
    b   Thinking distance
    c   Braking distance
2   a   i   6.0 m   ii   24.0 m   iii   30.0 m
    b   12 m
3   a   i   The thinking distance is proportional to the speed as the reaction
            time is constant.
        ii When the speed is twice as large and the braking force is
            constant, the braking time is greater, so the braking distance more
            than doubles (think about the area under the velocity–time graph).
    b   Yes; the braking distance divided by the square of the speed is the
        same for all three speeds. So the braking distance is proportional to
        the square of the speed.

2.7
1   a   The initial resultant force is equal to its weight.
    b   The frictional force is less than the weight.
    c   Zero   d   Zero
2   a   i   10 m/s²   ii   Zero
    b   The initial acceleration is zero because the drag force is zero initially
        so the resultant force is due to gravity only. As the object falls its
        acceleration gradually decreases because the drag force increases so
        the resultant force decreases. The acceleration becomes zero when
        the drag force is equal to the weight and the resultant force is then
        zero.
3   a   The frictional force due to the parachute increases with speed so the
        resultant force on the parachutist decreases. When the frictional force
        becomes equal to the weight, the resultant force becomes zero and
        the parachutist moves at terminal velocity.
    b   i   900 N   ii   900 N upwards
4   a   5.2 m/s² (i.e., the gradient at 0.10 s)
    b   For an object of mass m, the resultant force on it at 0.10 s = mass
        m acceleration a. Its acceleration = 5.2 m/s² = 0.52 g. Because the
        resultant force = weight − drag force, the drag force = mg − ma = mg
        − 0.52 mg = 0.48 mg which is approximately equal to half its weight.

Answers to end of chapter summary questions
1   Use the stopwatch to measure the time taken to complete 10 complete
    laps. Repeat the timing several times to obtain a mean value, then divide
    this mean value by 10 to give the time for one complete lap.
    Use the tape measure to measure the diameter by ⊠ of the circle made by
    the centre line of the track then multiply the diameter to obtain a value for
    the circumference (or lay a long string along the centre line of the track

exactly once round the track then mark the start and end of the string on
the track; the distance between the two marks can be measured using
the tape measure. This equals the circumference.
The speed of the car is the circumference divided by the time for one lap.
2   a   The speed decreased then became constant.
    b   i   17 m/s
        ii 10 m/s
3   a

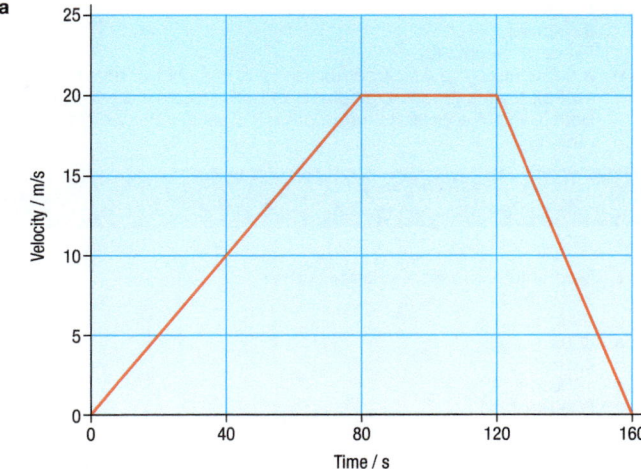

    b   0.25 m/s², 0, −0.50 m/s²
    c   2000 m
    d   Average speed = $\dfrac{\text{distance}}{\text{time}}$ = $\dfrac{2000}{160}$ = 12.5 m/s.

4   a   960 N
    b   i   1.1 m/s²   ii   883 N, 77 N
5   a   For an object of mass m, the gravitational force on it = m × g. Since
        this is the only force acting on the object, the resultant force on the
        object = m × g. The acceleration of the object = resultant force/mass
        = m × g/m = g.
    b   17 m/s
    c   i   The acceleration of X is constant and equal to 10 m/s²
        ii Object accelerates at first. The frictional force on it increases with
            speed so the resultant force on it and its acceleration decreases.
            When the frictional force is equal to the weight of the object, the
            resultant force is zero. The acceleration is then zero so the velocity
            is constant.
6   a   i   225 N   ii 450 N
    b   The cyclist exerts a constant force driving her forward. Crouching
        reduces the force of air resistance (the frictional force). The frictional
        force increases with speed. So the cyclist can get to a higher speed
        before the frictional force becomes equal to the driving force.
    c   a = (v − u)/t = (0 − 6.6 m/s)/3.7 s = − 1.8 m s⁻² [1]. (Magnitude of) the
        resultant force F = ma = 45 kg (−)1.8 m s⁻² = (−) 81 N. [1] Assuming
        other frictional forces are negligible, the resultant force is due to the
        braking force only therefore the braking force = 81

Answers to end of chapter practice questions
1   a   i   Deceleration/slowing down
        ii Constant speed
        iii Accelerating/speeding up (3)
    b   5 drops = 5 seconds
        Speed   = 100/5 = 20 m/s (3)
    c   Change in velocity = 5 − 25 = −20 m/s
        Acceleration   = −20/10 = −2 m/s²
        (1 mark for the correct unit; 1 mark for the minus sign) (5)
2   a   i   stationary
        ii travelling at constant speed (2)
    b   i   0.7 s (1)   ii   Speed   = 24 m/s
            Distance   = speed × time = 24 × 0.7 = 16.8 m (3)
    c   i   Change in velocity = 24 − 0 = 24 m/s
            Time   = 4.7 − 0.7 = 4 s
            Deceleration   = 24/4 = 6 m/s² (4)
        ii Distance = area = ½ base × height
                           = ½ × 4 × 24 = 48 m (3)
3   a   i   Measure out a distance; one student at either end; student gives
            signal when a vehicle passes; second student starts timing when
            sees the signal; stops timing when vehicle passes; calculate speed
            as distance/time.
            Equipment: metric tape or trundle wheel; stopwatch (6)
        ii any three of: as large a distance as possible; extra students timing;
            take average time; as many vehicles as possible; use sensors and
            electronic timers. (3)

b i bar chart
  pie chart
  One variable categoric; one variable discrete; can calculate percentage of total number at each speed. (5)
 ii Time needs to be long enough to give a large sample ideally a morning/afternoon/day; a bar chart would need 'the number of vehicles per ….' (2)

4 a i acceleration
 ii speed
 iii velocity–time (3)
 b X starts after Y (0.4 s); X completes race in a shorter time (X 9.6 s – Y 10.8 s); X wins the race; gradients of both graphs increase throughout; both X and Y accelerate throughout; X's acceleration > Y's; 100 m race (6)

3 Momentum and force

3.1

1 a Momentum = mass × velocity, kg m/s.
 b 240 kg m/s
 c 0.48 m/s
2 a 400 kg m/s
 b 0.5 m/s
 c 1000 m/s
3 a 5000 kg m/s
 b 2.0 m/s

3.2

1 a They exert equal and opposite forces on each other.
 b They have equal and opposite momentum just after they separate.
 c Just after they separate, the velocity of the 80 kg skater is three-quarters the velocity of the 60 kg skater and in the opposite direction.
 d Their total momentum is zero just after they separate.
2 a 120 kg m/s
 b 1.5 m/s
3 a 25 m/s
 b It would have been less.

3.3

1 a The seat belt increases the time taken to stop the person so the change of momentum per second is less, and therefore the force on the person is less.
 b −7200 N
2 a i 4000 N
 ii 800 N
 b Force = change of momentum divided by time taken. The change of momentum would be the same but the time taken would be much less. So the force would be much greater.
3 a Initial momentum = 2000 kg × 12 m/s = 24 000 kg m/s = final momentum. This is equal to the total mass × the velocity after impact which is therefore equal to 24 000 kg m/s ÷ 12 000 kg = 2 m/s.
 b i −33 m/s²
 ii −20 000 kg m/s
 iii 67 000 N

3.4

1 A cycle helmet protects the cyclist's head in a collision or if the cyclist falls off the cycle and his/her head hits the ground. The helmet increases the time taken to decelerate the head so it reduces the change of momentum per second and therefore reduces the impact force.
2 a The air bag increases the time taken to stop the person it acts on. This reduces the force of the impact. Also, the force is spread out across the chest by the air bag so its effect is lessened again.
 b In an accident where the car suddenly stopped, the child would press against the back of the car seat spreading out the force. This would prevent the child from being thrown forwards.
3 a 21600 kg m/s
 b 35 m/s
 c Yes

Answers to end of chapter summary questions

1 a i 45 000 kg m/s
 ii 3750 N
 b It is reduced to zero.
 c The car would probably have skidded as there is an upper limit on how much friction the road can exert on the tyres.
2 a i 115 N
 ii The force acts horizontally backwards from the seat belt.
 b The child would not have stopped when the car stops and would hit the back of the front seats or be thrown through the windscreen.
3 a The force of the student's foot on the boat pushes the boat away.
 b i 37.5 kg m/s
 ii 0.75 m/s

4 a i 36 000 kg m/s
 ii 20 000 kg m/s
 b i 16 000 kg m/s
 ii 13.3 m/s
5 a i Acceleration = $\dfrac{\text{change in velocity}}{\text{time}}$,
  since weight starts at rest.
  Speed = 10 m/s² × 0.63 s = 6.3 m/s.
 ii 44 kg m/s
 b 20 000 N
 c The impact force would probably have been the same because the increase of the speed and therefore the momentum of the weight would have been over a longer impact time.
6 a 8.4 kg m/s
 b Impact force = $\dfrac{\Delta p}{t} = \dfrac{8.4}{0.0384} = 218.75 = 220\,N$ (2 s.f.)

Answers to end of chapter practice questions

1 a Momentum is mass × velocity; velocity is speed in a given direction; so momentum has magnitude and direction. (3)
 b i Momentum of P = mass × velocity OR 20 000 × 14
  = 280 000 kg m/s
  (1 mark for answer; 1 mark for correct unit) (3)
 ii momentum of Q = 600 000 kg m/s (1)
 iii total momentum before collision = 880 000 kg m/s
  momentum after collision = 880 000 kg m/s
  = 50 000 × v
  v = 17.6 m/s (4)
2 a i kg m/s (1)
 ii **Similarities:** both have same *magnitude* of momentum.
  Differences: they have momentum in opposite directions and since the coach has a larger mass, he has a smaller velocity. (3)
 b momentum of dancer = 50 × 1.5 = 75 kg m/s
  momentum of coach = 75 kg m/s
  = 90 × v
  v = 0.83 m/s (3)
3 a Seatbelts stop drivers/passengers continuing forward when car stops suddenly; preventing them hitting the windscreen/seat in front; time of impact is increased; so deceleration is reduced; therefore the force of impact is reduced; seatbelts also spread the force; airbags inflate on impact; spread the force; increase time of impact; reducing the deceleration; also reducing the force. (6)
 b The batsman wants the impact force to be large; therefore the impact time should be as short as possible; the fielder wants the force on his hands to be as small as possible; therefore the impact time should be long. (4)
4 a The crumple zone 'gives'; making the impact time longer than it would be without the crumple zone; this reduces the deceleration; therefore reducing the force and the damage to the car. (4)
 b Acceleration = F/m OR 60/0.2
  = 300 m/s²
  = change in velocity/time OR change in velocity/0.008 s
  change in velocity = 300 × 0.008 = 2.4 m/s (5)
5 a i 8 m/s; 9600 kg m/s (2)
 ii Impact time = 2 milliseconds (1);
  Impact force = $\dfrac{1200 \times -8}{0.002}$ = −4 800 000 N = −4800 kN (3)
 b The decrease in velocity does not take place at a constant rate; as shown by the graph being a curve/line has a changing gradient; so momentum is not lost at a steady rate; therefore the impact force is not constant and is least at the end of the impact. (4)

4 Moments

4.1

1 The point in the object where all its mass can be thought to be concentrated.
2 a and c The centre of mass is where the two diagonal lines from the corners cross.
 b The centre of mass is found by drawing two diametric lines at right angles. The centre of mass is where the two line cross.
3 The centre of mass of the child is then directly below the midpoint M of the points of suspension of the swing. At this position, the moment of the child about M is zero.
4 See the practical instructions and Figure 4 on p 51.

4.2

1 a i Increased
 ii Unchanged
 iii Reduced to a quarter
 b 18 N m

2 a Anticlockwise
b i Increased
ii Decreased
3 a The moment of the applied force about the pivot is greater the longer the handle is, so a greater force can be exerted on the nail.
b The rust on the hinge increases the frictional forces in the hinge, so a greater moment and therefore a greater force must be applied to the door to overcome the moment of the frictional forces at the hinge.
4 72 N

4.3

1 a i 3 N
ii 1.2 N
b In both examples, the line of action of the effort is at a greater perpendicular distance from the pivot than the corresponding distance for the load. A smaller effort therefore gives an equal and opposite moment about the pivot to the load's moment.
2 a Dawn
b 340 N, 1.84 m
c Dawn needs to move 0.66 m towards the pivot so she is 1.84 m from the pivot.
3 1.5 N

4.4

1 a i It would be less stable as its centre of mass would be higher.
ii Without stabilisers, when the child leans to one side, the moment of the rider's weight (about the line between the points where the cycle wheels are on the ground) makes the bicycle fall over. The stabiliser wheel on that side touches to ground and an upward force from the ground acts on it. This upward force provides a moment (about the line between the points where the cycle wheels are on the ground) which counterbalances the moment of the rider's weight and stops the bicycle falling over.
b A supermarket trolley, a tall electric kettle, etc.
2 a The chair would topple over if the baby in the chair leans too far sideways.
b The lower the centre of mass, the harder it is to topple it over.
3 a When it is empty, its centre of mass is approximately halfway up the bottle. When it is standing upright and is less than half full, its centre of mass is approximately halfway between its base and the water level. This position of the centre of mass will always be lower than the position when it is empty.
b The cone has a wide base which is attached to a heavy square board. The centre of mass of the cone is therefore much lower than it would be if the base was narrow, and therefore more stable.

Answers to end of chapter summary questions

1 a The effort acts further from the fulcrum than does the force of the bottle opener on the top. So the force of the bottle opener on the top is greater than the effort.
b 75 N; the effort is 3 times further from the fulcrum than edge of the cap is. So an effort of 25 N causes three times as much force to be exerted on the edge of the cap.
2 a It would be less stable as it would be easier to disturb.
b 0.06 N
3 a The distance from the wheel axle to force F is much greater than the distance from the axle to the centre of mass. The weight of the sand and the wheelbarrow causes a certain moment about the wheel axle. To lift the wheelbarrow legs off the ground, force F must create a greater moment about the wheel axle. Because the distance from the wheel axle to force F is much greater than the distance from the axle to the centre of mass, force F can be much smaller than the weight to give an greater moment than that of the weight.
b i 149 N **ii** 84 N
4 a i 0.84 N m **ii** 3.5 N m
b The ruler would turn clockwise about the pivot and then fall off the pivot. This would happen because moving A towards the pivot would reduce its moment about the pivot. Therefore, the moment of B (which is unchanged) would be greater than the moment of A. So the ruler would no longer be balanced on the pivot and would turn clockwise and fall off the pivot.
5 a The centre of mass of the ruler is not directly above the pivot so the weight of the ruler causes a non-zero moment about the pivot. The distance from weight W_1 to the pivot is such that the moment of W_1 about the pivot is equal and opposite to the moment of the ruler about the pivot. Therefore the ruler is balanced and does not tip off the pivot.
b i The distance (d) from the pivot to the centre of mass of the ruler needs to be measured. The distance (x) from the pivot to the point where the thread supporting W_1 is tied to the pivot also needs to be measured.
ii The moment of the ruler about the pivot = $W_o d$. The moment of the ruler about the pivot = $W_1 x$ and is in the opposite direction

to the moment of the ruler about the pivot. Because the ruler is balanced, these two moments are equal and opposite therefore $W_o d = W_1 x$. Rearranging this equation gives $W_o = (W_1 x)/d$ Substituting the measured values of x and d and the known weight of W_1 gives W_o

Answers to end of chapter practice questions

1 a Points to be made: put a pin through one hole; suspend the sheet from the pin; hang a plumb-line from the pin; draw a vertical line along the plumb-line; repeat hanging the sheet from the other hole; where the two lines cross is the centre of mass. (6)
b i Moment = force × (perpendicular) distance from pivot OR = 2.5 × 30 (1) = 75 N cm (1)
ii Anticlockwise moment less (1); since perpendicular distance is less (1); therefore clockwise moment is less (1); so tension in spring is less. (1)
2 a i Two from: track; wheel base; position of centre of mass (2)
ii The government test should be unbiased/manufacturers' test could be biased. (1)
b i Weight = 12 000 N (1) moment = weight × distance from pivot OR = 12 000 × 0.5 (1) = 6 000 Nm (1 for answer; 1 for the unit).
ii Low centre of mass (1); clockwise moment returns vehicle to the horizontal (1); line of action of weight passes through base. (1)
iii Centre of mass raised (1); vehicle tilts less (1); before line of action of weight passes outside base. (1)
3 a The moment due to the applied force is equal to the moment of the cable force about the pivot.(1) The moment of each force is equal to the force the perpendicular distance from the pivot to its line of action. (1) The applied force acts further from the pivot that the cable force does . So the the cable force is much greater than the applied force. (1)
b The applied force is about 6 times further from the pivot than the cable force. (1) So the cable force is about 6 times bigger than the cable force. (1) The cable force is about 210 N (= 6 × 35 N). (1)
c The pivot is in equilibrium and is acted on by the applied force, the cable force and the force on it due to the steel bracket. (1) The resultant of the cable force and the applied force is to the right (just) above the horizontal. (1) So the steel bracket force onthe pivot acts to the left and (just) below the horizontal (or in the opposite direction to the resultant of the cable force and the applied force.) (1)
4 a Moment of X about P = moment of W about P therefore weight of X XP = weight of W WP (1)
weight of X 0.270 m = 3.0 N 0.400 m (1)
weight of X = (3.0 N 0.400 m)/0.260 m = 4.6 N (1)
b Using the same method as above, weight of X XP = 3.0 N 0.480 m (1)
XP = (3.0 N 0.480 m)/4.6 N = 313 mm (1)
so X would need to be moved by 53 mm away from P (1)

5 Forces and energy

5.1

1 a i Energy is transferred into kinetic energy of the boat and the water and thermal energy of the surroundings.
ii Energy is transferred into gravitational potential energy of the barrier and thermal energy due to friction and sound energy.
b 80 000 J
2 a The kinetic energy of the car is transferred by heating to the disc pads by friction.
b 140 000 J
3 a i 96 J **ii** 96 J
b 200 N

5.2

1 a i A mains-connected filament lamp.
ii 10 000 W electric cooker
b 600 000 kW (= 600 MW)
2 a 800 J
b 800 J
c 160 W
3 a i 1800 m
ii 9.0 MJ
b Force = $\dfrac{\text{work done}}{\text{distance}}$ = $\dfrac{9\,000\,000}{1800}$ = 5000 N

5.3

1 a On descent, gravitational potential energy of the ball is transferred to kinetic energy of the ball. On impact, the kinetic energy of the ball is transferred into elastic energy of the ball and some of the elastic energy is transferred back to kinetic energy as it rebounds. After the impact, the kinetic energy of the ball is transferred to gravitational potential energy of the ball as it rises.

b i 1.1 J

ii Energy transfer to the surroundings due to air resistance as the ball moves through the air; energy transfer by heating to the ball when the ball is deformed.

2 a 90 J

b 4500 J

3 a 450 J

b 375 J

4 Energy must be supplied to keep the biceps muscle in the arm contracted. No work is done on the object, because it doesn't move. The energy supplied heats the muscles and is transferred by heating to the surroundings.

5.4

1 a The brake pads becomes hot due to friction. Energy transfer from the brake pads to the surroundings by heating. Sound waves created by braking also transfers energy to the surroundings.

b Kinetic energy of the roller coaster is transferred to gravitational potential energy of the roller coaster and to kinetic energy of the air as the roller coaster goes up the hill. Gravitational potential energy is transferred to kinetic energy and the air as the roller coaster descends.

2 a **On descent:** Gravitational potential energy α kinetic energy + energy heating the surroundings due to air resistance.
On impact: Kinetic energy α elastic energy of trampoline + energy heating the surroundings due to impact + sound.
On ascent: Elastic energy of trampoline α kinetic energy α gravitational potential energy + energy heating the surroundings due to air resistance.

b The ball has less energy at the top of its bounce than at the point of release.

c Use a clamp to hold a metre ruler vertically over the middle of the trampoline surface. Hold the ball next to the metre ruler with its lowest point level with the top of the ruler. Release the ball so it rebounds vertically and observe the highest level of the bottom of the ball against the metre ruler after the rebound. Repeat the same test several times to obtain the average rebound position of the ball. Repeat the test with the same ball for the other two trampolines. The one with the highest rebound position is the bounciest.

3 Elastic energy of the rubber straps is transferred to kinetic energy of the capsule. This kinetic energy is transferred to gravitational potential energy as the capsule rises to the top of its flight etc.

5.5

1 a i 36 kJ

ii 88 J

b 17 m/s

2 a i Work done by the muscles transfer chemical energy from the muscles to elastic potential energy of the catapult.

ii Elastic potential energy of the catapult is transferred to kinetic energy of the object.

b i 10 J

ii 10 m/s

3 a 3600 N

b 800 kg

4 5.5 J

5.6

1 a Wasted: sound, kinetic energy of the air.

b Useful: light and sound. Wasted: heat.

c Useful: boils the water. Wasted: heat loss through surfaces, sound.

d Useful: sound. Wasted: heat loss.

2 a The gear box would heat up due to energy transfer through friction between the gears. The hotter the gear box gets, the less efficient the gears will work.

b The shoes would heat up due to energy transfer by conduction and infrared radiation from the feet. The feet would transfer less energy as the shoes warm up so the feet would become hotter.

c The drill would heat up and smoke if it burns the wood.

d The discs would heat up due to energy transfer by friction between the discs and the brake pads from the kinetic energy of the moving parts of the car.

3 a As the pendulum swings towards the middle, its gravitational potential energy decreases and its kinetic energy increases. As it moves from the middle to the highest position on the opposite side, its kinetic energy transfers back to gravitational potential energy. Air resistance acting causes some of its kinetic energy to be transferred to the surrounding as heat.

b Air resistance causes friction as the pendulum swings. This produces heat and so the pendulum transfers energy to the surroundings and stops.

5.7

1 a 85 J

b It is transferred by conduction to the surroundings

c

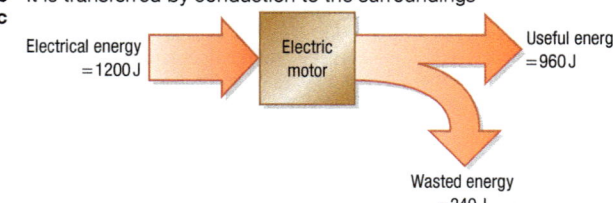

2 a 36 J

b 40%

3 a 800 J

b 50 W

Answers to end of chapter summary questions

1 a i 210 MJ

ii 6900 m

iii Resistive force = $\dfrac{\text{work done}}{\text{distance}}$ = $\dfrac{210\,000\,000}{6900}$ = 30 345 = 30 000 N (2 s.f.)

iv The acceleration of the train is zero so the resultant force on it is zero. Therefore, the driving force and the resistive force must be equal and opposite to each other.

b The train gains gravitational energy as it travels up the incline. The rate at which it transfer energy to the surroundings is unchanged as its speed is the same and the resistive forces acting on it are unchanged. So the output power of the engine needs to be greater as energy must be transferred to the train as gravitational potential energy as well as to the surroundings.

2 a 180 J

b work done = force × distance in the direction of the force = 11 N × 20 m = 220 J.

c The trolley did not gain kinetic energy as its speed was constant. The trolley gained 180 J of gravitational potential energy. Resistive forces such as friction at the trolley wheels must have transferred 40 J of energy to the surroundings as waste energy.

3 a i 960 J

ii 240 J

iii

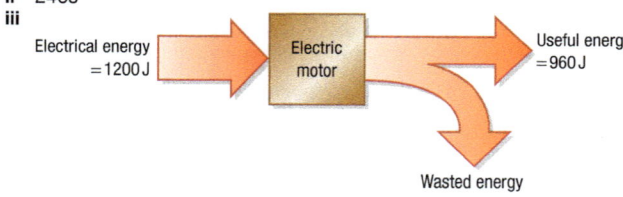

b 5040 J

4 a i 1500 J   **ii** 13 500 J   **iii** 10%

b i Apply oil to the bearings of the motor and the pulley to reduce friction.

ii Friction or air resistance can never be completely eliminated from the motor. In addition, the motor becomes warm due to the heating effect of the electric current passing through it.

5 a Energy is initially stored in the stretched cord as elastic potential energy. This energy is transferred to the arrow as kinetic energy when the cord is released. As the stone gains height, its kinetic energy decreases and its gravitational potential energy increases. At maximum height, the arrow has maximum gravitational potential energy and minimum kinetic energy. As it travels through the air, some of its energy is transferred to the air due to air resistance.

b i 4.7 J

ii 3.0 J

iii 15 m/s

6 a 135 kJ

b i 940 J

ii 675 kJ

c 810 kJ

Answers to end of chapter practice questions

1 a joule; kilojoule; (2)

b i W = m × g OR W = 58 × 10 (1)
= 580 N (1)

ii Work = weight × height OR work = 580 × 12 (1)
= 6960 J (1)

iii Power = work done/time taken OR power = 6960/120 (1)
= 58 W (1 mark for 58, 1 mark for the unit)

2 a i chemical energy from food (1); transferred to kinetic as the jumper runs (1); to gravitational potential as he rises above ground (1)

ii Gravitational potential energy = m × g × h = 65 × 10 × 1.25 (1)
= 812.5 J (1)

b i Kinetic energy = 812.5 = ½ × m × v² (1)
v² = 2 × 812.5/65 (1)
= 25 (1)
v = 5 m/s (1)

ii Energy is lost (1); due to drag along the track (1); and in rising (1)

3 a Equipment needed: stopwatch; scales/balance; tape measure. They need to measure: their mass; the vertical height of the stairs (they may measure the rise and number of steps and multiply together – in which case a tape measure would not be needed, just a ruler); and the time taken to climb the stairs – this would need co-operation and some form of signalling to indicate starting and stopping of the stopwatch. (6)

b Work done = 450 × 4 (1)
power = work/time = 1800/2.5 (1)
= 720 W (1)

4 a i gravitational potential (1)

ii Loss in height = 0.5 m (1)
percentage loss = 0.5/2.0 × 100 (1)
= 25% (1)

iii Transferred to thermal; warms the ground; spreads out. (3)

b i Equal to GPE = m × g × h (1)
= 0.2 × 10 × 2 (1)
= 4 J (1)

ii ½ × m × v² = 4 (1)
= 8/0.2 (1)
= 40 (1)
v = 6.3 m/s (1)

iii Reaction force of the floor changes the motion (1); speed changes (1); direction is reversed (1); acceleration is the rate of change of the vector velocity. (1)

5 a acceleration a = (v − u)/t = (1.8 m/s − 0)/5.0 = 0.36 m/s² [1]. Using F = ma, resultant force = 220 kg 0.36 m/s² = 79 N (1)

b i Distance moved in 1 s = 1.8 m (1) so work done = force × distance = 79 N × 1.8 m = 143 J (1)

ii % efficiency = (143 J/500 J) × 100% (1) = 29% (1)

6 Energy resources

6.1

1 a i Coal, oil and gas-fired power station

ii A nuclear power station

b Nuclear fuel produces radioactive waste when it is used and radioactive waste must be stored for many years until it becomes non-radioactive.

2 a i advantage – no radioactive waste; disadvantage – produces greenhouse gases

ii advantage – can be started quicker; disadvantage – gas supplies will run out before coal supplies

b 10 000 kg (= 300 000 MJ/kg 30 MJ/kg)

3 a A biofuel is any fuel obtained from living or recent organisms such as animal waste or woodchip.

b The carbon dioxide released when it is burnt is balanced by the carbon that was taken in as carbon dioxide from the atmosphere when it grows.

4 Energy used per person per year = 500 million million million J/6000 million = 83 000 J
Energy used per person per second = 83000 J/(365 days × 24 hours × 3600 seconds/hour) = 2600 J/s

6.2

1 a A renewable energy resource is a natural source of energy that can never be used up

b i tidal power   **ii** wind power

2 a i 1000   **ii** 25 km

b from top to bottom; Hilly or coastal areas, estuaries, coastline, mountain areas

3 a Tidal schemes use sea water flowing through turbines in barriers built across estuaries. Hydroelectric schemes involve less construction because it uses rainwater trapped in upland reservoirs (OR Tidal power uses sea water trapped by tidal flow in an estuary by a long barrier which is periodic not constant. Hydroelectric schemes use water flowing continuously from upland reservoirs. Hydroelectric power is therefore continuous whereas tidal power is produced for only part of each tidal cycle.

b i Hydroelectricity   **ii** It is only possible in hilly areas with significant amounts of rainfall. It is not possible in flat dry areas.

4 a A pumped storage scheme uses electricity from power stations at time of low demand to pump water to an upland reservoir. At times of high demand, the flow is reversed and the water in the reservoir is used to generate electricity.

b Most power stations are not switched off when demand is low, because they can't be restarted quickly if demand suddenly rises. But the electricity generated at times of low demand is used to pump water to uphill reservoirs instead of being wasted. The water pumped uphill can be used later at times of high demand to generate electricity.

6.3

1 a Geothermal energy comes from the energy released by radioactive substances deep underground

b Solar energy is not available at night, but geothermal energy is released all the time (or the output of a solar panel is reduced by cloud cover, but geothermal energy is unaffected by cloud cover).

2 a 1500 (= 300 W/0.2 W per cell)

b To supply electricity when the solar panels are in darkness

3 a 4800 kW (= 200 kW 48 hours)

b Advantage; geothermal energy is energy released by radioactive substances inside the Earth and the rate of release does not vary whereas wind energy is energy generated by wind turbines which varies according to weather conditions.
Disadvantage; geothermal power stations can only operate in those areas where the flow of geothermal energy from within the Earth is significant.

4 a In 1 second, ΔE = mcΔθ = 0.010 kg × 4200 J/kg/°C × (35−14)°C [1] = 880 J. Thermal energy transferred by second = 880 J/s. [1]

b In 1 second, the mass of water that is heated = 0.017 kg and the thermal energy transferred = 880 J
Substituting these values into E = m c Δθ gives
880 J = 0.017 kg × 4200 J/kg/°C × Δθ [1].
Therefore, $\Delta\theta = \dfrac{880\ \text{J}}{0.017\ \text{kg} \times 4200\ \text{J/kg/°C}}$ [1] = 30 °C [1] which
means the output temperature is 44 °C [1].

6.4

1 a Gas

b Increase of carbon dioxide in the atmosphere; acid rain

c Advantages – they will never run out; they do not release greenhouse gases/carbon dioxide into the atmosphere; they do not produce radioactive waste. Disadvantages – they take up large areas ; they affect the habitats of plants and animals.

2 a A   **b** D   **c** C   **d** B

3 a Solar energy, wave energy, wind energy,

b Nuclear, geothermal, tidal

4 Your answer in essay form should discuss the three types in terms of reliability and environmental effects including use of land, effect on natural habitats, pollution and waste.

Answers to end of chapter summary questions

1 a i Renewable energy is energy from any source that is replenished by natural processes

ii Renewable fuels are fuels that are replenished by natural processes. For example, wood is replenished when trees grow.

b Although renewable fuels release carbon dioxide when they are burnt, they are produced directly or indirectly from plants and all plants take up carbon dioxide from the atmosphere when they grow, So renewable fuels are carbon-neutral because the plants that produce them take as much carbon dioxide from the atmosphere and is release when the fuel is burnt.

2 a Similarities 1. Both rely on water flow, 2. Both transfer energy from gravitational potential energy stored in water to electrical energy Differences 1. Tidal power stations are in estuaries whereas hydroelectric power stations are in hilly areas 2. Tidal power stations use seawater whereas hydroelectric power stations use rain water OR tidal power stations produce more power than hydroelectric power stations. **b i** wind **ii** hydroeletric **iii** tidal

3 a i geothermal   **ii** hydroelectric
iii coal-fired   **iv** nuclear

b i non-renewable (because a tree does not grow fast enough to replace wood from another tree that is cut down and burnt),
ii non-fossil

4 a Nuclear power stations are cheaper to build and run than wind turbines for the amount of energy produced.

b Reliability; Nuclear power is more reliable than hydroelectric power as hydroelectric power stations need rainwater. Wind turbines and solar power are much less reliable than hydroelectric power because they are both weather-dependant and solar power is not available at night.

5 a 1 wind   **2** hydroelectric   **3** solar
  b i hydroelectricity   **ii** solar
  Environmental effects; Nuclear power stations produce radioactive waste that must be stored safely for many years. Nuclear power stations do not produce greenhouse gases whereas gas and coal-fired power stations produce carbon dioxide when the fuel is burnt and carbon dioxide is a greenhouse gas; the amount of greenhouse gases in the atmosphere is increasing and this increase is causing increased global warming. Coal-fired power stations also release gases such as sulfur dioxide which cause acid rain.)
6 a 160 MJ/s   **b** 40 m³/s

Answers to end of chapter practice questions

1 a B. renewable energy cannot generate electricity all the time (1)
    D. renewable energy cannot supply electricity to millions of homes (1)
2 a C – E – D – B (4) **b** The lagoon may not provide enough power when it is operating. The houses need power during the other 10 hours of the day.
  c The system cannot generate electricity for 24 hours a day. (1)
  The system may need to shut down for maintenance or breakdown. (1)
  Damage to the environment, harm fishes etc. (1)
3 a Independent is direction of solar cell. Dependent is voltage (2)
  b Point solar cells towards the south (1)
  c No because distance higher on roof is negligible compared to distance from Sun (2)
  d Same time of day, same weather conditions. (1)
4 a Reasons could include: reducing impact of greenhouse gases; reducing dependence on fossil fuels; reducing reliance on imported fuel; increased energy security; preparation of infrastructure for eventual scarcity of fossil fuels.
  b Advantages 2 from:
  Nuclear energy power stations can work all the time.
  Nuclear power stations produce massive amounts of energy from a small amount of fuel.
  Nuclear power stations can provide the base load for electricity demand during the day.
  Nuclear power stations do not produce carbon dioxide.
  Disadvantages 1 from:
  Nuclear power stations produce nuclear waste which is difficult to dispose of.
  Nuclear power stations have high commissioning and decommissioning costs
  Nuclear power stations have a long start up time.
5 Plants, such as sugar cane, for biofuels use up carbon dioxide.
  Using sugar cane ethanol reduces the amount of fossil fuels burnt.
  Burning fossil fuels causes environmental problems such as global warming and acid rain
  Brazil benefits economically from export of sugar cane ethanol.
  Provides employment for farmers in Brazil (5)

7 General properties of waves

7.1
1 a The oscillations in a transverse wave are perpendicular to the direction of energy transfer. The oscillations of a longitudinal wave are parallel to the direction of energy transfer.
  b i An electromagnetic wave or waves on a stretched string or wire.
  ii Sound waves.
  c The particles are displaced so they are closer together.
2 a Transverse
  b i Along the rope from one end to the other.
  ii It oscillates in a direction perpendicular to the energy transfer.
3 a Stretch the slinky out with each person holding one end. To send transverse waves along the slinky, move one end so it oscillates at right angles to the slinky. To send longitudinal waves along the slinky, move one end so it oscillates parallel to the slinky.
  b The red coil moves to and fro in a direction parallel to the slinky axis about a fixed point along the axis.

7.2
1 a amplitude = 9 mm, wavelength = 37 mm
  b The number of wavecrests passing a point in one second or the number of cycles of the waves that pass a point in one second.
2 a i and **iii** see 7.2 Figure 1.
  b Point P will oscillate at right angles to the wave moving from a maximum as the wave peak passes down to a minimum as the wave trough passes.

3 a 6.0 m/s
  b i 6.0 m
  ii 360 m

7.3
1 a They are equal.
  b The angle of each refracted wavefront to the boundary becomes greater than the angle of each incident wave front to the boundary.
2 See 7.3 Figure 3
3 a The slopes prevent reflection of the waves at the sides of the tank.
  b Reflection of waves from the sides would occur and these reflected waves would spoil the pattern of the waves in the tank.

7.4
1 a The wavelength is unchanged.
  b i The waves spread out (or diffract) more.
  ii The waves spread out (or diffract) less.
2 a Diffraction is the spreading of waves when and after they pass through a gap or pass by an obstacle.
  b i The radio waves carrying the TV signal from the transmitter are short compared with obstacles such as the hills so the waves do not diffracted much. Therefore, fewer radio waves would reach a TV receiver on the other side of a hill to the transmitter.
  ii The longer wavelength radio waves are diffracted more than the waves carrying the TV signal and so they spread out more when the pass over a hill and can reach radio receivers at some locations which TV signals cannot reach.
3 The sound waves from the radio diffract when they reach the doorway so they spread out and travel along the corridor.

Answers to end of chapter summary questions

1 a See 7.2 Figure 1
  b In a transverse wave, the particles oscillate at right angles to the direction in which the wave travels. In a longitudinal wave, the particles oscillate along the direction in which the wave travels.
  c Transverse wave; waves on a string or a rope, electromagnetic waves Longitudinal waves; sound waves
2 a 0.05 Hz
  b 0.40 m/s
  c 8.0 m
3 a i Decreases
  ii Unchanged
  b i Unchanged
  ii Unchanged
  iii Unchanged
4 a See Figure 1a
  b See Figure 1b

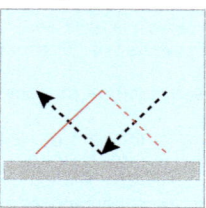

Figure 1a   **Figure 1b**

5 a 100 MHz
  b The waves would travel across the top of a hill without diffracting if their wavelength is much less than the height of the hill. As a result, no waves would spread down the hill so a receiver on the far side of the hill would not receive a signal.
6 a

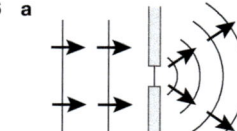

  i Diffraction
  ii The waves would be shorter (at the same distance apart), diffracted less.
7 a There is no change in any of these 3 quantities.
  b i See Topic 7.4 Figure 2 (Diffraction at an edge)
  ii The waves are diffracted less so they do not spread as far into the region behind the obstacle.

Answers to end of chapter practice questions

1 a i C   **ii** B
  b i Reflection of waves   **ii** B
2 a i X – transverse (1)
    ii Y – longitudinal (1)
    iii Perpendicular to the direction of energy transfer. (1)
    iv Parallel to the direction of energy transfer. (1)
  b Similarities: all travel in the form of waves; all transfer energy; light & radio have the same wave speed.
    Differences: light & radio are transverse waves/sound is longitudinal; wavelength of radio longer than that of light; sound has a much lower wave speed than light and radio; sound cannot travel through a vacuum/needs a medium. (6 – for full marks, there must be at least 2 similarities)
3 a i R
    ii Q (2)
  b $\lambda = 0.03\,\text{m}$ (1)
    $f = c/\lambda$ OR $f = 3 \times 10^8/0.03$ (1)
    $= 1 \times 10^{10}$ (1) Hz/hertz (1)
  c i Microwaves can not pass through metal so they can not reach M
    ii The microwaves from the transmitter reflect off the plate and N is in the path of the reflected the reflected waves.
    iii The reading at L would be less than at N because the microwaves spread out from the transmitter
4 a Diffraction increases with decreasing gap width; greatest diffraction when gap width = wavelength. (2)
  b Sound waves diffract through doorway; door width similar to wavelength; light waves have much shorter wavelengths and so do not diffract significantly through doorway. (3)
  c i Radio 4 carrier wave has a much greater wavelength than BBC1 television carrier wave; it is 3000 times as great. (2)
    ii Radio 4 carrier waves diffract round the hill; they have a similar wavelength to the width of the hill. (2)
    iii BBC1 television carrier wave has a much smaller wavelength; the carrier wave does not diffract round the hill. (2)

8 Electromagnetic waves

8.1

1 a Radio waves
  b The speed is the same for all electromagnetic waves.
  c X-rays
  d Microwaves
2 radio microwaves infrared visible light ultraviolet X-rays gamma rays
3 a 0.50 m
  b 1000 MHz
4 a Gamma radiation
  b All electromagnetic waves travel at the same speed in space. Since both types of waves travel the same distance and they are emitted at the same time, they reach the Earth at the same time.

8.2

1 a i Radio waves
    ii Light waves
  b i Microwaves
    ii Radio waves
2 a Mobile phone calls would not be clear as the phones would detect handset signals as well as mobile phone signals.
  b If other radio wave users operated in the same wavelength range as the emergency services, their signals would be detected by the emergency services and might 'mask' the emergency services signals making conversations carried by the emergency services signals difficult to listen to.
3 a i Radio waves
    ii Microwaves
    iii Infrared radiation
    iv Light
  b 0.125 m

8.3

1 a i Radio waves or microwaves
    ii Infrared radiation
  b i The two signals would interfere where they cross and there would be points where they cancel each other
    ii The signals cannot escape from the fibre so they cannot be detected except by the detector at the receiver end. Radio signals travel through the air so can be detected by any radio detector in the path of the waves.

2 a The skull of a child is thinner than that of an adult which means that mobile phone radiation can pass more easily through the skull of a child into the brain and cause a greater heating effect
  b Light waves have a much higher frequency and a much smaller wavelength in air than radio waves so can carry many more pulses per second than radio waves can.
3 a Microwaves are absorbed less by the atmosphere than radio waves and are not diffracted as much so they spread out less making them suitable for satellite TV. Terrestrial TV uses radio waves as diffraction as it helps to prevent signal problems.
  b Microwaves from such a transmitter dish are directed in a beam at the other dish so they travel in a straight line towards the other dish. If the other dish is visible from the first transmitter, it can detect the microwaves directed at it. The atmosphere does not absorb the beam much so the received signal is strong enough to be detected.

8.4

1 a X-rays pass through the crack but not through the surrounding metal. On the X-ray picture, the crack appears as a break in the shadow of the metal object.
  b Yes
  c A metal case would stop the X-rays so the X-rays would not reach the film inside the case.
2 a It harms the skin and can cause skin cancer. It damages the eyes and can cause blindness.
  b i It absorbs most of the ultraviolet radiation from the Sun.
    ii Ultraviolet radiation causes sunburn. Suncreams stop UV radiation reaching the skin. Suncreams absorbs the UV radiation that passes through the ozone layer.
3 a X-rays and gamma rays
  b Lead
  c i X-rays and gamma rays
    ii Ultraviolet radiation, X-rays and gamma rays

8.5

1 a The contrast medium absorbs X-rays. Without it, the X-rays would pass through the stomach so no image of the stomach will be seen on the X-ray photograph.
  b X-ray therapy can be used to destroy cancerous tissue.
2 a Dense material such as bone in the patient absorb X-rays from the tube and stop them reaching the film cassette. X-rays that reach the film cassette blacken the film and do not pass through such absorbing materials. When the film is developed, clear images of the bones and other absorbing materials in the patient are seen on the film because X-rays did not reach these areas.
  b If the film was not in a lightproof cassette, light from the room would blacken the entire film.
  c X-rays ionise substances they pass through. Ionisation in healthy cells can damage or kill the cells or cause cell mutation and cancerous growth. The shielding prevents X-rays from reaching and damaging cells in parts of the patient not under investigation.
3 a X-rays
  b Advantage = distinguishes between different types of soft tissue or gives a 3D image; disadvantage = higher cost or higher dose

Answers to end of chapter summary questions

1 a D E A B C
  b i Microwaves
    ii Gamma rays
    iii Infrared rays
2 i speed = wavelength × frequency
  ii 103 MHz
  b As the radio waves travel away from the transmitter, their amplitude is gradually reduced due to absorption by the air. At a certain distance d, the amplitude is just large enough for the waves to be detected. Reducing the power of the transmitter reduces the amplitude at that distance so the range is reduced.
3 a Mobile phones signals are carried by microwaves. Microwaves can heat substances which absorb them. If a mobile phone emits too much microwave radiation, the radiation absorbed by tissues in the head (e.g., brain tissue) may be adversely damaged by the heating effect of the microwaves.
  b A; it emits less microwave energy per second so it would not affect the organs in the head (e.g., the brain or the ear) as much.
  c Microwave radiation penetrates their skulls more than older skills because their skulls are thinner. Also, smaller heads heat up more easily than bigger heads.
4 a i Bone absorbs X-rays so a 'shadow' image is formed on the film.
    ii X-rays pass through the fracture but not through the bones.

b **i** Barium absorbs X-rays so an image of the stomach is formed on the film.
  ii The stomach movements would blur the images.
  iii X-rays ionise substances they pass through. The amount of ionising radiation the patient is exposed to is reduced by stopping the X-rays reaching the patient. The quality of the image is unaffected because the low-energy X-rays would not reach the film anyway.
c Ultrasound waves from a scanner do not harm the baby because they are non-ionising. X-rays are ionising and would harm the baby.

5 **a** Ionisation is the process of creating ions, which are charged atoms from uncharged atoms.
b X-rays, gamma rays
c Ionising radiation can damage or kill living cells and can cause cell mutation and cancerous growth.

6 **a** When ultraviolet radiation is directed at invisible ink, the radiation is absorbed by atoms in the ink which then emit light so the ink glows and becomes visible.
b The further infrared radiation travels through air, the more it is absorbed by air molecules so a beam of infrared radiation would be too weak to detect after more than a few metres.
c Local radio signals are carried by electromagnetic waves of a much longer wavelength than microwaves so they spread out much more from a transmitter and the amplitude decreases with distance more than for microwaves which spread out enough to reach a wide area but not so much as to become too weak to detect. In addition, radio waves are absorbed more by the atmosphere than microwaves because microwaves are much higher in frequency and are not affected as much by air molecules.

Answers to end of chapter practice questions

1 **a** **i** X-rays (1)
  ii radio waves (1)
  iii X-rays (1)
  iv ultraviolet (1)
b **i** They penetrate tissue (1); and clothing (1); but not bone (1); or metal. (1)
  ii They carry a lot of energy (1); they cause ionisation (1); can cause cancer (1); and kill cells. (1)

2 **a** Gamma: Prolonging the shelf-life of fruit (1);
  Infrared: In a TV remote control (1);
  Radio: Carrying TV programmes (1);
  Ultraviolet: Security marking of TV sets. (1)
b **i** They can pass through the atmosphere. (1)
  ii Points to be made: phones use microwaves; microwaves deliver energy to cells; can heat them up; phones are held near the brain; skull is thinner in young children; brain cells may be damaged. (6)

3 **a** **i** Can distinguish between different types of tissue (1); produce 3D images. (1)
  ii Larger doses of radiation are used (1); much more expensive. (1)
b X-rays can damage cells (1); foetus has developing cells which can mutate. (1)
c Any six from: the overall dose has more than doubled (1); dose from background radiation is unchanged (1) dose from medical uses has increased by a large amount (1); other man-made radiation has also increased (1) more medical procedures involving radiation in 2010 (1); some, such as CT scans, deliver large doses (1); other man-made due to nuclear accidents (1); and weapons testing. (1)

4 **a** Infrared (1); light. (1)
b **i** Any wavelength greater than 100 m. (1)
  ii Any wavelength between 1 m and 100 m. (1)
  iii Any wavelength less than 1 m. (1)
c Frequency = 2.4 × 10⁹ Hz (1)
  λ = c/f OR = 3 × 10⁸/2.4 × 10⁹ (1) = 0.125 m (1)

9 Sound and ultrasound

9.1

1 **a** Thick felt or a similar material.
b The bushes absorb some sound and also scatter it. Sound from traffic that reaches the fence panels is reflected back towards the motorway by the panels.

2 **a** 20 000 Hz
b When the whistle is blown, the ball inside the whistle revolves at high speed pushing the air in and out of the gaps in the case of the whistle. The vibrations of the air at the gaps causes sound waves to spread out from the whistle. The frequency of the sound is constant because the ball revolves at constant speed inside the whistle.

3 **a** **i** The cliff face would reflect sound from the horn and create an echo.
  ii distance = 340 m/s × 5.0 s/2 = 850 m
b The person creates a sound which spreads out in all directions. The echoes are due to sound reflected from different parts of the cavern walls.

9.2

1 **a** The amplitude decreases; the frequency does not change.
b The loudness of the sound decreases because the amplitude of the sound waves decreases. This happens because the amplitude of the vibrations of the wire decreases.

2 **a** The amplitude of the waves would be taller but the horizontal spacing between the peaks and troughs would be unchanged.
b The horizontal spacing between the peaks and troughs would be greater but the amplitude of the waves would be unchanged.

3 **a** **i** and **ii** The pitch or frequency is raised.
  iii The pitch is lowered.
b The sound of a violin (played correctly) lasts as long as the violin bow is in contact with a string. The sound of a drum dies away after the drum skin has been struck. A drum note is less rhythmical than a violin note.
c The vibrating tuning fork makes the table surface vibrate. The vibrating table surface creates sound waves in a much greater volume of air than the tips of the vibrating tuning fork does.

9.3

1 **a** The organs have a different density to the surrounding tissue. So ultrasound is reflected at the tissue/organ boundaries.
b Ultrasound is not ionising radiation whereas X-rays are. Ionising radiation is harmful to living tissue. Ultrasound is reflected at the boundaries between different types of tissue whereas X-rays are not.
c The reflected pulses would be weaker if there is more diffraction because the waves would spread out more and become weaker. Also, the pulses would be reflected from a wider area of each boundary so the position of the boundary would be more difficult to locate.

2 **a** 3 if the last pulse is due to the other side of the body.
b **i** 96 millionths of a second
  ii 0.144 m

3 **a** 12 mm
b ±2–3 mm

Answers to end of chapter summary questions

1 **a** **i** As the surface of the object vibrates, it alternately pushes air particles away as it moves into the air then allows them to return as it retreats from the air, in effect pulling them back. The air particles pushed away from the surface push on other air particles further away then allow them to retreat as they retreat. These further particles in turn alternately push and pull on particles further away. In this way, waves of compressions and rarefactions pass through the air.
  ii The sound waves spread out as they travel away from the loudspeaker so the sound becomes fainter and the amplitude becomes smaller.
b **i** The waves on the screen become taller (higher amplitude).
  ii The waves on the screen become more stretched out across the screen so fewer waves appear on the screen.
c The signal generator is connected to the loudspeaker and to the oscilloscope. The signal generator should be adjusted so the subject can hear the sound from the loudspeaker comfortably. The frequency of the signal generator should then be increased gradually until the subject can no longer hear the sound. The frequency at this point can then be determined by measuring the time period of the waveform on the oscilloscope screen. The upper limit frequency is equal to 1/the measured time period.

2 **i** reflected, smooth
  ii rough, scattered
  iii soft, absorbed

3 **a** Approximately 20 000 Hz.
b Keep the frequency and the loudness of the sound from the loudspeaker the same throughout. Keep the loudspeaker, the board or cushion and the sound meter in the same positions throughout. With the board in position, measure the sound meter reading. Replace the board with the cushion and measure the sound meter reading again. If the reading for the board is higher than the reading for the cushion, the board reflects more sound than the cushion.

4 **a** Sound waves created by clapping her hands together travel through the air to the wall where they reflect from it. Some of the reflected waves travel back to the person who hears the echo when the reflected sound waves reach her.
b 51 m

5 270 m

6 **a** 8.5 mm
b If the bat detects echoes directly ahead, the time delay between each pulse being emitted and being detected enables the bat to sense

the distance to the reflecting object. The difference in the intensity of the echo at each ear enables the bat to sense the direction of the reflecting object.

7 a 0.75 mm

  b i They could not be detected because they would be absorbed by the tissue.

    ii The reflected pulses would be much weaker. Also, if the pulses spread out, it is difficult to tell which boundary a reflected pulse is from.

  c Ultrasound as it is non-ionising unlike X-rays. Ionising radiation can damage or kill cells in living tissue.

Answers to end of chapter practice questions

1 a i Signal generator (1); loudspeaker (1); microphone (1); cathode ray oscilloscope. (1)

    ii A is louder than B (1); A is lower in pitch than B. (1)

  b i An echo. (1)

    ii Measure the distance to the wall (1); start a stopwatch when they hear the first sound (1); stop the stopwatch when they hear the echo (1); double the distance to the wall to find total distance travelled (1); divide the total distance by the time. (1)

    iii Each student should measure the time and then find the average. (1)

2 a i longitudinal; transverse (1) (both correct for mark)

    ii slower (1)

    iii cannot; can (1) (both correct for mark)

  b i 1 cm (1)

    ii $f = 1/T = 1/0.002$ (1) $= 500\,\text{Hz}$ (1)

    iii $\lambda = c/f = 340/500$ (1) $= 0.68\,\text{m}$ (1)

    iv Diffraction most marked when gap width similar to wavelength (1); 0.68 m is of the same order as a doorway (1); so waves would be diffracted. (1)

3 a i Longitudinal wave (1); series of compressions in water (1); and rarefactions. (1)

    ii There is nothing to be compressed. (1)

    iii $\lambda = c/f = 1400/2000$ (1) $= 0.7\,\text{m}$ (1)

  b i Sound/longitudinal waves (1); frequency above the range of the human ear. (1)

    ii Distance = speed × time = 1400 × 0.1 (1) = 140 m (1) This is twice the depth – answer 70 m (1)

    iii One from: prenatal scans; dispersing kidney stones; detecting flaws in metals; removing plaque; cleaning surgical instruments (1)

4 a 30 000 Hz (1)

  b Points to be made: transducer transmits ultrasound pulses into body; ultrasound can travel through solids and liquids; ultrasound travels through the body; partially reflected at boundaries between different types of tissue/baby/foetus; reflected ultrasound returns to the transducer; where it is detected; the computer processes the information to give a visual image on the screen. (6)

10 Reflection and refraction of light

10.1

1 a i 20°   **ii** 40°

  b 42°

2 a, b i

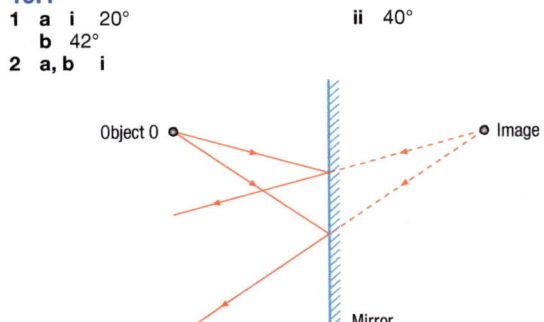

    ii Use a millimetre ruler to measure the perpendicular distance from O to the mirror and from the image to the mirror.

3 a

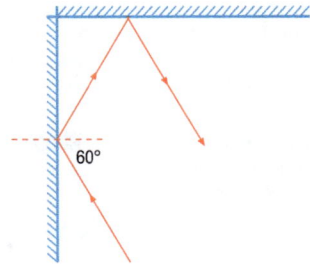

  b i 180°

    ii As the two mirrors are perpendicular to each other, the angle of incidence for the second reflection is always 90° – first angle of incidence. Therefore, adding both angles of incidences always equals 90° and therefore both reflected rays also equal 90° giving a total of 180°.

10.2

1 a Decrease

  b Zero

  c Smaller

2 a

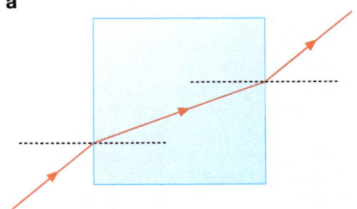 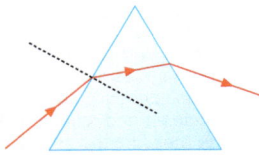

  b i Blue

    ii In glass, blue light travels slower than red light.

3 a

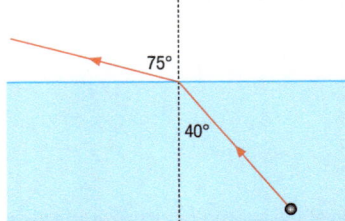

  b All the light rays from a point on the bottom of the pool that refract at the surface appear to travel straight from a point above the bottom.

10.3

1 a 1.54

  b 1.53

2 a 1.47

  b The measurement in Q2 was a single pair of measurements and three pairs of measurements were made in Q1 so the result in Q2 is less reliable than that in Q1; the refractive index of glass of the second block might have been different.

3 a 25.5°

  b 70.1°

  c 1.52

10.4

1 a The angle of incidence must be greater than the critical angle.

  b 90°

  c 1.47

2 a

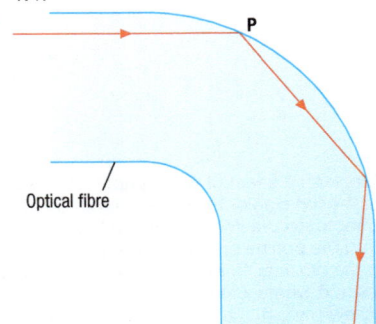

b Any two advantages:
1. The endoscope uses light which is non-ionising (unlike X-rays).
2. Movement of the fragments can be seen with an endoscope.
3. Fragments may be hidden by other fragments on an X-ray picture.

3 a 48.8°
  b i 1.49
    ii Range of critical angle = 41.5° – 42.5°. At 41.5°, the refractive index = 1.51. At 42.5°, the refractive index = 1.48. Therefore, the largest difference to the answer in part **i** is 0.02.

Answers to end of chapter summary questions

1 a i They are the same.
    ii

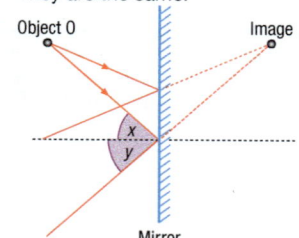

    iii They are the same.
  b Draw a straight line XY on a sheet of white paper and use a protractor to draw a 'normal' line perpendicular to XY. Place the mirror exactly on XY and use the ray box to direct a light ray at the point P where the normal intersects XY. Adjust the direction of the light ray so the angle of incidence is about 10°. Use a pencil to mark the direction of the incident and reflected rays. Remove the mirror and use a protractor to measure the angles of incidence and reflection. Repeat the text for several different angles of incidence. Record all your results in a table. Plot a straight-line graph of the angle of reflection against the angle of incidence. The line should show that the angles of incidence and reflection are always equal to one another.

2 a i

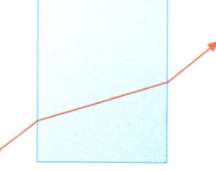

    ii The change of direction at the second refraction is exactly equal and opposite to the change at the first refraction. Since the opposite sides of the block are parallel, the light ray that emerges is therefore exactly parallel to the incidence light ray.
  b i

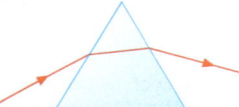

    ii A continuous spectrum of colour is seen on the screen with blue light refracted most and red light least. This happens because the refractive index of the glass varies with the wavelength of the light and is greatest for blue light and least for red light.

3 a i 1.53
    ii 196 000 km/s
  b i 19.5°
    ii 80°

4 a

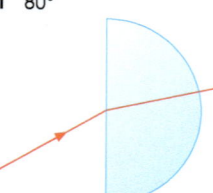

  b Draw an outline of the block on a sheet of white paper and use a ruler to locate the centre C of the flat side of the outline. Mark point C on the sheet and use a protractor to draw the normal to the flat side at C. Place the block exactly on the outline and direct a light ray at C so the light ray passes into the block at C. Mark the path of the light ray before it enters the block and where it leave the block as point D. Remove the block and draw the path of the incident light ray and the path of the light ray from C to D. Use a protractor to measure the angle of incidence and the angle of refraction at C.

Calculate the refractive index n using the formula $n = \sin i/\sin c$ from C to D. Repeat the test for different angles of incidence and calculate a mean value for the refractive index from the individual values.

5 a i Total internal reflection is the total reflection of a light ray in a transparent medium at a boundary between the medium and a less refractive medium.
    ii The angle of incidence at the boundary should be greater than the critical angle of the boundary.
    iii 1.56
  b i 41°
    ii

    iii One bundle takes light into the cavity. The other bundle is used to observe an image formed by a lens near the end of this bundle in the cavity.

6 a $\sin r = \dfrac{\sin i}{n}$; $\sin r = \dfrac{\sin 40}{1.59} = 0.404$; $r = 24°$
  b i 39°
    ii The angles in the triangle formed by the two normals at P and Q and the line PQ add up to 180°. The angle between the two normals is 90°. If x is the angle of incidence at Q, $x + 24° + 90° = 180°$ so $x = 66°$.
    iii As angle x is greater than the critical angle of the block, the light ray undergoes total internal reflection at Q so it does not enter the air at Q.

Answers to end of chapter practice questions

1 a i Rays only appear to come from it/rays do not pass through it (1); cannot be formed on a screen. (1)
    ii It is 'the wrong way round' or words to that effect BUT NOT upside down. (1)
  b i Image 1 is 1 m behind mirror A (1); image 1 is 2 m behind mirror B (1); distance = 6 m. (1)
    ii Image 1 in mirror B acts as virtual object for mirror A (1); image 1 in mirror A acts as virtual object for mirror B, whose image acts as object for mirror A etc. (1)

2 a i Light waves slow down on entering the other medium (1); the end X of the wavefront reaches the boundary first and so slows down first (1); this changes the direction of the wavefront downwards/towards the normal. (1)
    ii If the wavefront is parallel to the interface. (1)
  b i Normal correctly drawn (1); angle between normal and incident ray marked angle of incidence (1); angle between normal and refracted ray marked angle of refraction. (1)
    ii A protractor (1)
    iii Calculate the sines of all the angles (1); divide sin(i) by sin(r) for each pair of values (1); take the average of sin(i)/sin(r). (1)
    iv Refractive index = sin 35/sin 22 (1)
          = 1.53 (1)
        sin C = 1/1.53 (1)
           C = 41 degrees (1)

3 a i total internal (1); reflection (1)
    ii endoscope; in phone cables; etc. (2)
  b i The normal (1)
    ii Ray bends away from the normal (1); with arrow to show direction. (1)
    iii Critical angle (1)
    iv Total internal reflection with angle of reflection equal to angle of incidence, judged by eye. (1)

4 a Correct refraction at the first surface – towards the normal (1); correct refraction at the second surface – away from the normal. (1)
  b Light must be travelling into a less dense medium (1); and the angle of incidence must be greater than the critical angle. (1)
  c Points to be made: two bundles of optical fibres; inserted through the throat; one bundle to shine light into the stomach; the other to carry light back to the observer; the lens forms an image on the end of the fibres; the process involved is total internal reflection; surgery is invasive; X-rays pose health risks. (6)

11 Lenses and the eye

11.1

1 a A real image is formed where light rays from an object meet. A virtual image is formed where light rays from an object appear to originate from.
  b i A real image
    ii A virtual image

2 a Upright, enlarged and virtual.
  b i Inverted, magnified and real.
    ii The slide must be moved towards the screen.
  c i ×3
    ii The magnification would increase until the flower is at the focal point of the lens when no image is seen.
3 a The image is real, inverted and enlarged. Magnification ×3.
  b The image would be smaller than it was and would still be upside down.

11.2

1 a

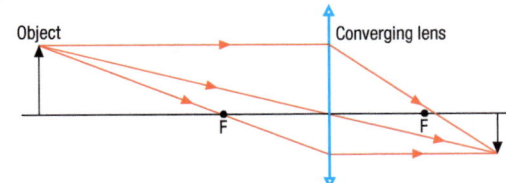

  b i Real
    ii Diminished
    iii Inverted
2 a

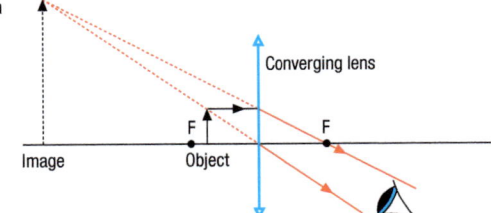

  b i Virtual
    ii Magnified
    iii Upright
  c A diverging lens produces images smaller than the object.
3 a

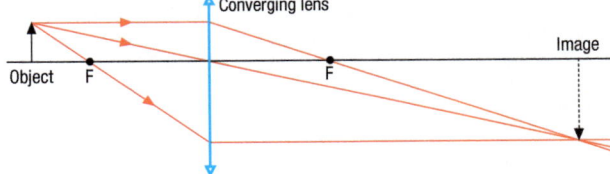

  b $f = 1.8$ cm
  c The nearer the object is to the focal point, the further the image would be from the lens and the larger it would be. At the focal point, no image would be formed (because the rays from it that pass through the lens would be parallel).

11.3

1 a i Alters the thickness of the eye lens to alter its power.
    ii Focuses light onto the retina.
    iii Allows light to enter the eye after passing through the cornea.
    iv Joins the eye lens to the ciliary muscles.
    v Protects the front of the eye and refracts light.
    vi Controls the width of the pupil so controlling the amount of light passing through the eye lens.
    vii Layer of light-sensitive cells on which the image is formed.
  b It widens in darkness to allow as much light as possible to pass through the eye lens.
2 a Each eye lens becomes thinner.
  b The focal length of each eye lens increases.
3 a The iris becomes narrower so the pupil is not as wide and less light passes through to the retina.
  b The aperture stop is made narrow so less light passes through it to the camera film or CCD.

11.4

1 a Short sight.
  b A diverging lens.
2 a The sight defect is long sight. It may be caused by an eyeball that is too short or an eye lens that is not strong enough.
  b The eye lens does not have enough focusing power to focus a near object on the retina. Making the cornea flatter would decrease the effective focusing power of the lens. This would worsen the sight defect.

3 a i The near point is the least distance from the eye to the position of an object that can be seen clearly in front of the eye
    ii The far point is the furthest distance to the position of an object that can be seen clearly in front of the eye
  b i Short sight
    ii A concave lens which has a suitable focal length needs to be used as a spectacle lens

Answers to end of chapter summary questions

1 a i converging
    ii

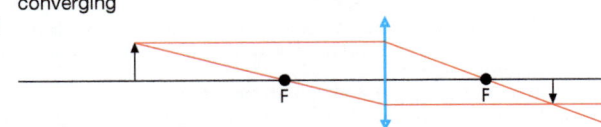

  b Real, inverted and smaller than the object; camera.
2 a

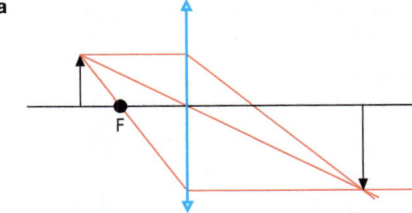

  b i Real
    ii Inverted
  c 1.7
3 a See 10.2 Figure 4
  b i Virtual
    ii Upright
4 a

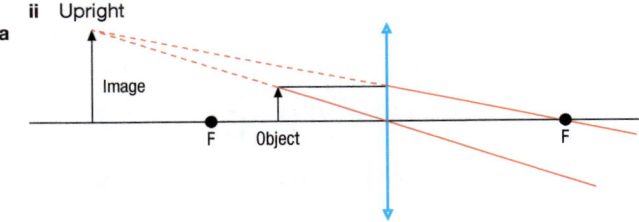

  b about 2.7
  c i Virtual
    ii Upright
5 a i Short sight is where a distant object can not be seen clearly but a nearby object can.
    ii

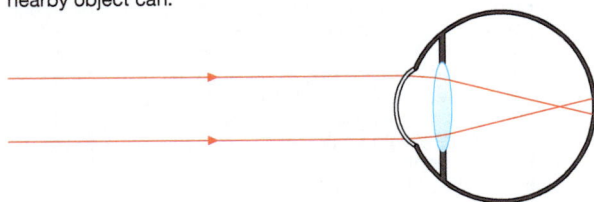

  b Diverging
  c i The cornea. It is made slightly thinner so it becomes flatter at the front surface.
    ii Laser light entering the eye damages the retina.
6 a i

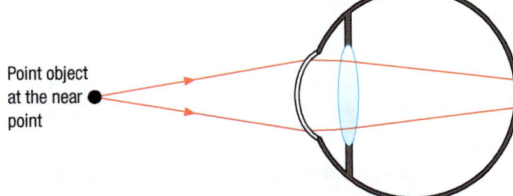

    ii When a long-sighted eye views a near object, the eye lens is unable to focus the light rays on the retina. The image would be formed behind the retina. The eye lens is too weak to focus light from the object onto the retina. A blurred image is therefore seen.
  b i A converging lens is needed
    ii The eye lens on its own is unable to focus the light from the object onto the retina because the lens is either too weak or the eyeball is too short. The convex lens makes the light rays from the object diverge less so that the eye lens can then focus them onto the retina.

Answers to end of chapter practice questions

1 a i Diagram showing rays parallel to the principal axis (1); brought to a point on the axis (1); distance from point to the lens marked 'focal length'. (1)

ii Diagram showing rays parallel to the principal axis (1); diverging from a point behind the lens (1); distance from point to lens marked 'focal length'. (1)

b i The ray parallel to principal axis passing through F (1); ray from top of object to centre of lens going straight on (1); both rays dotted back to where they meet (1); line from this point vertically down to axis labelled 'image'. (1)

ii M = height of image/height of object OR = image distance/object distance (1) = 2 OR their height of image/10 OR their image distance/5 (1)

iii magnified (1); distance from lens (1); virtual (1)

c i Two (1 mark each) from: both have light-sensitive detectors; both form real images; both can focus objects at different distances; both form diminished images; both have inverted images.

ii Camera moves the lens (1) away from the film for near objects (1); eye changes the curvature/focal length/power of lens (1); more curved/shorter focal length/more powerful for near objects. (1)

iii Two (1 mark each) from: camera has a film or CCD ; eye has light-sensitive retina; camera has an adjustable aperture; eye has an iris.

2 a

Device	Nature of image
Camera	real; magnified; inverted
Magnifying glass	virtual; magnified; upright
Projector	real; diminished; inverted

(3)

b i 0.1 cm (1)

ii

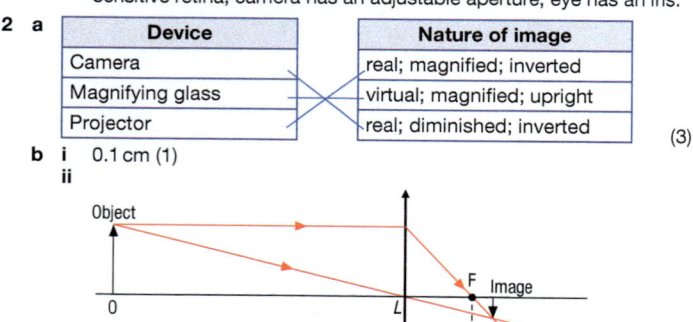

$f = (LF/OL) \times 40\,\text{cm} = 10.0\,\text{cm}$

iii Draw a ray diagram for each pair of object and image distances to find the focal length from each diagram [1] and average the values [1]

iv Object inside F (1); virtual image formed (1); cannot be formed on a screen/rays do not pass through the image. (1)

3 a A = iris (1); B = retina (1); C = optic nerve (1); D = cornea (1); E = pupil (1)

b Points to be made: iris controls the size of the pupil; in dark conditions, pupil has larger diameter OR the converse for light conditions; ciliary muscles alter the shape of the lens; making it more curved for nearer objects; increasing the power of the lens; making the focal length shorter; so that diverging rays from object are brought to a focus on the retina; eye at rest/muscles relaxed the eye lens can focus parallel light on the retina, i.e., light from a distant object. (6)

c i Short sight (1)

ii Concave/diverging (1)

12 Kinetic theory

12.1

1 a Vaporisation
b Freezing
c Melting

2 a Condensation
b Evaporation/vaporisation
c Melting
d Freezing

3 a The particles start to move about each other at random and are no longer in fixed positions.

b The particles in water vapour are not in contact with each other except when they collide and they move about at random. When water vapour condenses on a cold surface, the vapour particles lose energy when they collide with the surface and they stick to the surface as a film of liquid. The particles in film move about at random in contact with each other.

12.2

1 The small bucket warms up faster because the mass of water in it is much less than in the large bucket.

2 a Lead has a lower specific heat capacity than aluminium. Less energy is needed by lead for a given temperature rise.

b i 4500 J
ii 42 000 J
iii 46.5 kJ

c 25.6 MJ

3 A storage heater contains bricks or concrete that are heated by the heater element. An ordinary electric heater does not contain bricks or concrete. A storage heater transfers energy to the surroundings gradually. An ordinary electric heater transfers heat instantly.

12.3

1 Boiling takes place at a certain temperature whereas evaporation occurs from a liquid at any temperature; boiling takes place throughout a liquid whereas evaporation is from the surface only; evaporation can cause a liquid to cool whereas boiling can never have this effect.

2 a i

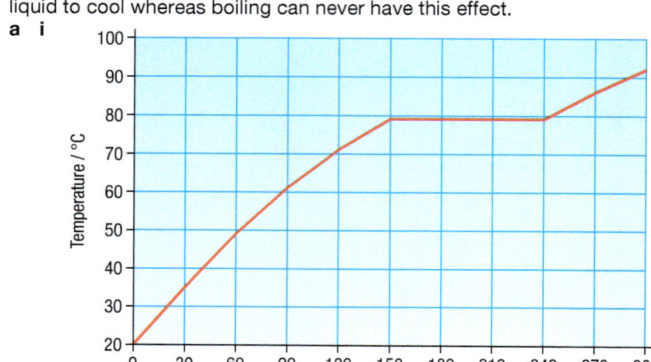

ii 79 °C

b At 60 °C the substance is solid. Once it reachers 79 °C it begins to melt. After 90 seconds it has all melted and the liquid then increases in temperature to 90 °C.

3 Salt and water on the road forms a solution which will not freeze on the road unless the temperature drops below the freezing point of the solution. So no ice forms on the road unless the temperature drops below the freezing point of the solution. If the solution does freeze, grit helps to stop vehicles sliding as it provides friction between the tyres and the ice.

12.4

1 a 0.044 kg
b 340 kJ/kg

2 a 2.3 MJ/kg
b Energy was transferred from beaker to the surroundings. Therefore not all the energy supplied to the heater was used to boil the water as some was transferred by heating to the surroundings. The specific latent heat value obtained from the data would have been less as the energy needed to boil away the water was less than 18 400 J.

3 a 3020 J
b $E = m\,c\,\Delta\theta = 0.008 \times 4200 \times 9 = 302.4\,\text{J} = 300\,\text{J}$ (2 s.f.)
c 340 kJ/kg

Answers to end of chapter summary questions

1 a i The substance was a solid from A to B and its temperature increased towards its melting point as it was supplied with energy.

ii The substance melted from B to C and its temperature did not changed until all of it had melted.

iii The substance was liquid from C to D and its temperature increased as it was supplied with energy.

b 78 °C

c From A to B the particles are in contact with each other and they vibrate about fixed positions. As the temperature increases the vibrations increase. From B to C, more and more of the particles break away from the fixed positions and move about at random. From C to D, the particles have all broken free and they move about at random in contact with each other.

2 a i 101 MJ
ii 2 MJ/s

b If the air were pumped at a slower rate, the temperature increase would be greater.

3 a 13 kJ
b The temperature of the air leaving the heater would decrease as the mass of air flowing through it each second would be greater and the energy supplied to it each second is unchanged.

4 a 6300 J
b 14 J/s

5 **a** 90 kJ
  b 0.039 kg
6 **a** $E_k = \frac{1}{2}mv^2 = \frac{1}{2} \times 1200 \times (30)^2 = 540\,000\,J = 0.54\,MJ$
  b 60 °C

Answers to end of chapter practice questions

1 **a** Boiling point (1)
  b Any three (1 mark each) from: some escapes into the air; some is used to warm the gauze; some is used to warm the beaker; it spreads out.
  c Points to be made: during OA the molecules gain energy; they vibrate more; some have sufficient energy to break free of the structure; the ice melts; during AB the molecules of liquid gain energy to move about faster; some gain enough energy to escape; the water evaporates; BC all the molecules have enough energy to escape; the water boils. (6)
  d Data is recorded continuously/frequently (1); greater resolution. (1)
2 **a** Energy needed depends on the mass (1); half the mass, half the energy needed/energy proportional to the mass. (1)
  b $E = m \times c \times \theta$ OR $= 8 \times 4200 \times 20$ (1)
      $= 672\,000$ (1)
      $= 672\,kJ$ (1)
3 **a** **i** large mass per unit volume (1); large specific heat capacity. (1)
      ii $E = m \times c \times \theta$ (1)
          $\theta = E/(m \times c)$ OR $= 540\,000/(12 \times 900)$. (1)
          $= 50\,°C$ (1)
      iii It will be less (1); because some energy is not absorbed by the bricks (1); some is lost to the surroundings. (1)
  b **i** It is lighter/portable (1)
      ii It stores more energy (1)
4 **a** **i** Mass evaporated $= 16\,g$ (1)
              $= 0.016\,kg$ (1)
          $L = E/m$ OR $= 36\,800/0.016$ (1)
              $= 2\,300\,000\,J/kg$ (1)
      ii Put lagging round the beaker. (1)

13 Energy transfer by heating

13.1

1 **a** Plastic and wood do not conduct by heating, so a plastic or wooden handle would not become hot when the pan was hot. A steel handle would become as hot as the pan as steel is a good conductor.
  b Felt because it contains fibres that trap layers of air and dry air is a good insulator.
2 **a** Felt or synthetic fur could be used, because they are good insulators.
  b Student's plan. Look for design of a fair test.
3 **a** The free electrons that gain kinetic energy diffuse through the metal quickly, passing on energy to other electrons and ions in the metal.
  b When part of an insulator is heated, the atoms there vibrate more than elsewhere and they make atoms in adjacent parts vibrate so these parts become hot. The atoms in these parts make the atoms in adjacent colder parts vibrate so these colder parts become hot. Energy is therefore transferred through the non-metal.

13.2

1 Hot gases from the flame heat the base of the pan. Energy is transferred by conduction through the base of the pan to heat the water in contact with the base. The water at the base rises because it becomes less dense when it is heated. The rising water makes the water throughout the pan circulate through convection currents and colder water sinks to the base where the colder water is heated so it rises and causes the circulation to continue until the water through out the pan is hot.
2 **a** It heats it reducing its density so it rises.
  b The hot air passes through the grille into the room.
  c Cold air flows into the heater at the bottom.
3 **a** Drop the crystal into a beaker of water through a tube. Heat gently under one corner. The colour rises above point of heating and travels across the top and falls at opposite side of beaker (where density of cooler water is greater). The colour then travels across the bottom of the beaker to replace lower density warmer water that rises above the Bunsen flame.
  b Air in contact with the radiator is heated so it rises and circulates in the room. The air in the room becomes warmer as a result. The rising air against the radiator is replaced by cold air so the cold air is then warmed and it circulates. All parts of the room therefore become warm. If the radiator was elsewhere, warm air from the radiator that circulates and reaches the top of the window would become cold and move down the window, becoming colder as it transferred energy by heating to the cold window. People near the window would feel colder than elsewhere in the room.

13.3

1 **a** Water particles in the surrounding air collide with atoms on the cold surface and lose energy as a result. They are held on the surface by the surface atoms to form a film of liquid.
  b The more energetic particles in the water escape from the liquid at the surface. The average energy of the remaining liquid particles becomes lower as a result. So the temperature of the liquid decreases.
2 **a** The air in the bus becomes damp as everyone breathes out warm water vapour. The water vapour condenses on the inside of the windows.
  b When the door is opened, warm air carrying water vapour enters the refrigerator. The water vapour now in the refrigerator condenses on the cold walls of the inside of the refrigerator and runs down the walls to the drip tray. If the door is opened too often, water vapour keeps entering the refrigerator and condensing so filling the drip tray.
3 **a** Water evaporates faster from the wet clothes on a hot day than on a cold day because more energy is transferred to the clothes by the warmer air leading to a faster rate of evaporation.
  b Evaporation of water from the wet clothes on a windy day transfers more energy from the skin, which makes the wearer colder than someone wearing dry clothes.

13.4

1 **a** Electromagnetic radiation emitted from the surface of objects due to their temperature.
  b The city is hotter than the rural areas surrounding it because the hotter an area of a surface is the more infrared radiation it emits. This may be due to the greater amount of energy being used in urban areas.
2 **a**

Object	Infrared	Light
A hot iron	✓	✗
A light bulb	✓	✓
A TV screen	✗	✓
The Sun	✓	✓

  b Put your hand near it and see if it gets warm due to radiation from the iron.
3 **a** They lose less heat through radiation when they huddle together because they radiate energy to each other.
  b The wavelength is short enough to enable the radiation to pass through the prism and lies just outside the red part of the visible spectrum.

13.5

1 **a** The sand grains provide a rougher and darker surface than ice and can therefore absorb more infrared radiation from the Sun. They therefore warm up and melt the surrounding ice.
  b A matt black surface absorbs infrared radiation from the Sun better than any other type of surface.
2 **a** The black surface absorbs more infrared radiation from the Sun than the silver surface.
  b As the cars are identical except for their colour, the black car would cool faster because it radiates more infrared radiation than the silver car. However, the temperature in both cars would probably decrease to the same value which would be the temperature of the surroundings.
3 **a** To make the test fair. The temperature recorded will differ at different distances from the cube as the radiation spreads out.
  b **i** B
      ii D
  c Greater accuracy, collects multiple sets of data at whatever time intervals you choose.

13.6

1 Energy transfer from the hot water to the outer surface of the radiator takes place due to conduction. Air in contact with the radiator surface is heated by infrared radiation and conduction. The hot air near the radiator rises and circulates causing energy transfer to the air in the room by convection.
2 **a** **i** To prevent the component overheating.
      ii Metal is a good conductor. The heat sink is plate-shaped to increase its surface area, so it transfers energy to the surrounding air as effectively as possible.
  b Plan must have a fair system that compares a single plate of glass to a pair of plates, ideally with a sealed air gap between.
3 **a** Student's explanation to include the role played by the plastic cap, double-walled plastic container, silvered inside surfaces, vacuum layer.

b Water vapour from the warm water condenses on the frosted surfaces in the freezer and releases energy in the process which transfers by conduction to the frost and causes it to melt.

13.7

1 When the oven heats up, the brass tube expands more than the Invar rod. When the oven overheats, the difference in the expansion of the brass tube and the Invar rod is sufficient to move the valve so it closes the large opening between the two parts of the chamber.

2 a Without expansion gaps between sections of concrete, adjacent sections would make contact and push against each other and the concrete would crack.
  b Rubber in the expansion gaps prevents rubble falling into the gap. If rubble fell into the gap, the concrete sections either side of the gap would be unable to expand and the sections would push via the rubble on each other.

3 a The bimetallic strip consists of two different metals fixed to each other. One of the metals expands more than the other when the temperature of the strip is increased. As a result the strip bends.
  b The bimetallic strip should be reversed so it bends away from the contact screw when its temperature increases instead of bending towards it. The contact screw should be adjusted so it remains in contact with bimetallic strip until the required 'switch off' temperature is reached.

Answers to end of chapter summary questions

1 a i The temperature variations causes the roofing material to expand and contract. If the temperature variations are great enough, this repeated expansion and contraction would cause small cracks to develop.
    ii A smooth shiny surface is better because it would reflect sunlight and would therefore not get as hot in sunlight. It would also radiate less energy to the surroundings at night.
  b The panel with the transparent cover would reflect sunlight and the fluid in the panel would absorb some radiation so not as much radiation would be absorbed by the matt black surface. However, the matt black surface would heat the fluid in the panel directly. The panel with the matt black cover would absorb sunlight very effectively so it would become warmer in sunlight than the base of the other panel. However, it would need to be an effective conductor to heat the fluid underneath effectively and it would also emit radiation into the surrounding air. The panel with the matt black cover would probably be more effective further form the equator as the Sun is lower in the sky.

2 a i By conduction through the plate.
    ii By radiation and convection in the air.
  b The fins increase the surface area of the heat sink. The larger the surface area, the greater the energy loss due to radiation and convection from the plate.
  c The greater the density of the material, the larger the mass of the heat sink will be. The greater the mass and specific heat capacity of the material, the lower the increase of temperature will be for a given amount of energy transferred by heating to the heat sink.

3 a i Energy transfer from the hot water in the radiator takes place due to convection. Energy transfer through the radiator metal takes place due to conduction. Energy transfer from the outside of the radiator takes place due to convection in the air and radiation from the radiator surface.
    ii Air between the panels becomes hotter than the air near the panels on the outside of the radiator. The hot air in the gap rises and is replaced by cooler air drawn in to the gap at the bottom of the panels. The hot air from the radiator circulates in the room.
  b Radiation from the heater element rapidly heats the outer surface of the bricks. The inside of the bricks slowly become warmer due to conduction and therefore store energy by heating. When the heater is switched off, the outer surface of the bricks emits radiation to the surroundings so it becomes cooler than the interior. Conduction from the hot interior to the surface takes place gradually so the interior gradually cools down.

4 a Wool fibres is a good insulator as the fibres are made of insulating material and they trap dry air which is a good insulator. The inside of the clothing becomes warm due to radiation from the body. The body inside stays warm because the clothing does not conduct energy away from the inside.
  b In cold weather, radiation from exposed skin causes energy transfer from the head. Hair is a insulator and it contains some trapped air so it does reduce some conduction. Wearing a hat cuts out radiation from

exposed parts of the scalp and reduces conduction by providing extra insulation.
  c The surface area of the ear is relatively large in relation to its mass. In normal conditions, radiation from your ears is balanced by radiation received from the surroundings by the ears. In very cold weather, energy received is significantly less so the ears radiate more energy than they receive. Because their surface area is so large in relation the their mass, they soon cool down.

5 a Their clothing becomes damp due to sweat and cooling by evaporation occurs so the clothing and therefore the skin becomes cold.
  b i Infrared radiation
    ii The reflective coating traps infrared radiation in the space between the body and the blanket so the space becomes warmer. The warm air in the space keeps the body warm. The blanket also prevents moving air from causing sweat to evaporate.

6 a i The water heated at the bottom rose to the top causing convection in the tube and melting the ice cube at the top.
    ii The water was warmed at the top and stayed there as it is less dense than cold water. Conduction through the water eventually made the water at the bottom warm, and then the ice cube melted.
  b 2 Energy transfer in water is mainly due to convection.

Answers to end of chapter practice questions

1 a i diameter of rods (1); length of rods. (1)
    ii Attach a pin to the end of each material with wax. Energy makes the wax melt (1); the first pin to drop was on the best conductor. (1)
  b i Points to be made: heated water expands; becomes less dense; heated water rises; the colder water at the top is denser; colder water falls; setting up a convection current (6)
    ii Any five of (1 mark each): metal is a good conductor; because there are free electrons in its structure; the free electrons near the inside surface gain KE; they move faster; and collide with electrons and ions nearer the outside; transferring some energy; outer surface warms up.
  c i For example in bridges/roads (1); by leaving a small gap for expansion to take place safely. (1)
    ii A fire causes the temperature of the strip to rise (1); the metals in the strip expand (1); brass expands more than steel (1); top of strip bends to the right (1); making contact at the contact screw (1); completes circuit/bell rings. (1)

2 a absorbers (1)
  b reflectors (1)
  c Large surface area; facing south; angled at 45 degrees to horizontal (1)

3 a i nature of the surface of the mug (1); surface area of the mug (1); temperature of the water (1); temperature of the surroundings. (1)
    ii The mass of the water (1); the specific heat capacity of water. (1)
  b The petrol evaporates (1); the most energetic molecules escape (1); the mean kinetic energy of the molecules is reduced (1); temperature is proportional to the mean kinetic energy. (1)
  c i Larger area (1); warmer temperature (1); both increase the rate of evaporation. (1)
    ii Particles of water experience forces of attraction (1); closer together than in vapour (1); have less energy than in the vapour. (1)

14 Electricity

14.1

1 a i Electrons transfer from the cloth to the polythene rod when the rod is rubbed with the cloth.
    ii Electrons transfer from the perspex rod to the cloth when the rod is rubbed with the cloth.
  b Glass loses electrons.

2 a Attraction
  b Attraction
  c Repulsion

3 a X and Y have the same type of charge.
  b Suspend R horizontally on the end of a thread and then charge R by rubbing it with a dry cloth. Charge X and hold it near R. If it repels R, X is also positive. If X attracts R, X is negative. Y has the same type of charge to X.

14.2

1 1 = cell; 2 = switch; 3 = indicator; 4 = fuse

2 a

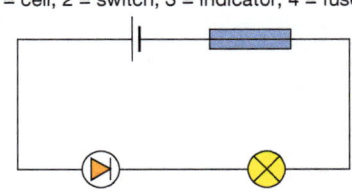

b A variable resistor.

c 15 C

3 a A light-emitting diode is a diode that emits light when current passes through it.

b A variable resistor is used to change the current in a circuit.

14.3

1 a 8.0 Ω

b 10.0 Ω

2 W: 6.0 Ω; X: 80 V; Y: 2.0 A

3 a 50 Ω

b i 18 C

ii 54 J

14.4

1 a i Thermistor

ii Diode

iii Filament bulb

b i 5 Ω

ii 10 Ω

2 a 15 Ω

b The ammeter reading increases because the resistance of the thermistor decreases.

3 a When the LDR is covered, its resistance increases. The current decreases because the resistance of the LDR increases and the potential difference across the LDR is still 3.0 V.

b i The current is zero until the potential difference is about 0.7 V then the current increases rapidly.

ii The resistance is very large until about 0.7 V then it decreases rapidly.

14.5

1 a 0.4 V

b 0.20 A, 0.5 V

2 a

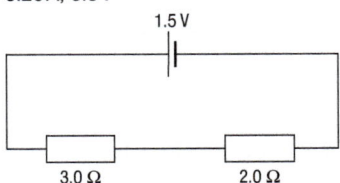

b i 5.0 Ω

ii 0.3 A

3 a i 12 Ω

ii 3.0 V

b $\dfrac{3\,V}{12\,\Omega} = 0.25\,A$

c P = 0.5 V, Q = 2.5 V

d i 15 Ω

ii 0.20 A

iii P 0.40 V, Q 2.00 V, R 0.60 V

14.6

1 a 0.30 A

b The 3 Ω resistor.

c 0.60 Ω

2 a

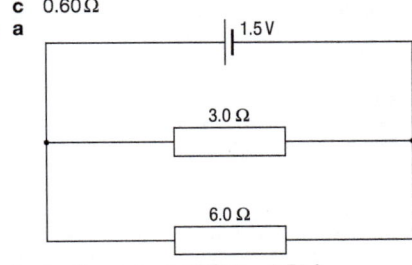

b i Current = 1.5 V/3 Ω = 0.50 A

ii Current = 1.5 V/6 Ω = 0.25 A

c Cell current = 0.5 + 0.25 = 0.75 A

3 a i I_1 = 3.0 A; I_2 = 2.0 A; I_3 = 1.0 A

ii Current through the battery = 6.0 A

b 6.5 A

14.7

1 a i See Figure 1

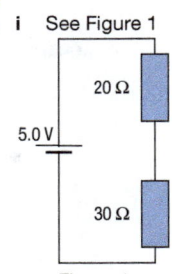

Figure 1

ii The current is the same in both resistors because they are in series. The p.d. across each resistor is equal to its resistance × the current. So the p.d. across the bigger resistance is greater than the p.d. across the other resistance.

b 0.10 A, 3.0 V

2 When the LDR is covered, the LDR resistance increases so total circuit resistance increases and the current therefore decreases. The output p.d. is the p.d. across R so it decreases because the current through R decreases.

3 a See Figure 2

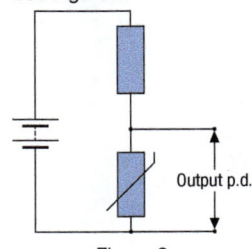

Figure 2

b When the thermistor temperature is reduced, the thermistor resistance increases so total circuit resistance increases and the current therefore decreases. The p.d. across R decreases because the current through R decreases. As the sum of the output p.d. and the p.d. across R is equal to the battery p.d., the output p.d. therefore decreases.

4 a A certain p.d. is needed to switch the buzzer on. The variable resistor is used to alter the circuit so a higher or lower temperature is needed to reach the p.d. needed to operate the buzzer. If the resistance of the variable resistor is increased, the thermistor temperature will need to be higher before the buzzer is switched on.

b i The thermistor needs be replaced by an LDR and the positions of the LDR and the variable resistor swapped so the output p.d. is across the LDR. In darkness, the LDR resistance will be high so the output p.d. will be high and therefore the lamp will be switched on.

ii See Figure 3

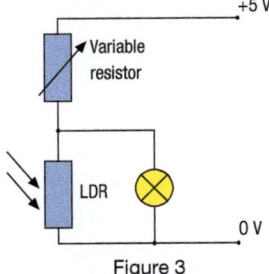

Figure 3

Answers to end of chapter summary questions

1 The resistance increases when the current is increased. This is because the increase of current makes the bulb hotter. As a result, the metal ions of the filament vibrate more, so they resist the passage of electrons through the filament more.

2 a Filament bulb

b Resistor

c Thermistor

d Diode

3 a

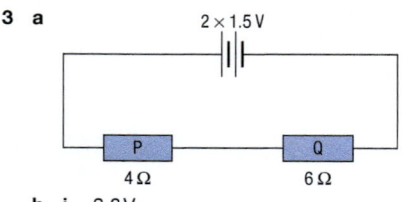

2 × 1.5 V

P 4 Ω Q 6 Ω

b i 3.0 V
ii 10 Ω
iii 0.3 A
iv P: 1.2 V; Q: 1.8 V

4 a

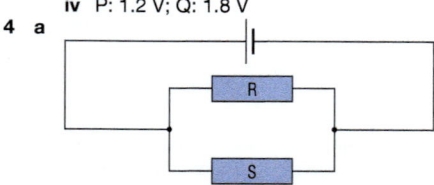

R

S

b i 1.0 A
ii 0.5 A
iii 1.5 A

5 a i 2.0 V
ii Potential difference across LDR = 3.0 − 2.0 = 1.0 V
iii 100 Ω
b i Ammeter reading decreases when LDR is covered as resistance increases.
ii The potential difference across the LDR = $V - (I \times R)$ so the LDR resistance = potential difference across LDR ÷ current I

6 a

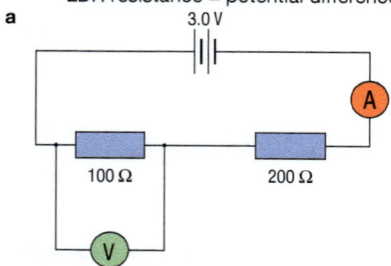

3.0 V

A

100 Ω 200 Ω

V

b i 300 Ω
ii 0.01 A
iii 1.0 V
iv 100 Ω 1.0 V; 200 Ω 2.0 V

7 a i The battery p.d. of 3.0 V is shared between the LED and the resistor. Since the potential difference across the LED is 0.6 V when it emits light, the potential difference across the resistor is 3.0 − 0.6 = 2.4 V.
ii 0.0024 A
b 160 Ω
c The current would be (almost) zero as the 'reverse' resistance of the LED is very high. The total resistance of the LED and the resistor would therefore be much greater than it was when the LED was in its 'forward' direction so the current would be much less than 0.0024 A.

8 a

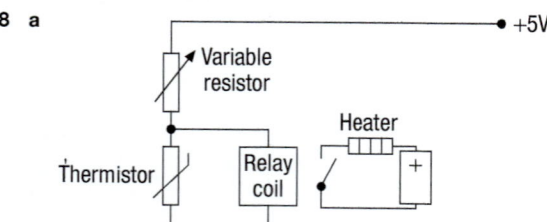

+5V

Variable resistor

Heater

Thermistor

Relay coil

0V

b When the temperature decreases, the thermistor resistance increases so the thermistor, and therefore the relay, take a greater share of the supply p.d. When the temperature drops below a certain value, the p.d. across the relay is large enough to operate which then switches the heater on.

Answers to end of chapter practice questions

1 a i negative
ii uncharged
iii positive
b Electrons are transferred to surface atoms of ruler from the duster (1); ruler has an excess of negative charge (1); duster has a deficiency of electrons (1); so is positively charged. (1)

c Points to be made: friction at the nozzle/connection to + terminal of a supply; causes paint molecules to lose electrons; paint droplets have the same type of charge; so repel each other; causing droplets to spread out forming a fine spray; the opposite charge on the car door; attracts the droplets; to give an even coating of paint. (6)

2 a i Correct symbols for battery, switch, variable resistor, resistor (piece of wire), ammeter (2); in series (1); with correct symbol for voltmeter in parallel with resistance wire. (1)
ii So that current only flows when readings are to be taken (1); if current flows all the time, the piece of wire will heat up (1); increasing its resistance. (1)
iii So that a series of results can be obtained (1); and a mean value for resistance calculated/graph plotted. (1)
b i 60 Ω (1)
ii 115 V (1)
c i 230 V (1)
ii 30 Ω (1)
iii 15.4 A (1)

3 a 3 kΩ (1)
b Total R = 5.5 kΩ (1)
   I = V/R OR = 6/5500 (1)
     = 1.09 mA (1)
c i V = I × R OR = 0.00109 × 2500 (1)
     = 2.73 V (1)
ii R at 0 °C = 17.5 kΩ (1)
     V = (2.5/20) × 6 (1)
       = 0.75 V (1)
d The potential difference across R increases with temperature (1) so could be used to switch on/off a system when temperature rises (1) thermostat (1); e.g., for cooling system. (1)

15 Motors, generators, and transformers

15.1

1 a i N
ii S
b P is a N-pole; P repels X because it has like polarity and it attracts Y because it has unlike polarity.
2 a N
b S
c Unmagnetised
3 a See 15.1 Figure 4
b i X = N, Y = S
ii The needle of the compass would also rotate in the same direction as the bar magnet.

15.2

1 a See 15.2 Figure 2
b Although both iron and steel can be magnetised, steel does not lose its magnetism when the current is switched off. Iron does lose its magnetism when the current is switched off.
2 C; B; E; D; A
3 a The current through the electromagnet coil magnetises the core of the electromagnet. The armature is pulled on to the core. This opens the make-and-break switch which cuts the current. The electromagnet loses its magnetism and the make-and-break switch closes so the cycle repeats itself.
b The armature of the buzzer has a much lower mass so it moves faster than the bell's armature and it therefore has a higher frequency of vibration.

15.3

1 a When a current passes through the coil of the electric motor, a force acts on each side of the coil due to the magnetic field of the magnet in the motor. The force on each side has a turning effect on the coil and because the current on each side is in opposite directions, the forces on each side are always in opposite directions so the motor turns. Each time the coil passes the position where the coil is at right angles to the magnetic field, the split-ring commutator reverses the connections to the battery so the current round the coil reverses direction. Without the split-ring commutator, the forces would reverse and so the coil would turn back. The action of the split-ring commutator allows the forces to continue to turn the coil in the same direction.
b The force on the loudspeaker coil would be in one direction only with a direct current so the coil would not vibrate.
2 a The current is in the opposite direction to what it would have been so the force on each side is in the opposite direction to what it would have been. The coil therefore rotates in the opposite direction.

b i Faster because the coil is lighter.

ii Faster because the field is much stronger due to the presence of the iron.

3 a The force decreases gradually as the wire is turned and becomes zero when the wire is at right angles to the field lines.

b A force acts on the coil when a current passes through it because it is in a magnetic field. An alternating current causes an alternating force to act on the coil. The coil therefore vibrates because it is acted on by an alternating force. The vibration of the coil makes the diaphragm vibrate which produces sound waves.

15.4

1 a A potential difference is induced in a wire when it cuts across the magnetic field lines of the magnet. The induced potential difference causes a current to pass through the wire and the ammeter whilst the wire is cutting the field lines.

b The ammeter would show a smaller reading in the opposite direction because the wire cuts across the field lines more slowly and in the opposite direction so the induced potential difference is less and in the reverse direction.

2 a There would be no deflection of the pointer.

b The pointer deflection would be bigger.

c The deflection would be in the opposite direction.

3 a The current in X creates a magnetic field which passes through coil Y. The increase of the magnetic field in Y induced a potential difference in Y.

b The magnetic field does not change. A potential difference can only be induced when the magnetic field is changing.

15.5

1 a A potential difference is induced in the coil when the sides of the coil cut across the field lines. The potential difference reverses direction each time the coil is at the position where its sides are moving parallel to the field lines. This happens every half-turn of the coil so one full turn of the coil corresponds to one full cycle of the alternating voltage.

b The alternating voltage would have a greater peak value (i.e., amplitude) and its time period would be less (i.e., its frequency would be greater).

2 a The peak value would be smaller. The waves would be stretched more across the screen.

b The peak value is less because the sides of the coil cut more slowly across the field lines so the potential difference at any position of the coil is less than when the coil spins faster. The waves are more stretched out across the screen because the time for each cycle would be longer.

3 a The split-ring commutator reconnects the coil the opposite way round in the circuit every half-turn each time the coil is perpendicular to the magnetic field lines. As a result, the induced potential difference changes its polarity.

b See 15.5 Figure 4b

15.6

1 a An alternating current is passed through the primary coil. This coil creates an alternating magnetic field that passes through the secondary coil. As a result, an alternating potential difference is induced in the secondary coil.

b i The 4000-turn coil

ii A steel core would not be easily magnetised and demagnetised. When an alternating current passes through the primary coil, a steel core would not produce as strong a magnetic field as the iron core would, so the induced potential difference in the secondary coil would be much smaller.

2 a Direct current in the primary coil would not produce an alternating magnetic field so no potential difference would be induced in the secondary coil.

b The current would short-circuit across the wires instead of passing through them. This would cause the coil to overheat if it did not cause a fuse to blow.

c Iron is a magnetic material so it makes the magnetic field much stronger. It is easily magnetised and demagnetised when the current alternates.

3 a i If the mains supply fails, the battery takes over.

ii The transformer steps the potential difference down.

b It has a ferrite core, which is much lighter than an iron core of the same size.

15.7

1 a 1200 turns

b 1150 turns

c i 6.0 A **ii** 0.26 A

2 a 2000 turns

b i 3 A

ii 0.15 A

3 a The current in A is less than the current in B.

b The cables have the same resistance and the current in A is less than in B. So the heating effect of the current in A is less than that in B.

Answers to end of chapter summary questions

1 a i See 15.1

ii The compass points in a direction parallel to the axis of the magnets to the left or right according to whether the magnet with the N-pole at the gap is on the left or the right of the gap.

b i current, force

ii current, lines, field

2 a When a current passes through the coil of the electromagnet, the core of the electromagnet becomes magnetised and attracts the iron armature. The iron armature turns about the pivot and its lower end pushes against one side of the switch and makes the switch close.

b When the ignition switch is closed, the core of the relay coil becomes magnetised so the relay switch closes. The motor is switched on as a result of the relay switch closing.

3 a Upwards

b The force is zero.

c The force on the sides make the coil turn.

4 a i See 15.5 Figure 4

ii The split-ring commutator would need to be replaced by two separate slip rings with a connecting brush at each ring.

b Direct current through the primary coil does not produce an alternating magnetic field. No potential difference is induced in the secondary coil as the magnetic field through it does not change.

5 a i 10 A

ii 100 A

b The higher the potential difference, the less the current needed to transfer a certain amount of electric power. The smaller the current through the cables, the less power is wasted in the cables due to their resistance and the heating effect of the current.

6 a 12 V

b 0.5 A

c 5 A

7 a 150 turns

b 0.15 A

Answers to end of chapter practice questions

1 a i strength of magnetic field (1)

ii If you double or treble the current, you double or treble the force (1); the force is directly proportional to the current. (1)

b i Force on AD is out of the page (1); force on BC is into the page (1); so coil turns (1).

ii Forces on AD and BC still in same direction (1); so coil turns (1) in opposite direction. (1)

iii Reverse the current (1)

2 a Points to be made: the coil spins in a magnetic field; the coil cuts the magnetic field lines; the coil is effectively in a changing magnetic field; a potential difference is induced across the coil; since the coil is part of a circuit, a current flows; the slip rings carry the current to the brushes; which are connected to the external circuit. (6)

b i The current varies in magnitude (1); and reverses at regular intervals. (1)

ii One complete cycle in 0.04 s (1); $f = 1/t = 1/0.04$ (1); $f = 25\,Hz$ (1)

3 a iron (1)

b i 5000 (1)

ii $n_s/n_p = V_s/V_p$ OR $n_s/5000 = 110/230$ (1)

$n_s = (5000 \times 110)/230$ (1)

$n_s = 2391$ (1)

iii An alternating current passes through the primary coil (1); this produces an alternating magnetic field in the core (1); the secondary coil is therefore in a changing magnetic field (1); an alternating potential difference is induced across the secondary coil (1); this drives an alternating current through the shaver. (1)

c The car battery provides direct current (1); so there is no changing magnetic field. (1)

d $V_s \times I_s = V_p \times I_p$ OR $230 \times I_s = 48\,000$ (1)

$I_s = 48\,000/230$ (1)

$I_s = 209\,A$ (1)

16 Household electricity

16.1

1 a 12 V

b 230 V

c 1.5 V

d 325 V

2 The number of cycles on the screen would: **a** increase **b** decrease.

3 25 Hz

4 a Direct current is in one direction only. Alternating current repeatedly reverses.

b The diode only allows current to pass in one direction, its direction. So it rectifies the alternating current to direct current.

c i

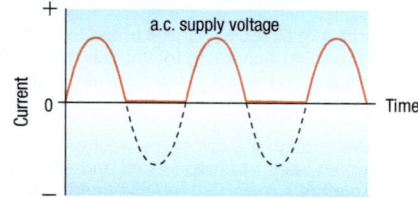

ii The peaks would not be as tall; the horizontal spacing would be unchanged

16.2

1 a The sockets are in parallel so each of the appliances connected to the sockets can be switched on and off without affecting the other appliances.

b Brass is a good conductor and doesn't oxidise like copper does. Brass is harder than copper and doesn't deform as easily as copper.

c The live wire could be exposed where the cable is worn away or damaged.

2 a 1 C; 2 D; 3 A; 4 B

b 1 Rubber is flexible and is an insulator.
2 Stiff plastic is an insulator, it doesn't wear and it can't be squashed.
3 Brass is a good conductor and doesn't deteriorate.
4 Copper is an excellent conductor and copper wires bend easily.

3 a The three wires must be insulated from each other otherwise there would be a dangerously large current in the cable due to the very low resistance between the live wire and the other wires where they touch.

b The earth wire of the cable is connected to a terminal fixed to the metal case. The other end of the earth wire is connected to the earth pin in the three-pin plug attached to the cable. When the plug is connected to a three-pin wall socket, the metal case is therefore connected via the earth wire to the ground.

c The cables to the wall sockets need to be thicker so their resistance is lower and more current passes through them than through the lighting cables. If they were not thicker, the heating effect of the current would be greater and the cables would overheat.

16.3

1 a A fuse protects an appliance or a circuit.

b So it cuts off the live wire if too much current passes through it.

c It is faster than a fuse and doesn't need to be replaced after it 'trips'.

2 a Yes.

b The element is live.

c

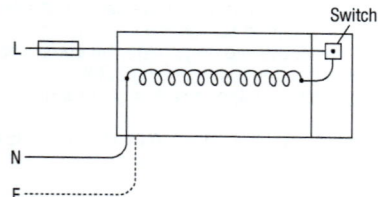

3 a A circuit breaker is an electromagnet in series with a switch that opens when too much current passes through it and cuts the current off. A fuse is a piece of resistance wire in a circuit. When too much current passes through the fuse, the fuse melts and cuts the current off.

b A circuit breaker works faster than a fuse and can be reset more quickly.

16.4

1 a 1 W

b 1150 W

c The current through the lamp in normal operation is 0.4 A. A 13 A fuse would not melt if a current greater than 0.4 A passed through it.

2 a i 36 W
ii 460 W

b i 3 A
ii 5 A

3 a The fuse would melt as the (normal) current would be 3.5 A (= 800 W/230 V)

b i 3.0 A
ii 36 W

c i 6.5 V, 169 W
ii 2.8%

16.5

1 a 150 C
b 120 J
c 180 J

2 a i 80 C
ii 720 C

b i 120 J
ii 300 J

c i 150 C
ii 12 J/C from the battery; 8 J/C to the lamp; 4 J/C to the variable resistor.
iii 1800 J from the battery =1200 J to the lamp + 600 J to the variable resistor.

3 a 12.0 Ω
b 0.50 A
c 30 C
d 4 Ω: 2.0 V; 8.0 Ω: 4.0 V
e 4 Ω: 60 J (= 30 C × 2.0 V); 8.0 Ω: 120 J (= 30 C × 4.0 V)
f 180 J

16.6

1 a 2.4 kWh
b 15 kJ
c £177.80

2 a i 1.5 kWh
ii 0.5 kWh
iii 0.8 kWh

b 39.2 p

3 a i 4 kW
ii 43.2 million joules (4000 × 3 × 60 × 60)

b i 390 W
ii 51 kWh
iii £6.10

16.7

1 An example of each possible electrical hazard is given in the table.

Appliance	Hazard
Electric drill	The drill might 'hit' a live wire in a cable in the wall.
Electric saw	The saw might cut the cable (or cut a limb).
Hairdryer	Anyone with wet hands using a hairdryer would be at risk.
Vacuum cleaner	The vacuum cleaner might run over and damage its cable.

2 a i The fault needs to be put right or the appliance replaced otherwise a new fuse will melt as soon as the appliance that caused the fault is switched on.
ii Three-core as an iron has a metal base.

b i 30
ii 11
iii 3000 kWh at 10p per kWh = £300
iv Cost of 30 bulbs + electricity = £315; cost of 11 LEDs + electricity = £61; saving = £254

3 a 4.8 A

b The metal case must be 'earthed'. The 3 core cable includes an earth wire for this purpose. A 2 core cable does not have an earth wire.

c i 5 A
ii If a fault develops and the current is much greater than 4.8 A, a higher current fuse such as a 13 A fuse would not blow if the current was less than the fuse rating. Too much current would pass through the appliance and the cables and either the appliance or the cables would overheat.

Answers to end of chapter summary questions

1 a i The neutral wire.
ii The live wire.

b i The waves would be taller as the amplitude would increase.
ii The waves would be closer together as the time for each cycle would be less.

2 If a live wire touches the case, the case would become live. Anyone touching the case would be electrocuted as an electric current would pass through their body to earth.

3 a i parallel
ii series, live

b i A fuse is a thin wire that overheats and melts if too much current passes through it; a circuit breaker is an electromagnetic switch that opens and stays open if too much current passes through it.
ii A circuit breaker works faster than a fuse and can be reset more quickly.

4 a i 11 A
      ii 13 A fuse
      iii 35 kWh
   b i The kettle.
      ii 64p
5 a

   b i 432 J
      ii 108 J
      iii 324 J
   c i 30 Ω   ii 0.4 A
      iii battery 4.8 W; 5 Ω resistor 0.8 W; 25 Ω resistor 4.0 W.
6 a i 3.0 A   ii 600 C
   b i $E = P \times t = 36\,W \times 200\,s = 7200\,J$
      ii 12 J/C
   c i 3.0 A through each bulb; 6.0 A through the battery.
      ii Energy supplied per second to each bulb = 36 W = 36 J/s; energy supplied per second by the battery = 12 V × 6.0 A = 72 J/s. Therefore energy supplied per second to the two bulbs = the energy supplied per second by the battery.
7 a 28.7 A
   b i D because the maximum safe current through D is greater than the current that would pass through it when the oven operates at full power. So D would not overheat. E would not overheat either but it would be more expensive than D.
      ii Cables A, B and C would overheat as their maximum safe current is less than the current that would pass through them when the oven is at full power. The overheated cable might cause a fire. Also, the cable insulation could melt and cause a short-circuit that may start a fire.
8 a i To change the voltage from the power station generator to a suitably high grid voltage and reducing the grid voltage to a suitable mains voltage for our homes.
      ii A step-up transformer.
   b i The grid voltage is much higher.
      ii The current supplied to the grid is much smaller.
      iii Power is wasted in the cables due to the heating effect of the current. The less the current, the less the power that is wasted.

Answers to end of chapter practice questions

1 a It needs an ac supply (1); needs a potential difference of 230 V (1); frequency 50 Hz (1); so mains (1); When on hottest setting, with maximum fan speed it takes 1200 joules of energy per second. (1)
   b i No. of kilowatt-hours used = 7198.5 − 6471.5 = 727.0 (1)
         Cost = 727 × 0.15 (1)
              = £109.05 (1)
      ii 2000 W = 2 kW (1)
         Cost = 2 × 3 × 0.15 (1)
              = £0.90 (1)
   c i P = V × I OR 60 = 230 × I
         I = 60/230 (1)
           = 0.26 A (1)
      ii 30 hours = 30 × 3600 s (1)
         Q = I × t OR = 0.26 × 30 × 3600 (1)
           = 28 080 (1) coulomb/C (1)
2 a i 4.3 A [1]
      ii The 5 A fuse. [1]
      iii The 3 A fuse would melt as soon as the heater is switched on. The 13 A fuse is unsuitable because it would only melt if the current exceeded 13 A. If a fault developed, the current could be much greater than the normal current without melting the fuse. [1] The heating effect of this large current could cause the wires or the heater element to overheat and catch fire. [1]
   b Water conducts electricity/wet hands have lower resistance than dry hands (1); water can provide a route from hand to live pin. (1)
   c Points to be made: hairdryer has a plastic case; it is double-insulated; if there is a fault and the live lead touches the metal case of the kettle, the case becomes live; anyone touching it will provide a route to earth; they will conduct current; will get an electric shock; the earth lead is connected to the metal case; if live lead touches the earthed kettle, a large current will flow to earth; melting the wire in the fuse; therefore breaking the circuit; and rendering the kettle unusable until fault rectified. (6)
3 a LED (1)
   b filament lamp (1)

c LED (1)
4 a Time period = 1 cm = 0.02 s (1)
          F = 1/T = 1/0.02 (1)
             = 50 Hz (1)
   b peak potential difference = 1.5 cm (1)
             = 3 V (1)

17 Nuclear physics

17.1
1 a Radiation from uranium consists of particles whereas the radiation from a lamp is electromagnetic waves; radiation from uranium is ionising whereas radiation from a lamp is non-ionising.
   b Radioactive atoms have unstable nuclei whereas the atoms in a lamp filament do not. The decay of a radioactive atom cannot be stopped whereas the atoms in a lamp filament stop emitting radiation when the filament current is switched off.
2 a Alpha radiation.
   b Beta or gamma radiation.
3 a There are atoms in the substance that have nuclei that are unstable. These nuclei become stable by emitting radiation.
   b Any two from radioactive isotopes in the air, the ground or in building materials; X-ray machines; cosmic radiation.
   c i The substance is radioactive.
      ii The Geiger counter continues to detect background radiation.

17.2
1 i The nucleus is much smaller than the atom.
   ii The nucleus is positively charged.
   iii The mass of an atom is concentrated in its nucleus.
   iv All the positive charge of an atom is concentrated in the nucleus of the atom.
2 a B b A is wrong because it is attracted by the nucleus; C is wrong because it is unaffected by the nucleus; D is wrong because it is repelled in the wrong direction by the nucleus.
   c The nuclear model explains why some of the alpha particles are scattered through large angles. This is because alpha particles are positively charged and if they approach the nucleus very closely they are repelled by the nucleus because it is also positively charged.
3 a A proton and a neutron have about the same mass (or they are both found in nuclear matter): A proton is charged whereas a neutron is uncharged.
   b Because the mass of a helium nucleus is four times as greater as the mass of a hydrogen nucleus, a helium nucleus must contain a total of 4 neutrons and protons whereas a hydrogen nucleus is a single proton. Two of the particles in the helium nucleus must be protons because the helium nucleus has twice as much charge as a hydrogen nucleus. The other two particles in a helium nucleus must therefore be neutrons

17.3
1 a 6 p + 6 n
   b 27 p + 33 n
   c 92 p + 143 n
   d 4 protons, 10 neutrons
2 a 92 p + 146 n
   b 90 p + 144 n
   c 91 p + 143 n
3 a i $^{235}_{92}X \rightarrow \,^{231}_{90}Th + \,^{4}_{2}\boxtimes$
      ii $^{64}_{29}Cu \rightarrow \,^{64}_{30}Zn + \,^{0}_{-1}\boxtimes$
   b i $^{210}_{83}Bi \rightarrow \,^{210}_{84}Po + \,^{0}_{-1}\boxtimes$
      ii $^{210}_{84}Po \rightarrow \,^{206}_{82}Pb + \,^{4}_{2}\boxtimes$

17.4
1 a To stop the radiation so it can't affect objects or people nearby.
   b Charged particles are deflected by an electric or magnetic field. Gamma radiation is not deflected by an electric or a magnetic field so gamma radiation is not made up of charged particles.
   c To keep the source out of range.
   d ⊠, ⊠ radiation
2 a i Gamma
      ii Alpha
      iii Beta
   b i The reading increased because the paper stopped alpha radiation reaching the detector but not beta or gamma radiation. So when the paper was removed, the alpha radiation was also detected by the counter.
      ii The reading decreased because the lead plate absorbed all the incident radiation from the source so the counter detected no radiation from the source.

iii The Geiger tube continued to detect background radiation so the counter continued to count.

3 a Radiation can knock electrons from atoms. This ionisation damages the genes in a cell which can be passed on if the cell generates more cells.

b Place the Geiger tube in a holder so the tube can be moved horizontally. Move the holder and tube so the end of the tube is close to the source and the Geiger counter detects radiation from the source. Move the tube and holder gradually away from the source until the count rate from the counter decreases significantly. The distance from the end of the tube to the source is the range of the α radiation from the source.

17.5

1 a The half-life is the average time it takes for the number of nuclei of the isotope in a sample to halve.

b 75 cpm

c 6.5 hours

2 a i 4 milligrams

ii 1 milligram

b About 65 hours (= just over 4 half-lives)

3 a i 160 million atoms

ii 10 million atoms

b Just less than 180 minutes (= just less than 4 half-lives)

17.6

1 a β; thin metal stops α radiation completely and does not stop γ radiation. The amount of β radiation passing through a thin metal sheet depends on the thickness of the sheet.

b γ; α radiation would be wholly absorbed by the body so could not be used. β radiation would be partly absorbed but γ radiation is absorbed much less so γ radiation is more reliable.

c γ; α radiation would be wholly absorbed by the pipe wall so could not be used. β radiation would be partly absorbed by the ground but γ radiation would much less affected so γ radiation is more reliable.

2 a γ radiation would hardly be absorbed by the foil as it would all pass straight through the foil.

b A stable isotope in the body (or elsewhere) would not be dangerous whereas an unstable isotope would be harmful as it is radioactive.

3 a It needs to be detectable outside the body, non-toxic, have a short half-life (1–24 hours) and decay into a stable product.

b 11 200 years old.

c The count rate measurements would be due to background radiation as well as the wood. The count rate due to background radiation is measured by measuring the count rate without the wood present. This is then subtracted from the count rate with the wood present to give the count rate due to the wood only.

17.7

1 a The nucleus splits into two fragments and releases energy and several neutrons in the process.

b The nucleus absorbs a neutron without undergoing fission.

2 a (In order) B, A, C, D, B …

b i The control rods absorb fission neutrons and keep the chain reaction under control, maintaining an even rate of fission.

ii More fission neutrons will be absorbed so the number of fission neutrons in the reactor core will decrease and the rate of release of energy due to fission will therefore decreases.

3 a i A and D

ii They have undergone fission and released neutrons and energy.

iii C and E

b i

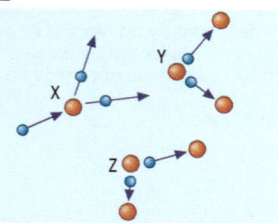

ii Either, the third neutron released by X was absorbed by a control rod, or slowed by the moderator, or absorbed by a non-fissionable nucleus, or escaped from the reactor.

17.8

1 a Nuclear fusion is the formation of a nucleus when two smaller nuclei collide and fuse together.

b A helium nucleus with 2 protons and a single neutron is formed, $^3_2$He.

2 a i So the nuclei have enough kinetic energy to overcome the force of repulsion between them and fuse.

ii The energy output would be less than the energy input so it would not produce any energy overall.

b *Advantage* – the fuel is readily available or, the reaction products are less harmful than fission products or, the reactions would stop if the plasma touches the sides of the reactor.

Disadvantage – the reactions cannot be maintained for long periods of time or, strong magnetic fields are needed to control the plasma.

3 a 1 proton and 1 neutron.

b $^2_1H + ^1_1p \rightarrow ^3_2He$

c $^3_2He + ^3_2He \rightarrow ^4_2He + ^1_1p + ^1_1p$

17.9

1 a i It needs to be stored securely because it is hazardous and would be a danger to people and animals if it escaped.

ii It needs to be stored for a long time because it contains radioactive isotopes with long half-lives.

b The α radiation from the source will be absorbed by the surrounding tissues and it could damage or kill cells in the body or cause cancer. Outside the body, it is less dangerous as α radiation has no penetrating power, but it can damage skin cells if within range of them or retinal cells if near the eye.

2 a Radon gas in a house may be more concentrated than outdoors and people in the house would breathe it in. The lungs would be exposed to α radiation from radon gas atoms that enter the lungs. The ionising effect of the α particles in the tissue cells would damage or kill the cells or cause cancer.

b Install pipes under the house and connect them to a suction pump to draw radon gases out of the ground before it seeps into the house. The top of the outlet pipe from the pump would need to be high up outside the house.

3 Benefits to building either type of reactor should include no greenhouse gas emissions, reliable and secure electricity supplies, and large-scale generation from small sites compared with renewable supplies that would take up much larger areas etc. Drawbacks should include long-term storage of nuclear waste, possible escape of radioactive substances into the environment, impracticality of fusion reactors, etc.

Answers to end of chapter summary questions

1 a i 6p + 8n

ii 90p + 138n

b i 7p + 7n

ii $^{14}_7N$

c i 88p + 136n

ii $^{224}_{88}Ra$

2

	α	β	γ
Identity	helium nuclei	electrons	electromagnetic radiation
Stopped by	paper	5 mm aluminium	thick lead
Range in air	about 5 cm	about 1 m	unlimited
Relative ionisation	very strong	strong	weak

3 a Student graph

b 1 hour 40 minutes

4 a i The activity of a radioactive source is the number of unstable atoms in the source that decay per second (or the rate of decay of the unstable atoms in the source).

ii 3 half-lives

b 16 800 years

5 a Background radioactivity.   **b** 356 cpm

c Beta radiation, because it penetrates thin foil and is stopped by an aluminium plate. Alpha radiation would be stopped by the foil. Gamma radiation would pass through the foil and the plate.

6 a They are positively charged so they are repelled by the nucleus because it is also positively charged.

b It doesn't approach the nucleus as closely as A does so the force on it is less.

c The nucleus is very small in size compared to the atom so most particles don't pass near enough to the nucleus to be affected by it.

7 a A chain reaction is a sequence of induced fission events in which each event causes further fission events which cause further fission events and so on. Each fission event occurs when the nucleus of an atom of a fissionable isotope absorbs a neutron and splits into two smaller fragment nuclei and releases releases several neutrons which can cause other fissionable nuclei to split.

b **i** The reactor core would overheat because energy is still released from the fuel rods as they are very hot.

ii The control rods would absorb more neutrons and so the number of fission events per second would decrease. As a result, the rate at which energy is released from the fuel rods would decrease.

8 **a** **i** Nuclear fusion is the process in which two small nuclei fuse together to form a single larger nucleus, releasing energy in the process.

ii There is a force of repulsion between any particles that have the same type of charge. All nuclei are positively charged. Therefore when two nuclei approach each other, they repel each other.

iii When two nuclei collide, they only fuse together if they have sufficient kinetic energy to overcome the force of repulsion that acts between them. So they need to collide at high speed to fuse together.

b In the reactor, the plasma containing the nuclei to be fused must be at a very high temperature so the nuclei are moving fast enough to fuse. In addition, strong magnetic fields are used to prevent the plasma touching the inside walls of its container and losing internal energy. The very high temperature needed is difficult to achieve and the plasma is difficult to control at such a high temperature.

Answers to end of chapter practice questions

1 **a** Smooth curve of best fit drawn with correct axes and labels (3)

b As the time increases the number of `heads' decreases but not directly proportional.(2)

c To control the variable and to make comparison. (1)

d Timing error or counting error (1)

e The numbers of isotopes that have decayed (1)

2 **a** 23% (1)

b Examples such as X-ray machines etc. (2)

c Each flight exposes them to additional radiation.
So 20 - 2.0 (from average background radiation) = 18millisieverts.
Therefore 18/0.3 = maximum of 60 flights (3)

3 **a** 86 (1)

b 218 (1)

c Different number of neutrons (1)

d 5 half-lives shown (1) 19 days (1)

4 **a** Alpha - same as helium nucleus
Beta - high speed electron
Gamma - electromagnetic radiation

b 88---alpha---222

c alpha and gamma

5 examples of typical answers:
Nuclear fusion reactor - only produces small amounts of energy but there are no radiation dangers and therefore no radioactive waste to dispose. They could be sited locally to provide energy to where it is needed. The fuel is readily available and easily extracted.
Nuclear fission reactors - produce large amounts of energy from small quantities of fuel but they are radioactive and the radioactive waste is difficult to dispose of. Decommissioning costs are high.

18 Space

18.1

1 **a** Comets are frozen rocks that orbit the Sun. When they get near the they heat up so much that they emit light and can then be seen. Meteors or shooting stars are small bits of rocks that burn up when they enter the Earth's atmosphere.

b Comets and asteroids both orbit the Sun. The orbits of comets are highly elliptical (ie non-circular) whereas the orbits of asteroids are circular or nearly so.

2 **a** **i** Jupiter

ii Mercury

b Water in the liquid state is necessary for life forms to exist. The planets nearer than the Earth to the Sun are to hot for liquid water to exist on the surface. The planets further than the Earth to the Sun are too cold for liquid water to exist.

3 The Earth would be frozen when it was far away from the Sun because it would receive much less energy from the Sun. When it was close to the Sun, water on its surface would evaporate if it travelled too near the Sun and its atmosphere would be heated so much its particles would into space.

4 **a** The Sun formed out of clouds of dust and gas where the particles in the clouds were pulled together by their own gravitational attraction. The clouds become more and more concentrated to form a protostar. As the protostar became denser,its particles collided more and its temperature increased until it became hot enough for the nuclei of

hydrogen atoms to fuse together, forming helium nuclei and releasing energy to make the protostar hot enough to emit light.

b A main sequence star is a star in which hydrogen nuclei in its core fuse together to form helium nuclei. This is the main stage in the life of a star as it can maintain its energy output for millions of years until there are no more hydrogen nuclei left to fuse together. The Sun is a main sequence star because most of its core consists of hydrogen nuclei in its core.

18.2

1 **a** b, d, c, a.

b **i** d
It will fade out and go cold to become a black dwarf.

2 **a** The force of attraction due to its gravity acting on its own mass.

ii The force of the radiation flowing outwards to its surface from its core.

b A white dwarf cools down and when it no longer emits light it has become a black dwarf because it can no longer be seen.

3 **a** **i** hydrogen

ii uranium

iii helium, iron

b A red giant star is much larger than a neutron star. A neutron star consists only of neutrons whereas a red giant star contains helium and other light elements in its core. A red giant star emits light and a neutron star does not.

4 **a** The Sun and the rest of the Solar System formed from the debris of a supernova. Much of the uranium-238 formed from the debris of the supernova still exists because has it has a half-life which is comparable with the age of the Earth.

b Some plutonium-239 would have been created in the supernova from which the Sun was formed. Since this event was at least 4500 million years ago, any plutonium-239 created then would have long since decayed into other elements so it is not found naturally now.

18.3

1 **a** **i** The force of gravity on a satellite in a circular orbit is directed towards the centre of the Earth.

ii The acceleration of the satellite is towards the centre of the circle

b The direction the satellite's velocity (ie its direction of motion) is changed by the force of gravity on it so it it continues to circle the Earth. Because the direction of its velocity is always at right angles to the direction of the force of gravity on it, no work is done on it so its speed does not change.

2 **a** Below; A geostationary satellite takes 24 hours to go round its orbit once. A GPS satellite goes round each orbit in half the time a geostationary satellite takes. The larger a satellite orbit is, the longer the satellite in it takes to go round the orbit once. So the GPS satellite must be in a lower orbit

b Above; A weather satellite takes less time to go round its orbit compared with a GPS satellite. The larger a satellite orbit is, the longer the satellite takes to go round the orbit once. So a weather satellite must be in a lower orbit than a GPS satellite.

3 **a** Jupiter is the slowest of the three; It takes about 11 times longer to go round its orbit than the Earth and its orbit is only about 5 times bigger so its speed must be less than that of the Earth. Mercury moves faster than the Earth because it only takes 0.24 years to go round its orbit yet its orbit is 0.39 times bigger than the Earth's orbit. Therefore Jupiter is the slowest of all three.

b As explained above, Jupiter is slower than the Earth and the Earth is slower than Mercury. So Mercury moves round faster than the Earth

OR

a For a circular orbit, the speed of a satellite in a circular orbit is equal to its circumference the time it takes to go round once. The circumference of a circular orbit is proportional to its radius. So the speed of a satellite in a circular orbit is proportional to the orbit radius the time the satellite takes to go round once. This ratio is equal to 1.6 for Mercury, 1 for Earth and 0.4 for Jupiter. So Jupiter is the slowest of the three because it has the lowest ratio.

b Mercury has the highest ratio in the analysis in a). So it is the fastest of the three.

4 50 km /s. The speed of a satellite in a circular orbit is proportional to the orbit radius the time the satellite takes to go round once. This ratio is equal to 1.6(3) for Mercury and 1.0(0) for Earth. So Mercury's speed is about 1.6 times that of Earth. Mercury's speed is therefore 1.6 30 km/s which gives 48 km/s. Note the data in the table is given to 2 decimal places so the ratios here are calculated to 3 sf. The ratios are then used to give the final answer which is rounded off to 1 significant figure because (although the Earth's speed is given to 2 significant figures) the question asks for an estimate.

18.4

1 a i Receding
  ii Approaching
 b The light from Andromeda shows a blue-shift which means Andromeda must be moving towards us.

2 a Earth, Sun, Andromeda galaxy, universe.
 b i Their red-shift is of the same order of magnitude as that of the distant galaxies. Since red-shift depends on distance, this means that quasars can be as far away as the distant galaxies.
  ii The brightness of a quasar seen from Earth is the same as that of billions of stars in a distant galaxy even though a quasar is much smaller object than a galaxy. So the power output of a quasar is about the same as billions of stars even though a quasar is much smaller.

3 a The light from a light source (e.g., galaxy) that is moving away from us is increased in wavelength due to the motion of the source moving away from us. This increase in wavelength is called a red-shift.
 b i Y
  ii X

4 These galaxies have different speeds because they have different red or blue shifts. The ones with red shifts are moving away from us. The ones with blue shifts are moving towards us. As well as towards or away from us, galaxies may be moving across but we can't tell because they are so far away.

18.5

1 a The Big Bang theory holds that the universe was created in a massive explosion about 13 billion years ago.
 b They had no evidence for a massive explosion and they could explain Hubble's finding that the universe is expanding by assuming that the universe has always existed and is expanding because matter is entering it and pushing the galaxies apart.
 c Cosmic background microwave radiation provided evidence that the universe was created in a massive explosion.

2 C D B A

3 a The distant galaxies are accelerating away from each other.
 b The universe would stop expanding and go into reverse, ending in a Big Crunch.

4 a About 6.8×10^9 or 6800 million light years $(150000 \text{ km/s} \div 22 \text{ km/s}) \times 10^6$ light years)
 b ~100000

Answers to end of chapter summary questions

1 a A, C, B, D, E.
 b i It will become a red giant.
  ii It will explode as a supernova, leaving a neutron star at its core. If the mass of the neutron star is large enough, it will be a black hole.
 c The velocity of the Earth changes because the direction of its velocity changes as it moves round the Sun. The force of gravity between the Sun and the Earth acts on the Earth and makes its velocity change direction so it experiences an acceleration towards the Sun.

2 a i A large star that explodes.
  ii A star that becomes a supernova suddenly becomes much brighter then it fades. A star like the Sun has a constant brightness.
 b i A massive object, from which nothing can escape.
  ii They would be pulled in by the force of gravity and then disappear.

iii A neutron star is composed entirely of neutrons. It is formed at the core of a supernova if there is not enough matter to form a black hole.

3 a i Helium
  ii Helium
 b i lead, uranium
  ii The two elements would have been formed in a supernova explosion.
  iii Heavy elements can only have formed in a supernova. The presence of heavy elements in the Earth tells us that the Solar System formed from the debris of a supernova.

4 a 3, 1, 4, 2
 b i The force of gravity between them.
  ii Gravitational potential energy is released and transferred into kinetic energy as dust and gas clouds pull together. As the clouds of gas and dust become denser and denser, the particles in the clouds move faster and faster and so the clouds heat up.
  iii The force of gravitational attraction towards the centre of the galaxy acts as the centripetal force to keep the stars revolving about the centre of the galaxy.

5 a i an increase.
  ii Red light has a longer wavelength than any other colour of light. The wavelength of the light is made longer by the motion of the galaxy and is shifted towards the red part of the spectrum.
  iii The galaxy is moving away from us. The speed of the galaxy can be deduced from the amount of the red-shift.
 b The galaxy is moving towards us.
 c i The universe is expanding.
  ii The discovery of cosmic microwave background radiation.

6 a i Galaxy A.
  ii Galaxy C is further away than galaxy A.
 b i It is expanding and all the galaxies are moving further away from each other.
  ii We are not in any special place.

7 a 0.05 metres per second.
 b 0.50 metres per second.

Answers to end of chapter practice questions

1 a Big bang b Red shift c All directions

2 a The distance traveled by light in one year. b Galaxies further away from the Earth move away faster. c Observing a larger number of galaxies; individual measurements more accurate due to newer techniques and equipment.

3 a Cloud of dust and gas accumulates; gravity draws most of the matter into the centre of the cloud; the loss of potential energy heats up mass in the middle of the cloud; high temperature and density of the forming star allows nuclear fusion to begin.
 b Fusion of hydrogen in the core, higher elements are not fused; forces within the star are balanced so the star keeps its shape.
 c The remaining core of a very heavy star its supernova stage; it is so dense that light cannot escape.
 d supernova; heavy elements; universe

4 Moons and artificial satellites both orbit a parent body; artificial satellites weigh far less than moons; moons weigh less than their parent planets; moons orbit more distant from the planet than artificial satellites; artificial satellites have very short orbital times compared to moons; artificial satellites orbit at a faster speed than moons;

Index